The Scholarly Publishing Office
the University of Michigan
University Library

CITY OF NEW YORK

# Board of Estimate and Apportionment

OFFICE OF THE CHIEF ENGINEER

# PRECISE LEVELING IN NEW YORK CITY

BY

FREDERICK W. KOOP

Assistant Engineer, Board of Estimate and Apportionment

EXECUTED 1909 TO 1914

CITY OF NEW YORK

# Board of Estimate and Apportionment

OFFICE OF THE CHIEF ENGINEER

# PRECISE LEVELING
IN
# NEW YORK CITY

BY

FREDERICK W. KOOP

Assistant Engineer, Board of Estimate and Apportionment

EXECUTED 1909 TO 1914.

J. J. Little & Ives Co. N.Y.
Printers and Bookbinders

4683-15-600

City of New York
Board of Estimate and Apportionment
Office of the Chief Engineer.

October 3, 1914.

To the Board of Estimate and Apportionment:

Gentlemen: On February 19, 1909, the Board of Estimate and Apportionment authorized the running of precise levels and the establishment of a series of permanent bench marks in all boroughs of the City of New York. It was expected that what was left of the organization which carried out the triangulation of the city would be used for this purpose. Mr. Frederick W. Koop was placed in charge of this work, and had at no time more than four, and part of the time three, assistants, one of these being an axeman and the others rodmen. The field work was completed during the first quarter of 1912, and since that time Mr. Koop has been engaged in the laborious process of computing, checking and preparing accurate descriptions of the various bench marks which were established. This work has now been completed, and I submit herewith Mr. Koop's final report.

A high standard of accuracy has been maintained, and I believe that this report will be one of the most valuable contributions yet made to the literature of precise surveying. In commenting editorially upon this work as described in a paper read by Mr. Koop before the Municipal Engineers of the City of New York the *Engineering News,* in its issue of September 4, 1913, said: "The work which the City of New York has recently completed in establishing a precise level net embracing the entire area of the city is worthy of the attention of every city engineer. Aside from the marking of street lines with sufficient care and accuracy to prevent encroachments which in business streets are expensive to remove, there is perhaps no part of city survey work more important and more appreciated by private engineers as well as municipal employees than a really accurate and systematic net of bench marks referred to some definite datum plane." In commenting upon the small cost at which this work has been accomplished the editorial notes that such accurate results are desirable when not achieved at excessive cost, and that the New York City work offers an excellent example of how this can be done.

The report herewith submitted contains: First, a description of the instruments and methods which were used, which will be of interest and value to those engaged in similar work, and especially to the topographical bureaus of the several boroughs of the city. This is followed by a history of the datum planes in use in the City of New York, the number of which are bewildering and the existence of which prompted action by the Municipal Engineers of the City of New York, to which organization must be credited the initiative in securing this new system of precise levels. This section of the report contains a chart illustrating the relation between the various datum planes, the differences having been found to vary somewhat from those heretofore published. The report also contains a valuable chapter on the subject of coastal subsidence, a phenomenon which has been the cause of much trouble and considerable controversy. Then follows a description of each of the bench marks, of which 215 are located in the Borough of Manhattan, 327 in the Borough of Brooklyn, 234 in the Borough of The Bronx and some nearby points in Westchester County, 233 in the Borough of Queens and in Nassau County, and 177 in the Borough of Richmond, including a few in New Jersey, or a total of 1,186 bench marks.

Some of the circuits were very long, one of them covering 73 miles, and the closing error of this circuit was only 0.0125 foot. Several other circuits were between 35 and 40 miles in extent, and the closing errors in all of them were very small.

In accordance with authority given by the Board, a list of the bench marks for each borough is to be published separately, and copies will be made available not only for the topographical bureaus of these boroughs, but for all city departments and surveyors engaged in private work.

I feel confident that the report will prove to be of great general value.

Respectfully,

NELSON P. LEWIS,

*Chief Engineer.*

CITY OF NEW YORK

BOARD OF ESTIMATE AND APPORTIONMENT

OFFICE OF THE CHIEF ENGINEER.

July 1, 1914.

NELSON P. LEWIS, *Chief Engineer,*

Board of Estimate and Apportionment.

SIR:—I beg to submit herewith the final report on Precise Leveling in New York City. In the report is embraced an account of the methods of precise leveling in large cities and tables and formulas for the reduction of observations are likewise incorporated.

The 1,186 bench marks established during the work are described as to character and location and their elevations in meters and feet are recorded.

On the work done under your direction the United States Standard Datum has been adopted as the plane of reference, and the report contains a modern datum plane chart by means of which the various departmental datum planes in the several boroughs can readily be reduced to this universal standard. The history of the more important datum planes at present in use in the city has been briefly outlined.

During the progress of the work the writer has frequently been confronted with the problem of the supposed evidence of coastal subsidence in New York City and vicinity. As this question is one of considerable importance when running precise levels on or near the sea coast, the results of the writer's investigation of this matter conducted during the progress of the work have been set forth in some detail.

The tabulation of the data concerning the bench marks established has been arranged by boroughs to facilitate reference on the part of those interested.

Respectfully,

FREDERICK W. KOOP,

*Assistant Engineer.*

# CONTENTS.

CONTENTS—(*Continued*)

# ILLUSTRATIONS.

# BOARD OF ESTIMATE AND APPORTIONMENT

OFFICE OF THE CHIEF ENGINEER

# PRECISE LEVELING IN NEW YORK CITY

By Frederick W. Koop

Assistant Engineer, Board of Estimate and Apportionment.

## General Statement.

At a meeting of the Board of Estimate and Apportionment of the City of New York held on February 19, 1909, a resolution was adopted appropriating the sum of $10,000 for the running of precise levels and the establishment of bench marks in connection with the topographical work of all boroughs, and on March 2 this resolution was approved by the Board of Aldermen. On February 16, 1911, an additional $5,000 was voted for the completion of this work, making a total appropriation of $15,000. From this sum were paid the salaries of the men employed on the work, office rentals, cost of instruments, railroad fares and incidentals. The salary of the writer, as Assistant Engineer in charge, was paid by the Board of Estimate and Apportionment.

When the initial appropriation was secured the writer was directed to carry out the work and to follow in its execution a plan previously presented by him to the Chief Engineer of the Board of Estimate and Apportionment. The plan proposed was as follows:

(1) To call on the United States Coast and Geodetic Survey for a list of bench marks in and around New York City, with their locations and elevations. These benches had all been adjusted, referred to mean sea level at Sandy Hook and, though few in number, were to form the basis of future leveling in New York City and vicinity.

(2) To make a reconnaissance of each borough, and select the level lines necessary to give a good level net in each; also to locate suitable points for bench marks along these lines and to establish others, so that there would be at least one permanent bench mark to each mile of level line.

(3) To call on the topographical bureaus and other engineering departments of the city for a list of bench marks along the proposed level lines, with the datum planes to which they were referred and the history concerning the adoption of such planes, and to ascertain if possible what other departments or organizations made a practice of referring their elevations to these datum planes.

(4) To adopt the methods now in use by the Coast Survey, both in the field operations and in the reduction of the notes and office computations, in order to secure the best results for the city, and to attain a degree of precision in the leveling at least equal to that attained by the Coast Survey in their precise leveling during recent years.

(5) To refer all elevations to mean sea level at Sandy Hook, N. J., which is the Standard Datum of the United States.

(6) To publish a list of all elevations determined, both in meters and in feet, together with descriptions of all bench marks.

### Instruments Used.

On March 16, 1909, an order was placed with the Bausch & Lomb Optical Company of Rochester, N. Y., for two precise levels with tripods and three metric leveling rods with foot plates and foot pins, all of which were to be of the latest Coast Survey type and to be made from specifications furnished by the Instrument Division of the Coast Survey.

The rods ordered were the standard Coast Survey metric rods. All differences of elevation are therefore in meters, and the horizontal distances in kilometers.

A Locke hand-level and a strong buff colored canvas umbrella 60 inches in diameter completed the outfit.

### Recovery of Coast Survey Bench Marks.

During 1886-87 the United States Coast Survey established a series of bench marks along a portion of the waterfront of New York City, the resulting elevations being referred to mean sea level as determined by tidal observations made at Sandy Hook during 1876-81. While awaiting the completion of the instrumental outfit a search was begun to recover as many of these bench marks as possible. Comparatively few were found, however, the majority of them having been destroyed.

### Reconnaissance.

A reconnaissance was then made over the Boroughs of Brooklyn, Queens and Richmond, level routes were selected and a level net planned for each. These level lines included all the Coast Survey bench marks recovered and the bench marks of the various city departments, about one to each mile. A reconnaissance of the Boroughs of Manhattan and The Bronx was made during the following winter.

### Inspection of Instruments.

In June, 1909, the precise levels, rods, etc., were delivered by the makers to the Instrument Division of the Coast Survey for inspection, such inspection and approval being a condition of the contract. The instruments and rods were tested as to material and workmanship in the writer's presence, and the leveling rods were then sent to the United States Bureau of Standards to have their absolute lengths determined; these determinations were also witnessed.

The instruments and rods were approved and passed by the Coast Survey, but the tripods, foot plates and foot pins were returned to the maker for alteration. The outfit was finally delivered complete in September, 1909.

### Field Party.

A field party was then organized, consisting of the Assistant Engineer in charge, who acted as observer, three rodmen, one of whom acted as recorder, and one man whose duty it was to shade the instrument from the sun with the umbrella and to prepare bench marks. This made a total of five men in the party.

The levels and rods were tested in the field and on October 13, 1909, the actual running of the level lines was commenced. Except during the winter months when little precise leveling could be done, this work continued uninterruptedly until May 1, 1912, when the field party was disbanded. All of the field work, covering each of the boroughs with primary level lines and connecting one with the other, was accomplished with this one level party of five men, which very often was reduced to four.

## The Precise Level.

On November 29, 1896, a committee of four was appointed by the Superintendent of the United States Coast and Geodetic Survey to consider the subject of precise levels and in particular to investigate "the accuracy of various methods, their relative freedom from systematic errors and their relative quickness, cheapness and facility for the reduction of the observations." * This committee met almost daily and submitted a very complete report on February 9, 1899.

When it became necessary to provide instruments for carrying out the geodetic leveling planned for the season of 1899 the recommendations of this committee were acted upon, and several of the older Coast Survey levels were reconstructed. An examination of these instruments shows that several of the changes were designed primarily to eliminate systematic errors due to temperature changes, and the instruments in general exhibit many of the essential features of the precise level of the present Coast Survey type.

A departure from previous practice was made in the reconstruction of these instruments, in that nickel alloys were freely used for the first time for such a purpose. With the modification of these instruments the method of observation was also radically changed from that formerly in use.

A careful study of the results obtained with these instruments proved conclusively that the use of the new alloy of iron and nickel and the reduction of the distance between the level and the line of collimation were decided improvements and practically eliminated errors due to temperature effects.

Keeping these facts in view, the new or present type of precise level was designed by Mr. E. G. Fischer, Chief of the Instrument Division, United States Coast and Geodetic Survey. This type of instrument entered into field service in the summer of 1900 and has been used continuously ever since. The justification of the present instrument and of the method of observation adopted since its introduction is found in the accuracy, rapidity and cheapness of the leveling done since 1900.

The Board of Estimate precise level of the latest Coast Survey type is shown in illustrations Nos. 1 and 2. The principal statistics of the Board of Estimate levels of this type are as follows:

| | | |
|---|---|---|
| Telescope inverting, orthoscopic ocular. | | |
| Diameter of objective | | 44 mm |
| Effective aperture of objective | | 42 mm |
| Focal length of objective | | 410 mm |
| Length of telescope | | 438 mm |
| Length of telescope focused | | 476 mm |
| Magnifying power of telescope | | 30 |
| Angular value of one division (2 mm) of level vial | | 2.16 sec. |
| Angular value of one division of micrometer head | | 2.6 sec. |
| Weight of instrument alone | 6 kg. | = 13.2 lbs. |
| Weight of tripod | 9 kg. | = 19.8 lbs. |

Reticle contains one vertical and three horizontal spider threads. The horizontal threads are equidistant, and the upper and lower threads intercept an interval of 300 millimeters on the rod at a distance of 100 meters.

The instrument is essentially a dumpy level, as the telescope does not rest in wyes, cannot be removed from its supports and can neither be inverted nor reversed:

* This committee consisted of John F. Hayford, Expert Computer and Geodesist; Isaac Winston, Assistant; J. J. Gilbert, Assistant, and A. L. Baldwin, Assistant.

No. 1.—The Coast and Geodetic Survey Precise Level of 1900. Right Side View.

No. 2.—The Coast and Geodetic Survey Precise Level of 1900. Left Side View.

the level vial is fixed relatively to the telescope and is placed as near as possible to the line of collimation, being in fact countersunk into the barrel of the telescope. In this position the parallelism of the two is not disturbed by local temperature changes. The nickel-iron alloy has been used almost entirely in the construction of the telescope and adjacent parts.

The following description of the Coast Survey level is reproduced, somewhat condensed, from a description written by Mr. E. G. Fischer, designer of the instrument.

### The Material in the Level.

For all those parts upon which depend the constancy of the relation between the line of collimation and the level—the telescope, the tube incasing the level vial, the draw-tube, reticle ring and the supporting cylinder—the material selected is a nickel-iron alloy. As a slight variation in the proportions of this alloy causes a change of several units in the sixth place of the expansion coefficient, a number of alloys were made of different proportions, of which one of 66⅔ parts of a medium-grained cast iron, furnished by the Brown & Sharpe Manufacturing Company, of Providence, R. I., and 33⅓ parts of what is called "grain nickel" was finally adopted. It can be cast free from sand and blowholes, and has a coefficient of 0.000004.

No thorough tests as to strength, etc., were made of this alloy, but so far as shop practice reveals its physical properties it can be said to be rather brittle, easily worked in the lathe and with the file, entirely unmalleable, and behaving practically like the better and softer grades of cast iron. It can readily be brazed and soldered, and, unlike cast iron, very easily takes an exceptionally fine polish, resembling that of nickel.

The pointed screws pivoting the telescope, the screws holding in place the level tube and by which the level is adjusted, the screws holding and adjusting the reticle ring, and the fine-motion micrometer screw, upon all of which depends the constancy of the relation between the line of sight and the plane tangent to the middle point of the level vial, and which require to be made of a material much harder than the casting above described, are made of nickel-steel, with a coefficient of 0.000001, obtained from the Société Anonyme de Commentry-Fourchambault, 26 Place Vendôme, Paris.

A longitudinal section of this instrument is shown in illustration No. 3. The material used in the construction of other portions of the instrument will be named in the description of those parts.

### The Tripod of the Level.

The tripod is of the usual form, and those for the Board of Estimate levels are made of a fine-grained white ash. The three legs separating some distance above the feet into two rectangular rods, pivot in the head by means of bolts about $1^{cm}$ diameter at points forming a regular hexagon. The feet consist of pointed hollow sockets about $14^{cm}$ long and $3^{cm}.5$ diameter at the top, fitted and fastened by screws to the legs. They are made of 10 per cent. aluminum bronze, an alloy but little inferior to steel in hardness and toughness. The two rods forming the leg are $2^{cm}$ by $3^{cm}.5$ and are fastened together at two points by braces which are screwed between them. The tops of the legs are brass-bound to guard against the splitting out of the bolt holes. In obtaining the length of the legs, which should be made to suit the observer's height, their normal angle with the ground was taken to be 60°, the vertical distance between the bolt holes in the head of the stand and the line of collimation being $13^{cm}$. The head of the stand is $4^{cm}.5$ thick and carries sunk into its upper surface the three V-grooved plates forming the supports for the foot screws of the instrument. In

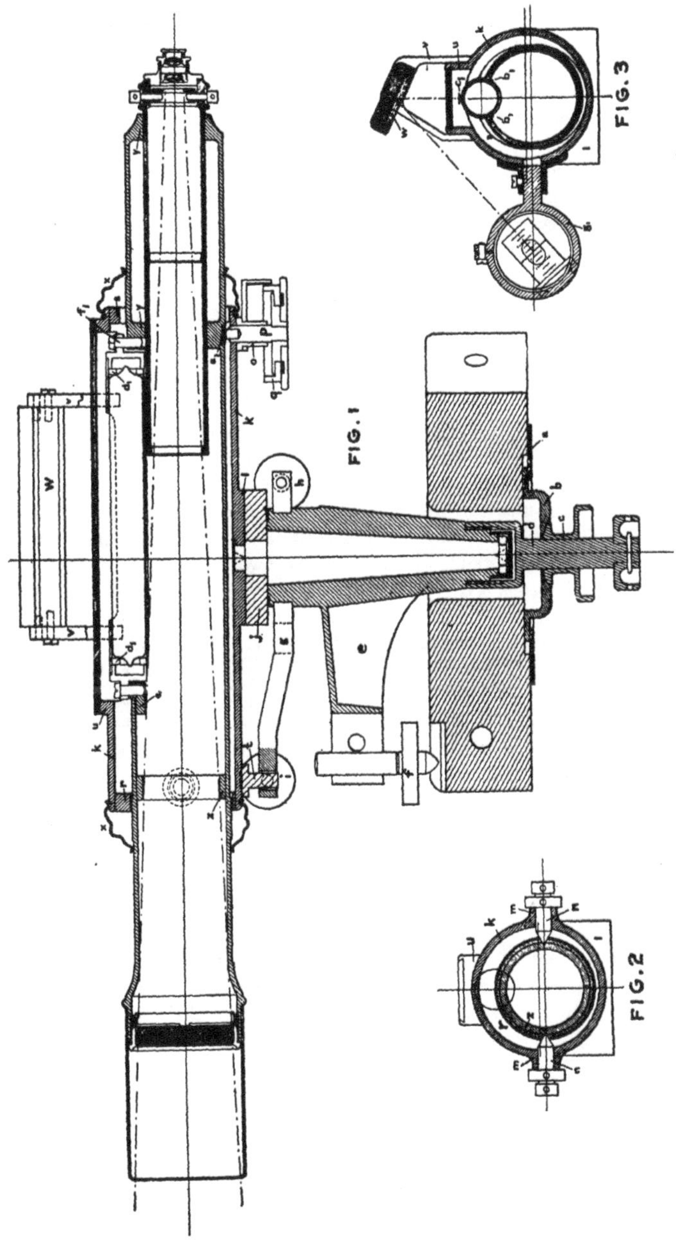


No. 3.—The Coast and Geodetic Survey Precise Level of 1900. Longitudinal Section.

a circular recess in the lower surface is held by a ring-shaped plate marked *a* in Fig. 1, a washer, *b*, shaped so as to form a seat for the convex shoulder of the nut *c*, which is threaded on the screw *d*. This screw, $1^{cm}.3$ in diameter and of a pitch of 8 threads per centimeter, enlarges at its upper end to a cup-shaped nut, which can be screwed upon the lower end of the center socket of the instrument. The washer *b* is not confined in its recess so closely but that it can move laterally and adapt itself to any position the vertical axis may assume in relation to the head of the stand. When the instrument is set upon the stand the lower end of the center socket will come to rest upon the cup-shaped nut before the foot screws can touch their supports, thus leaving it in an unstable position and making it practically impossible for the observer to forget to secure the instrument to the stand before it is carried to the place of work. The nut *c* is loosened before observing, and tightened only when the instrument and stand are to be carried from station to station.

## The Instrument Base and Center.

The instrument base, designated by *e* on the diagram, is a single piece of hard and fine-grained cast iron. In its three legs, at a radial distance of $9^{cm}$, are threaded the foot screws *f*, of $9^{mm}.5$ diameter and 15 threads per centimeter, and having a bearing of $2^{cm}.3$. The screws are of such length as to permit a motion of $6^{mm}$ above and below the normal position, thus allowing the instrument to be leveled even under unusual inclination of the head of the tripod. No position of the foot screws can prevent the fine-motion or micrometer screw from freely passing over them. The ends of the legs are split in the usual manner and provided with milled-head screws for clamping the foot screws. The clamp arm *g*, with its clamp screw *h*, is fitted into a groove near the top of the center socket, and carries at the outer end the fine-motion screw *i* for moving the telescope horizontally in azimuth. The central portion of the instrument base is bored out conically and affords a bearing throughout its length for the unusually long center ($10^{cm}$), which is made of the hardest grade tool steel, Sanderson's No. 6. It is secured against being withdrawn by a small nut screwed to its lower end. Upon its upper end is fastened permanently, by screwing and riveting a disk or flange *j*, of $5^{cm}.8$ diameter, made of hard cast iron, which forms the base of the supporting cylinder.

## The Supporting Cylinder of the Telescope.

This, indicated in Figs. 1, 2, and 3 by *k*, is a nickel-iron casting, as stated above. Its length is $21^{cm}.6$, its outer diameter $5^{cm}.9$, its inner diameter $5^{cm}.4$, leaving a thickness of wall of $2^{mm}.5$. At its middle point is a cylindrical base or hub (*l* in Figs. 1, 2, and 3), of the same diameter as the flange of the center, to which it is firmly fastened by four steel screws. Two lugs, *m* in Fig. 2, are threaded to receive the pivoting screws *n*, which are made of nickel-steel, and, with their points $2^{mm}.6$ below the center of the supporting cylinder, form a horizontal axis for the telescope. At a distance of $1^{cm}.2$ from the rear end and below is fastened, by two screws, the nut *o* (Fig. 1), made of nickel-iron, which carries the fine motion or micrometer screw *p*. This latter, of 39 threads per centimeter nearly (100 per inch) and $7^{mm}$ diameter, is provided at its upper end with a small tip of glass hard steel and carries, below, an adjustable micrometer head of white zylonite *q*, which is $4^{cm}.1$ in diameter and is divided into 100 parts. A hard rubber disk with milled edge, projecting beyond the micrometer head, not only protects the graduation from the fingers, but, by reason of its large diameter, facilitates the setting of the sensitive level. An index for reading the micrometer head is provided.

The supporting cylinder carries a movable eccentric ring *r* inserted into its forward end (Figs. 1 and 2), of which the inner diameter is such as to permit the telescope pivoted between the screws *n* to rotate slightly without touching. A similar ring *s* (Figs. 1 and 3) at the rear end, however, is cut out so as to permit the telescope to move up and down, above and below the normal or horizontal position, by about 2 mm., while the sides of the ring permit of no lateral play, but form a guide for that amount of vertical motion.

Directly in front of the micrometer screw is fastened to the supporting cylinder a small case holding an eccentric which can be rotated by a lever handle at the right side of the instrument. When the lever handle is turned up the eccentric pushes against the telescope, lifts its weight off the micrometer screw, and presses it gently against a spring sunk into the upper part of the ring *s*. In this position the instrument can be carried without the risk of jarring the telescope and thereby disturbing the level adjustment. This device is not shown in the diagrams but can be seen in the photographic view (illustration No. 1).

Against the hub *l*, on the right side of the instrument, is fastened a bracket carrying a small universal level, which is easily observed from the eye end of the telescope by means of a mirror mounted above it at an angle of 45° (see illustration No. 1).

At the forward end of the supporting cylinder and below is mounted a post *t* (Fig. 1), reaching downward between the horizontal pointing screw *i* and the spring case of the clamp arm *g*.

The upper part of the supporting cylinder has cast into it a rectangular opening with a framing *u* surrounding it. A piece of plate glass, fitted into this framing by dovetail grooves, closes the opening against dust or air currents, but can quickly be moved forward for the purpose of adjusting the level by loosening a small milled head screw (see illustration No. 1), and turning up a hinged locking piece. Over this opening and against the sides of the framing is mounted by brass arms *v* the glass mirror *w*, arranged so as to permit of a small rotary adjustment for the purpose of adapting the level reading device to individual observers. It may be stated here that the opening in the supporting cylinder was placed as near as possible to its rear end and away from the middle of the instrument, because the level could at that place be put closer to the line of collimation without entering the cone formed by the apertures of the objective and the reticle ring.

Small grooves around the ends of the supporting cylinder afford the means of fastening, by wire rings or narrow metal bands, the leather cones *x* (Fig. 1). They are fastened to the telescope in a similar manner, and effectively shut out dust and air currents without in the slightest degree preventing the telescope from assuming the position determined by the pivoting screws at one end and the micrometer screw at the other end of the supporting cylinder.

## The Telescope of the Precise Level.

The tube with the objective head and drawtube bearings, forming the telescope, is cast of nickel-iron in one piece and bored and turned in the lathe. Its outer diameter being $4^{cm}.37$ and the inner $4^{cm}.05$, giving a thickness of wall of $1^{mm}.6$. Immediately at the eye end and at a distance of $9^{cm}$ from it are two constrictions forming the bearings *y* for the drawtube. A ring *z* is fitted and soldered into the telescope at the place where the 60° points of the pivoting screws *n* are bored into it ($28^{cm}.9$ from the eye end) for the purpose of strengthening it to resist strains caused by undue tightening of these screws.

The drawtube, cast solid of nickel-iron and bored out, is fitted closely into its bearings, and carries within an enlargement at its outer end, by means of four nickel-steel screws, the nickel-iron reticle ring. Great care was taken to fit the threads of these screws very tightly to insure, as much as possible, invariability of the position of the reticle. One vertical and three horizontal spider threads of the finest grade obtainable are mounted upon the reticle ring. The horizontal threads are equidistant and the upper and lower embrace a space of 30 centimeters at a distance of 100 meters. Two Steinheil eyepieces, of $12^{mm}.5$ and $9^{mm}.5$ (one-half inch and three-eighths inch), equivalent focus, to suit different weather conditions, are supplied. The objective lens is mounted in a cell cast of nickel-iron. It is held in place by a spring ring, fastened with three small screws, in such manner as to hold it firmly in position, without restraining it from expanding and contracting with changes of temperature.

The drawtube is moved into focal distance in the usual way, by means of a rack and pinion, and has sufficient range to enable the observer to point on an object as near as 3.5 meters.

Just within or under the leather cones $x$ the telescope carries two enlargements or collars, which are turned to equal diameters, and serve the purpose of placing the pointing line into the geometric axis of the telescope. This adjustment is made in the shop permanently. It is done by laying the telescope, with these collars, upon two metal wye supports provided with leveling foot screws. Pointing on an object and rotating the telescope in the wyes reveals any want of parallelism between the axis of the two collars and the line connecting the intersection of the vertical and middle horizontal threads and the optical center of the objective. This is corrected by means of the four screws holding the reticle ring. Since the spider threads move with the drawtube, it also must move in a line parallel to the axis of the collars, in order to preserve true collimation in any position required by focusing upon the rod at different distances. To insure this parallelism, great care was taken in making the telescopes. The objectives were centered with special care, and the collars were turned true at the same chucking under which the drawtube bearings were bored. Inasmuch as any error of collimation enters into the result of leveling only to the small amount due to differences between backsights and foresights, it may be said that these instruments, as far as collimation error is concerned, are practically faultless.

In the same wye supports above mentioned, the level attached to the telescope is adjusted so that its axis is parallel to the vertical plane containing the line of collimation. This adjustment eliminates what is commonly called the "wind" of the level, and cannot readily be made in the field.

The position of the forward drawtube bearing and that of the micrometer screw were selected with the view to sufficient rigidity of that part of the telescope which rests upon the micrometer screw. The point of contact with the hardened tip of the screw is a small hardened steel plate $a_1$ fastened into the telescope at the forward drawtube bearing. The distance between the axis of the micrometer screw and the axis of rotation formed by the pivoting screws $n$ is $19^{cm}.15$ nearly, which, with the screw pitch of 39 threads per centimeter, gives a value of about $2''.6$ per division of micrometer head. The distance between the axis of rotation of the telescope and the vertical center is $9^{cm}.8$.

## The Level Vial.

The level vial was made by A. Pessler, of Freyburg, Saxony, and is of the chambered type. It is $11^{cm}.5$ long, $1^{cm}.5$ in diameter, and carries a graduation 8 centimeters long in 2 millimeter spaces. The length of the bubble used is about 25 divisions, or 5 centimeters. The mounting of the vial has been attended to with special care, with the aim of securing the greatest possible constancy of adjustment. The glass vial rests within a tube of nickel-iron upon the ends of four small screws, $b_1$, piercing the tube, two at each end of the vial, 120° apart. A small tip $c_1$, at the end of a flat spring fastened to the tube and also piercing it, presses with sufficient force upon the vial at each end, exactly over the supporting screws, to hold it firmly in place and yet permit it to expand and contract independently of the tube. Longitudinally the vial is confined by two cork rings $d_1$, one at each end, which, however, leave a small clearance so that the vial is free also in that direction. The level tube, with the vial thus supported, is secured to the telescope, sunk through an oblong opening close to the cone formed by the apertures of the objective and the reticle ring. At the forward end it is held by a screw holding it down to a rounded support $e_1$, screwed to the telescope, upon which it can be moved laterally by two opposing screws for adjusting the "wind." The other end is made adjustable in the vertical for the purpose of keeping the level parallel to the line of collimation. This is the only adjustment required on the part of the observer in the field. A square-headed vertical screw $f_1$, of about 27 threads per centimeter and fitting closely in the level tube end, is threaded tightly into that part of the telescope forming the forward drawtube bearing. Two strong helical steel springs, one on each side, press the level tube tightly upward against the shoulder of the screw $f_1$. A socket wrench, with a lever arm $7^{cm}.5$ long, permits of applying rotary force to the screw without exerting any other pressure against the instrument and thereby displacing the pointing of the telescope, as is the case when using a screw driver or capstan bar, so that this delicate adjustment is made quickly and with ease, and seldom requires to be repeated. As already stated, the adjustment of the reticle is made permanently in the shop, the observer having no means of testing it in the field. It is of great importance, therefore, that the reticle ring should not be disturbed, but that, when necessary, the level be moved into parallelism with the pointing line.* The observer, carrying the instrument from station to station, readily learns to hold it in such position as to prevent any change of the length of the bubble by establishing communication between the chamber and the interior of the vial. In the vials used for these instruments the openings in the chambers are not at the bottom, but slightly to the side.

## The Level Reading Device.

The operation of reading the position of three fine lines, the spider threads, projected upon the graduation of the level rod, is a trying one under the best conditions, and subjects the observer, when the air is hazy or unsteady, to severe strains. As it is of the highest importance that these readings be taken only at the instant when the level indicates horizontality of the line of sight, the instrument should be designed particularly with a view to the observer's comfort, so as to enable him to observe the rod and the level as nearly as possible simultaneously. It is thought that the level-reading device provided for these instruments fulfills all requirements, since only the time required for transferring mental attention from one object to another need elapse between the two observations. Two clamp rings $g_1$, Fig. 3, support an aluminum tube with an eye end reaching back to a point even with the eyepiece of the telescope when focused for an average distance. Against this tube is screwed a dovetail bar $h_1$, (illustrations Nos. 4 and 5) upon which move, within the tube, two slides $i_1$ and $j_1$, carrying the prisms $k_1$ and $l_1$. These slides are connected by arms with a lever mounted upon a stem with a milled head $m_1$, the rotation of which moves the prisms equally toward or away from a central point between them. This motion is provided to adjust the distance of the prisms accu-

* In the instruments as originally constructed the heads of the four screws holding the reticle ring were exposed in the ordinary manner, as shown in Illustration No. 1. Since then the design has been so improved that the heads of these screws are completely inclosed in a protecting case and are inaccessible to the observer in the field.

rately to the length of the bubble which, during the day's work, may vary by reason of temperature changes. Those faces of the prisms which are directed toward the eye are ground to such curvatures as, with the aid of a lens mounted between them and the eye end, to reduce to that of distinct vision of the normal eye the distance from the eye end of the bubble to the eye, by way of the mirror *w*, the reflected faces of the prisms $k_1$ and $l_1$ and the lens. For the benefit of the observer required to use glasses the eye cap of the level-reading tube may be arranged to hold such a lens as he may require to enable him to observe without spectacles.

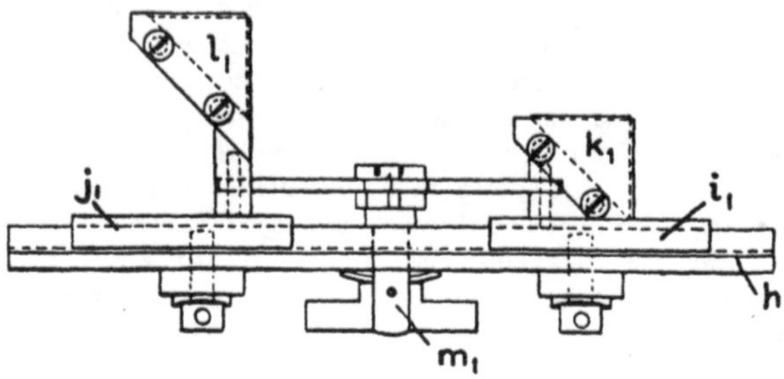


No. 4.—Prisms of Level Reading Device. Vertical Section.

The distance between the level-reading tube and the telescope can be changed to suit each individual observer, and provision is made for the rotary adjustment of the prisms and the mirror necessary in consequence of any such change, as can be seen in Fig. 3. The prisms are put in such position by means of the milled head $m_1$, that the ends of the bubble and the graduation marks above them are brought into view, appearing as if the bubble were very short. The lines forming the gradu-

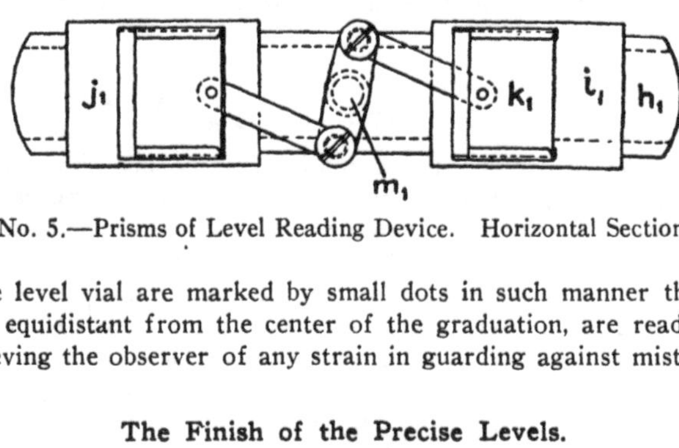


No. 5.—Prisms of Level Reading Device. Horizontal Section.

ation upon the level vial are marked by small dots in such manner that symmetrical lines, or lines equidistant from the center of the graduation, are readily distinguishable, thus relieving the observer of any strain in guarding against mistakes.

## The Finish of the Precise Levels.

The telescope throughout its length—with the exception of the eye end of the drawtube and the two collars turned to equal diameters—the supporting cylinder, and the level tube were covered with a heavy coating of cloth dust of a bluish-gray color. This coating is put on by painting the parts with a mixture of Japan varnish, turpentine, and linseed oil, which is colored with white lead, lampblack, and ultra-marine blue, to the same shade as that of the cloth dust; the latter is sifted over the freshly varnished pieces through a hair screen and pressed in with the hand. After allowing it to dry for two days and brushing off all loose cloth dust, a coating of a dilute solution of bleached shellac in alcohol is applied. The finish has the appearance of a fine quality of cloth, and affords considerable protection against sudden and temporary changes of temperature.

Other parts of the instrument, as the instrument base, mirror frame, level-reading tube, etc., are finished in black enamel of the kind introduced so extensively through the bicycle industry. It is heavy, hard though elastic, and surpasses in appearance and durability any of the black lacquers heretofore used in the art of instrument making.

## The Board of Estimate Precise Levels.

These instruments are manipulated with great ease and rapidity. They are irreversible and as simple as possible, differing from the dumpy level mainly in being provided with a level-reading device and with a quick leveling screw under the eye end of the telescope. The instruments have an unusually long center or vertical axis and the center of gravity is very low, both features being conducive to unusual stability. Both instruments stand very low on their tripods and the area exposed to wind pressure is reduced to a minimum, being less than for wye levels of corresponding optical power. This makes the instruments stable in heavy winds.

The weight of the instrument is $6^{kg}$ or $13^{lbs}.2$. This, no doubt, can be considerably reduced when tubing made of nickel-steel shall be obtainable in the market. The thickness of the cast tubing—about twice as great as would be necessary if wrought metal could be substituted—is considered as small as safety permits of in view of the loose texture of the alloy.

The weight of the tripod, $9^{kg}$ or $19^{lbs}.8$, is somewhat greater than that of stands formerly used for the same class of instruments. This is due to the much great length required in order to enable the observer to stand erect, which is considered of sufficient advantage to more than offset the slight addition in weight.

The magnifying power of these levels (30 diameters) is somewhat less than that of most of the Coast Survey levels of the same type. This power, however, was found to be entirely adequate for precise leveling in a built-up city, not only for the reason that in such a locality short sights are the rule, but also because under conditions where a great amount of vibration due to heavy traffic is encountered, instruments of unusually high power are not well adapted to accurate and economical work.

The chambered level vial is filled with pure sulphuric ether, which attains its equilibrium with greater rapidity and exactness than alcohol. This type of vial enables the observer to adjust the bubble to its most convenient length. In precise leveling operations, the chambered level vial is a necessity, and by a little manipulation an interchange of air and fluid may be accomplished and the length of the bubble easily regulated. When running the precise level lines, if the bubble happened to be too long, it was shortened by simply carrying the instrument part of the distance from one station to another with the eye end of the telescope pointed toward the ground. If the bubble were too short, it was lengthened by carrying the level with the objective pointed toward the ground. These manipulations were sometimes necessary several times a day on account of the sensitiveness of the ether to variable degrees of temperature. The leveling of the instrument is made approximately with the small universal level attached to the side of the telescope, after which it is perfected with the level proper.

The value of the level vials, about 2 seconds per division of 2 millimeters, is so much smaller than that usually selected for field instruments, that the behavior of the bubble is at first quite startling. When the instrument is leveled up and the bubble brought to the center of the vial with the micrometer screw, and the telescope then revolved 180° in azimuth, it is seen that the bubble does not remain in the center. When it is considered that the small value of the level vial is but little greater than that employed on the largest astronomical instruments which are mounted on massive

No. 6.—Precise Leveling Rod, Lower Portion, with Foot Plate and Foot Pin.

stone structures, it can readily be seen that it is not necessar to adjust the vertical axis so accurately that the bubble wi remain in the center in all positions of the telescope.

It is one of the principal features of the instrument tha the advantages of so sensitive a vial can be obtained withou loss of time for the making of accurate adjustments. Th length of the bubble as used on the Board of Estimate level ing was about 25 divisions, this being found to be a very con venient and practical length. A short bubble has a ver sluggish movement even in the most sensitive vials and mus be avoided. A great deal has been said for and against th sensitive bubble; the fact is, however, that it is useless t construct an accurate and very sensitive instrument unless i is to be equipped with a correspondingly sensitive bubble tha will reveal the fine qualities of the instrument.

The results obtained on the Board of Estimate levelin have convinced the writer that these instruments are not onl superior to any former precise levels, but that they can com pete successfully with the wye level type on leveling of an degree of accuracy. A similar conclusion has been reache by others who have made use of this type of instrument. Th instrument is well adapted for general engineering work.

## The Leveling Rods.

The rods used on the Board of Estimate leveling ar shown in illustration No. 6. They are direct-reading rods o the Coast Survey type, and measure 3.27 meters over al The cross section of each rod is a cross (+) of symmetrica proportions. The rods are made of well seasoned white pin treated with paraffin for the purpose of making their length independent of the hygrometric state of the atmosphere. Th paraffining process is applied as follows:

The leveling rods are made up nearly to dimensions an treated in a vat of boiling paraffin. A complete saturation i never quite accomplished even if the rods are boiled in th paraffin an entire day, and it is found that a penetration o ¼-inch, or thereabouts, accomplishes every desired purpose After treatment the rods are carefully dressed down to siz and any warping incidental to the paraffin treatment trued up The amount of paraffin ordinarily absorbed will add abou 5% or 6% to the weight. Rods thus treated are found to b sensibly as constant in length under varying hygrometri conditions as rods containing 20% or more of their weight i paraffin. The graduation, which is read directly from th telescope, is in black and white squares, 1 centimeter on side. Silver plugs, 5 millimeters in diameter, are inserted a the 0.1, 1.1, 2.1, and 3.1 meter points. A fine graduation o these plugs is used to determine the exact length of the rod and to study their changes of length while in the field. Fre quent measurements with the same steel tape during the sea son's work enable the observer to keep informed as to th nature and time of occurrence of any changes taking place i the field. The zero of the graduation is at the foot of th rod, which is brought forward so as to be in the plane of graduation on the rod instead of being in the axis of the rod prolonged.

The shoe of the rod is made of the hardest bell metal, with a rounded or hemispherical shaped bottom having a radius of $2^{cm}.6$ or about 1 inch. The meters, decimeters and centimeters are read directly and the millimeters are estimated. The rod is equipped with a handle, universal level, and Centigrade thermometer, and weighs $4^{kg}.7$ or $10^{lbs}.4$. The thermometer used to determine the temperature is sunk deep into the rod itself at the back and is covered by a wooden lid. It has seemed to the writer, that if the meter graduations were placed upon the rod horizontally instead of vertically, as at present, there would be less strain upon the observer in preventing the occurrence of erroneous rod readings. These graduations are shown at each decimeter (see illustration No. 6), beginning with the first meter above the foot of the rod. The reason for the suggestion above noted is that if the rod does not face the observer squarely for any one of several reasons, he is unable to tell if any of the upright meter graduations are cut off by the centimeter face of the rod. If the meter graduation were placed upon the rod horizontally, the observer would at least see the ends of all the graduations even though the rod were turned considerably away from him. The Board of Estimate rods were marked A, B, and C, respectively by a brass plug inserted in the back of the rod. A metric scale 1 meter in length was used in the case of bench marks upon which the rod could not be held exactly vertical. On the Board of Estimate leveling two rods were always used, but only a single line was run at one time.

## The Rod Lengths.

While in the field the rods were measured by means of a specially made steel tape 3 meters long, graduated to millimeters at both ends, used direct and reversed, and read with the aid of a pocket reading glass. This tape was used with a tension of 5 kilograms. The reading of the rod thermometer was always noted in connection with these measurements. Care was taken that the rods and tape should be at the temperature of the atmosphere so that reliable comparisons could be made. These measurements were made in order to detect, if possible, any change in the length of the rods during the progress of the leveling.

The results of these measures are given in the table on page 24. The long intervals between certain measures as, for instance, between January 5th and April 20th, and between July 25th and October 3rd, in 1910, indicate times when no leveling was in progress on account of snowstorms, vacations, time spent for reconnaissance, setting of bench marks, etc. The part of the rod measured was between the fine lines on the silver plugs at the first and thirty-first decimeter graduations. These measurements are made comparable by reducing the lengths as observed to the mean temperature of the measures, which is 20° C. In the reduction, both the coefficient of expansion of the tape and that of the rod are taken into account.

All three rods seem to have maintained their length with a fair degree of uniformity up to July 24, 1911. On September 19, 1911, rods A and C showed apparently a very decided change in length which remained practically constant until the close of the field work. The measures of April 16, 1912, show rod B to have been similarly affected. The total series is therefore divided into two distinct groups and for computation purposes, namely, correction for length of rod, a corresponding division was made. The lengths of the 3-meter intervals marked on each rod by the fine graduation on the silver plugs in its face, as determined by the National

FIELD MEASUREMENTS OF RODS—REDUCED TO 20° C.

| Date | | Rod A | Rod B | Rod C |
|---|---|---|---|---|
| 1909 | | m. | m. | |
| Nov. 15 .......... | | 3.0013 | 3.0013 | |
| Dec. 6 .......... | | 3.0012 | 3.0012 | |
| 1910 | | | | m. |
| Jan. 5 .......... | | 3.0014 | 3.0014 | 3.0015 |
| April 20 .......... | | 3.0012 | 3.0012 | |
| April 27 .......... | | 3.0013 | 3.0013 | |
| Mar. 4 .......... | | 3.0013 | 3.0013 | |
| May 10 .......... | | 3.0014 | 3.0014 | |
| May 20 .......... | | 3.0015 | 3.0013 | |
| June 1 .......... | | 3.0013 | 3.0013 | |
| June 14 .......... | | 3.0017 | 3.0016 | |
| July 13 .......... | First group | 3.0016 | 3.0016 | |
| July 25 .......... | | 3.0016 | 3.0015 | |
| Oct. 3 .......... | | 3.0016 | 3.0015 | |
| Oct. 14 .......... | | 3.0016 | 3.0016 | |
| Nov. 15 .......... | | 3.0012 | 3.0012 | |
| Nov. 28 .......... | | 3.0012 | 3.0012 | |
| 1911 | | | | |
| April 13 .......... | | 3.0013 | 3.0013 | |
| May 9 .......... | | 3.0012 | 3.0012 | |
| June 7 .......... | | 3.0014 | 3.0014 | 3.0013 |
| June 28 .......... | | 3.0014 | | 3.0014 |
| July 24 .......... | | 3.0014 | | 3.0014 |
| | Mean | 3.0014 | 3.0014 | 3.0014 |
| Sept. 19 .......... | | 3.0018 | | 3.0018 |
| Oct. 20 .......... | | 3.0018 | | 3.0018 |
| Nov. 9 .......... | | 3.0018 | | 3.0017 |
| Nov. 28 .......... | Second group | 3.0020 | | 3.0020 |
| 1912 | | | | |
| Mar. 18 .......... | | 3.0020 | | 3.0020 |
| April 16 .......... | | 3.0018 | 3.0018 | 3.0020 |
| | Mean | 3.0019 | 3.0018 | 3.0019 |

Bureau of Standards, before and after the field work and reduced to 0° Centigrade, are as follows:

| | Rod A | Rod B | Rod C |
|---|---|---|---|
| | m | m. | m. |
| June 19, 1909...................... | 2.9999 | 2.9999 | 3.0001 |
| June 8, 1912...................... | 3.0003 | 3.0002 | 3.0005 |

The coefficient of expansion used in the reduction to 0° C. was 0.000004 per Centigrade degree.

The absolute lengths at 0° Centigrade, as derived from the standardization of June 8, 1912, are a striking confirmation of the lengthening of the rods as shown by the tape measurements in the field.

As this lengthening appears to have occurred very suddenly, as indicated by the decided break in the field measurements shown in the two groups of the table on page 24, the rod lengths resulting from the standardization of June 19, 1909, were used in the computations up to July 24, 1911. The three values of the excess or deficiency of length per meter at 0° C. expressed as corrections for length of rod, are, respectively, —0.033, —0.033, and +0.033 millimeter per meter of difference of elevation between bench marks. From September 19, 1909, until the close of the work the rod lengths resulting from the standardization of June 8, 1912, were used in the computations. The three values of the excess of length per meter at 0° C. thus used in the computations were, respectively, +0.100, +0.067, and +0.167 millimeter per meter of difference of elevation between bench marks. The mean lengths of the rods for the two periods, together with the mean excess or deficiency of length per meter, are as follows:

JUNE, 1909, TO JULY, 1911.

| Rods | Mean Length at 0°C. | Length Corrections in Millimeters Per Meter. |
|---|---|---|
| | m. | mm. |
| A and B | 2.9999 | —0.033 |
| A and C | 3.0000 | 0.000 |
| B and C | 3.0000 | 0.000 |

SEPTEMBER, 1911, TO APRIL, 1912.

| Rods | Mean Length at 0°C. | Length Corrections in Millimeters per Meter. |
|---|---|---|
| | m. | mm. |
| A and B | 3.00025 | +0.083 |
| A and C | 3.00040 | +0.133 |
| B and C | 3.00035 | +0.117 |

The increase in the length of the rods is in accordance with previous experience with paraffined rods. The experience of the U. S. Coast and Geodetic Survey shows that the general tendency for paraffined rods is to increase slightly in length during the first season and to preserve a nearly constant length thereafter.

The field measurements of the Board of Estimate rods show them to have remained constant in length for a period of about two years before any change occurred.

The index correction for each of the rods at the beginning of the leveling was zero.

## Foot Plates and Foot Pins.

The foot plates and foot pins, or special metallic turning points as they are sometimes called, are shown in illustrations Nos. 6 and 7. The foot plate is a cast iron disk 6 inches in diameter, with a spherical cavity in the top to receive the foot of the rod. This cavity is turned to a radius of $3^{cm}.2$ (about 1¼ inches), or ¼-inch greater than the radius of the foot of the rod.

The foot plates were found to make the best turning points on dirt or macadam roads, or when crossing long stretches of sand, gravel, or farm lands. The method of using the foot plate is simply to set it down and press it firmly into the soil; when the rod is set on the plate, it is kept there until the rodman leaves his station; that is, he is not allowed to raise his rod, and then replace it.

The foot pin is made of hard steel, having a broad head with a spherical cavity for receiving the foot of the rod, the same as the foot plate. The pins are driven with a mallet and are very useful when running over grassy soils, through wooded country, and on very steep grades.

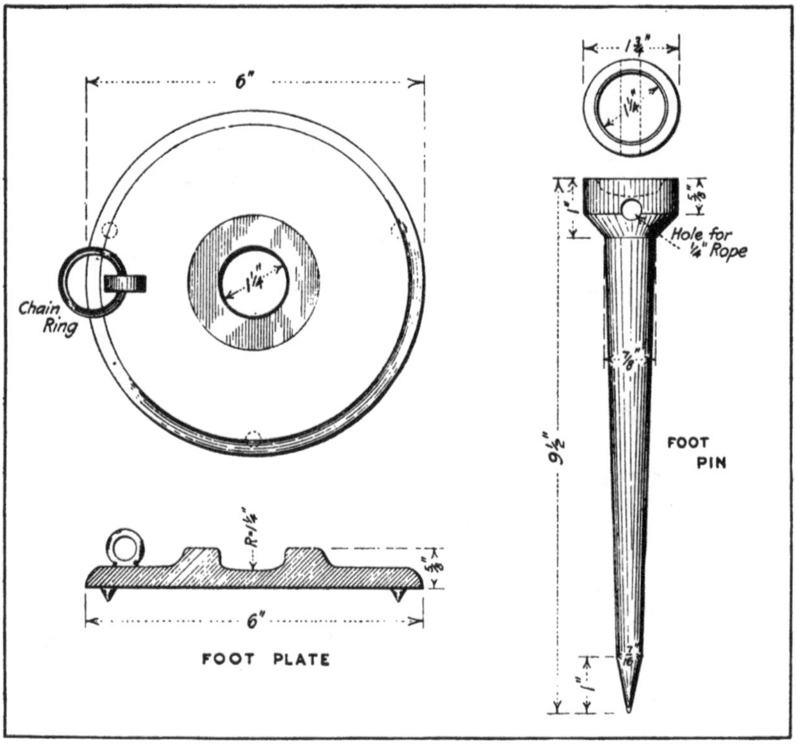


No. 7.—Foot Plate and Foot Pin.

The method used in taking and recording the observations on the Board of Estimate leveling is the one so successfuly applied by the U. S. Coast and Geodetic Survey since 1900. This method is defined by the instructions given to the precise leveling parties of the Coast Survey*; these instructions were, however, slightly modified to adapt them to the particular requirements of city work.

## General Instructions for Precise Leveling.

1. Except when specific instructions are given to proceed otherwise, all lines are to be leveled independently in both the forward and the backward direction.

* These instructions were prepared by Mr. J. F. Hayford, Expert Computer and Geodesist. Some minor changes have been made since 1900, making the instructions more complete.

2. The line of levels is to be broken up by bench marks into sections from 1 to 2 kilometers long, except where special conditions make shorter sections advisable.

3. All old bench marks are to be called by their old names and are to be fully described by quoting the old description, if one is available, and by making additions to it.

4. It is desirable that the backward measurement on each section should be made under different atmospheric conditions from those which occurred on the forward measurement. It is especially desirable to make the backward measurement in the afternoon if the forward measurement was made in the forenoon, and vice versa. The observer is to secure as much difference of conditions between the forward and backward measurements as is possible without materially delaying the work for that purpose.

5. On all sections upon which the forward and backward measures differ by more than $4^{mm}.0\sqrt{K}$ (in which K is the distance leveled between adjacent bench marks in kilometers), both the forward and backward measures are to be repeated until two such measures fall within the limit.

6. Whenever a blunder, such as a misreading of 1 decimeter or of 1 meter, or an interchange of sights (the backsight being recorded as a foresight) is discovered in any measure after its completion, and the necessary correction applied, such measure may be retained, provided there are at least two other measures over the same section which are not subject to any such uncertainty.

7. The programme of observation at each station is to be as follows: Set up and level the instrument. Read the three lines of the diaphragm as seen projected against the front (or rear) rod, each reading being taken to the nearest millimeter (estimated), the bubble being held continuously in the middle of the tube (i. e., both ends reading the same). As soon as possible thereafter read the three lines of the diaphragm, as seen projected against the rear (or front) rod, estimating to millimeters as before, and holding the bubble continuously in the middle of the tube.

8. At each rod station the rod thermometer is to be read to the nearest Centigrade degree and the temperature recorded.

9. At stations of odd numbers the backsight is to be taken before the foresight, and at even stations the foresight is to be taken before the backsight.

10. The maximum difference in length between a foresight and the corresponding backsight is to be 10 meters. The actual difference is to be made as small on each pair of sights as is feasible by the use of good judgment, without any expenditure of time for this particular purpose.

11. The recorder shall keep a record of the rod intervals subtended by the extreme lines of the diaphragm on each backsight, together with their continuous sum between bench marks. A similar record shall be kept for the foresights. The two continuous sums shall be kept as nearly equal as is feasible, without the expenditure of extra time for that purpose, by setting the instrument beyond (or short of) the middle point between the back and front rods. The two continuous sums shall not be allowed to differ by more than a quantity corresponding to a distance of 20 meters.

12. Once during each day of observation the error of the level should be determined in the regular course of the leveling and recorded in a separate opening of the record book, as follows:

The ordinary observations at an instrument station being completed, transcribe the last foresight reading as part of the error determination, call up the back rod and have it placed about 10 meters back of the instrument, read the rod, move the instrument to a position about 10 meters behind the front rod, read the front rod, and then the back rod. The rod readings must be taken with the bubble in the middle of the tube. The required constant C to be determined, namely, the ratio of the required correction to any rod reading to the corresponding subtended interval, is

$$C = \frac{(\text{sum of near rod readings}) - (\text{sum of distant rod readings})}{(\text{sum of distant rod intervals}) - (\text{sum of near rod intervals})}.$$

The total correction for curvature and refraction must be applied to the sum of the distant rod readings before using it in this formula. The level should not be adjusted if C is less than 0.005. If C is between 0.005 and 0.010 the observer is advised not to adjust the level, but if C exceeds 0.010 the adjustment must be made. If a new adjustment of the level is made, C should at once be redetermined. It is desirable to have the determination of level error made under the ordinary conditions as to length of sight, character of ground, elevation of line of sight above ground, etc. The level must be adjusted by moving the level vial, not by moving the reticle.

13. Notes for future use in studying leveling errors shall be inserted in the record, indicating the time of beginning and ending of the work for each section, the weather conditions especially as to cloudiness and wind, and whether each section of the line is run toward or away from the sun; and such other notes as may be of value in studying errors.

14. The instrument shall be shaded from the direct rays of the sun, both during the observations and the movement from station to station.

15. The maximum length of sight shall be 150 meters, and the maximum is to be attained only under the most favorable circumstances.

16. At the beginning and end of the season and at least twice each month during the progress of the leveling, the three-meter interval between metallic plugs on the face of each level rod shall be measured carefully with a steel tape, which shall be continuously kept in the party throughout the season for that purpose. The rod temperature at the time of each of these measures must be recorded. The purpose of these measures is to detect changes in the length of the rods rather than to determine the absolute lengths. The absolute lengths are determined at the office between field seasons. At least once each month, during the progress of the leveling, the adjustment of the rod levels shall be tested and a statement inserted in the record showing the manner in which the test was made, whether the error was found to be without the limit stated below, and whether an adjustment was made. The test must determine the inclination of the rod to the vertical, measured in the plane of sight from the instrument to the rod, as well as at right angles to that plane, when the bubble of the rod is held at the center. If the deviation from the vertical exceeds 10 millimeters on a three-meter length of the rod, the rod level must be adjusted.

## EXAMPLES OF RECORD AND COMPUTATION.

### The Determination of C.

The Coast Survey type of precise level being very simple in construction requires but one adjustment in the field and that adjustment should be examined daily. This is accomplished by a modification of the well known peg method and is called the "determination of C," or the error of the level.

The numerical value obtained for C gives a control of the non-parallelism of the axis of the bubble and the line of sight and enables proper corrections to be applied, when necessary, to results obtained when the line of sight is inclined to the vertical.

The determination of C as actually made in the field is shown in illustrations Nos. 8 and 9 and in the table on page 29. The unit of length in the table is the millimeter.

The instrument is set up at H, as shown in illustration No. 8, about 9 meters from rod A and about 66.7 meters from rod C. The distances are not measured, but should be approximately as indicated. A reading is taken on rod A and is placed in the backsight column, the mean reading in this instance being 1821.0. A reading is next taken on rod C and is placed in the foresight column, the mean being 0895.0. All three wires are read and recorded each time. The instrument is then moved to station H′, illustration No. 9, the rods remaining as before and the horizontal distances being about in the reverse order. A reading is taken on rod C and placed in

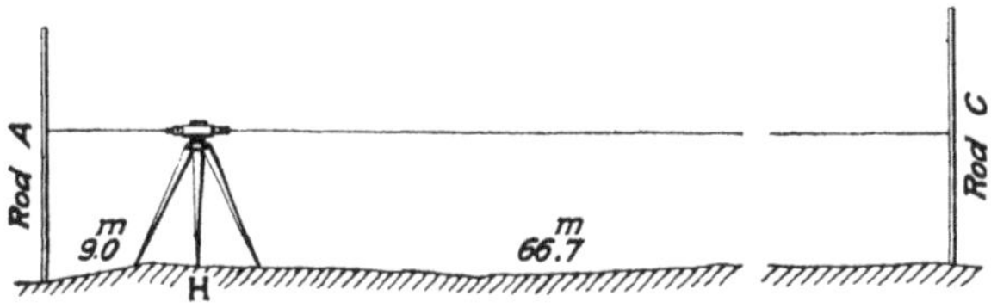


No. 8.—Method of Testing Precise Level; first position.

the backsight column, the mean being 1462.0. The two readings in the backsight column are called the "near rod readings." A reading is then taken on rod A and placed in the foresight column, the mean being 2389.0. The two readings in the foresight column are called the "distant rod readings."

The sum of the distant rod readings is then taken and a correction of —0.6

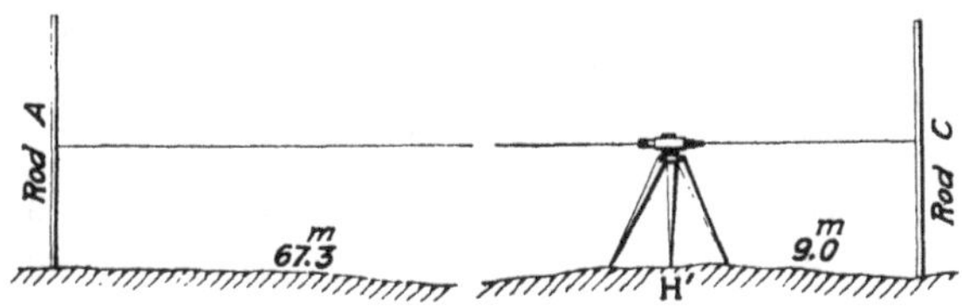


No. 9.—Method of Testing Precise Level; second position.

millimeter is applied for curvature and refraction, the two corrections for the two distant rods being combined. This corrected sum of the distant rod readings is then subtracted from the sum of the near rod readings, giving in this case a residual of —0.4. This residual is divided by the sum of the distant rod intervals minus the sum of the near rod intervals, which in this case is 346, and a quotient of —0.001 is obtained, which is the value of C. The rod readings must be taken with the bubble in the middle of its tube.

Determination of C, 9.00 A.M., Sept. 28, 1911.

**Left-hand page.** | **Right-hand page.**

| No. of Station. | Thread Reading Backsight. | Mean. | Thread Interval. | Rod and Temp. | Thread Reading Foresight. | Mean. | Thread Interval. |
|---|---|---|---|---|---|---|---|
| H | 1807 | | 14 | | 0795 | | 100 |
| | 1821 | 1821.0 | 14 | C | 0895 | 0895.0 | 100 |
| | 1835 | | 28 | | 0995 | | 200 |
| H′ | 1448 | | 14 | | 2288 | | 101 |
| | 1462 | 1462.0 | 14 | A | 2389 | 2389.0 | 101 |
| | 1476 | | 28 | | 2490 | | 202 |
| | | 3283.0 | 56 | | | 3284.0 | 402 |
| | | 3283.4 | | Çor. for Curv. & Ref. | | —0.6 | —56 |
| | 346) | −0.4 (C = −0.001 | | | | 3283.4 | 346 |

When C is less than 0.005 no adjustment of the instrument is necessary, and in fact it is not of much importance until C becomes equal to 0.01 or more. When the instrument must be adjusted, due to too large a value for C, it is done by raising or lowering one end of the level vial and *not by moving the reticle.*

## Adjustment of the Level.

The adjustment is made as follows: Point to the distant rod with the bubble in the middle of its tube and read the three horizontal wires. Move the telescope so as to raise the middle wire by an amount equal to C times the rod interval. While holding the telescope in this position bring the bubble to the middle of the tube by raising or lowering one end of the level vial. If C is negative the middle hair must be lowered on the rod and if plus it must be raised. After adjusting the instrument, a new determination of C must be made. The cross-hairs must never be disturbed as these have been permanently adjusted for collimation by the instrument maker.

A very high grade of workmanship was used in the construction of these instruments as is testified to by the constancy with which they retained their adjustment. Level No. 1, being used the most, was tested every day, but required adjusting only about once in three weeks. Level No. 2 was frequently used for months at a time and required adjusting about once in three months, and at no time was the angle between the line of collimation and the tangent to the level vial at its middle point greater than 4″.

As the elimination of errors due to the unequal expansion and contraction of various parts of the instrument is the most serious problem to be dealt with in precise leveling, it is of the utmost importance that the instrument be shaded from the direct rays of the sun, both during the observations and in moving from station to station. It is absolutely necessary to do this in order to keep the adjustments approximately constant. Even in ordinary leveling, the results are more accurate when the instrument is shaded, and on precise leveling every effort must be made to minimize heat effects. The instrument should also be protected from the wind whenever it is strong enough to disturb the level.

## Field Notes.

The form of record for the field notes is shown on page 31. The unit in the record is the millimeter. The first column contains the numbers of the instrument stations, the second the reading of each horizontal thread for the backsight, and the third the mean of the three readings. The mean of the three readings is treated in the computation in the same manner as the single target reading in wye leveling. The fourth column contains the thread intervals, and the fifth the sum of the intervals. The first column on the right-hand page contains the identifying letter of the rod and the temperature in degrees Centigrade for each foresight, the second the reading of each horizontal thread for the foresight, the third the mean, the fourth the thread interval, and the fifth the sum of the intervals.

The columns headed "Sum of Intervals" show the continuous sum of the total intervals, and as these values are proportional to the sums of the backsight distances and foresight distances, respectively, they enable the observer to keep these two sums nearly equal at all times, as required by the instructions, for the purpose of eliminating instrumental errors.

The temperature is read by the rear rodman just before he moves forward, and is called out to the recorder when the rodman passes. The explanation of the symbols used after the words "Sun" and "Wind" is printed on the bottom of the computation form shown on page 35.

**Left-hand page.** **Right-hand page.**

BOARD OF ESTIMATE AND APPORTIONMENT
CITY OF NEW YORK.

SPIRIT LEVELING.

From B.M.: 3. To B.M.: 4.

Date: *Dec. 10, 1911.* Forward—~~Backward~~ Wind: *C.* *10 A.M.*
Sun: *C.*

| Number of Station. | Thread Reading Backsight. | Mean. | Thread Interval. | Sum of Intervals. | Rod and Temp. | Thread Reading Foresight. | Mean. | Thread Interval. | Sum of Intervals. |
|---|---|---|---|---|---|---|---|---|---|
| 1<br>B.M. 3 at<br>141 B'way. | 1124<br>1190<br>1257 | 1190.3 | 66<br>67<br>133 | 133 | A<br>9 | 1984<br>2047<br>2110 | 2047.0 | 63<br>63<br>126 | 126 |
| 2 | 1386<br>1459<br>1532 | 1459.0 | 73<br>73<br>146 | 279 | C<br>10 | 1329<br>1405<br>1481 | 1405.0 | 76<br>76<br>152 | 278 |
| 3 | 1865<br>1930<br>1994 | 1929.7 | 65<br>64<br>129 | 408 | A<br>9 | 1626<br>1690<br>1755 | 1690.3 | 64<br>65<br>129 | 407 |
| 4 | 1902<br>1955<br>2009 | 1955.3 | 53<br>54<br>107 | 515 | C<br>9 | 1510<br>1562<br>1614 | 1562.0 | 52<br>52<br>104 | 513 |
| 5 | 1730<br>1790<br>1850 | 1790.0 | 60<br>60<br>120 | 635 | A<br>9 | 1240<br>1298<br>1356 | 1298.0 | 58<br>58<br>116 | 629 |
| 6 | 1833<br>1861<br>1890 | 1861.3 | 28<br>29<br>57 | 692 | C<br>9 | 1646<br>1671<br>1697 | 1671.3 | 25<br>26<br>51 | 680 |
| 7 | 1616<br>1633<br>1651 | 1633.3 | 17<br>18<br>35 | 727 | A<br>9 | 1445<br>1462<br>1478 | 1461.7 | 17<br>16<br>33 | B.M. 4 at N.W<br>cor. of G.P.O |
| | | 11818.9<br>+683.6 | | | | | 11135.3 | | 713 |

The instrument stations are numbered consecutively from bench mark to bench mark, this particular run being from B. M. No. 3 at 141 Broadway to B. M. No. 4 at the General Post Office. Such portions of the computation as are shown as forming a part of the record are kept up by the recorder as the work progresses. The principal checks the recorder has upon his notes are those given by the upper and lower thread intervals, which should be nearly equal, and those given by the total interval and sum of intervals.

It is desirable that as little time as possible should elapse between the reading of both backsight and foresight; the more rapidly both readings can be taken, consistent with careful manipulation, the more accurate the results will be.

It will be noticed that no column has been provided for the height of instrument. This is not necessary in precise leveling, as the work in the field is purely differential leveling and certain small corrections will have to be applied to the field results and adjustments made before the elevations can be considered.

## CORRECTIONS IN PRECISE SPIRIT-LEVELING.

Precise leveling differs from ordinary leveling not only in the type of instrument used and in the method of observation, but also in the application of certain corrections to the field results. These corrections are necessary on account of the high degree of accuracy required, but are not due to any peculiarity in the method of observing. They are corrections for errors which exist in the wye leveling as well as in the precise leveling. In no branch of surveying is a knowledge of the sources of errors and of the methods of eliminating the same more important than in precise spirit-leveling. However, with the present type of instrument and the present method of taking the observations, precise spirit-leveling has become a very simple matter and the corrections applied are few, but very important. The largest and most important corrections applied to the field results are the rod length and rod temperature corrections.

### Rod Length Correction.

The length correction is the amount the rod differs from the length which its graduations indicate, and its application corrects the results of the leveling for the erroneous length of the rod. In applying the length correction, the mean length of the two rods at 0° C. is determined and the excess or deficiency per meter of length is applied to each of the various differences of elevation determined by the leveling.

The correction is equal to the difference of elevation in meters times the excess or deficiency per meter. It is plus when the mean rod is too long at 0° C. and minus when it is too short. Any other temperature may be used as a standard. The following illustration will make the reason for this correction more clear, the principles being applicable also to ordinary leveling.

In Fig. 1, Illustration No. 10, the line of sight AB is exactly 3 meters above the foot of the rod D, and the absolute length of the rod at 0° C. is 3.001 meters. The rod being longer than its graduations indicate, the 3-meter mark is above the line of sight and the actual reading is less than it should be. The difference of elevation is therefore in error by the amount that the rod is long, which in this case is 1 millimeter, or 0.333 millimeter per meter of difference of elevation; this is the length correction, and it will have the same sign as the measured difference of elevation. In other words when the length correction is plus, its application will *increase* the measured difference of elevation.

In Fig. 2, Illustration No. 10, the rod E is 1 millimeter short at 0° C., and the reading will therefore be too great by that amount. In this case, the length correction will have the opposite sign to the measured difference of elevation, and its application will *decrease* the measured difference of elevation.

If these particular rods were used together their mean length at 0° C. would be taken, which in this case is exactly 3 meters. The length correction would then be zero. Usually, however, leveling rods are longer than their graduations indicate, and the length corrections are very often as much as + 0.5 millimeter per meter of difference of elevation.

### The Temperature Correction.

In graduating the Board of Estimate leveling rods the unit of length chosen is the meter. As the material of which the rods are made has a coefficient of expansion of 0.000004 per degree Centigrade, the rods will have their true length as indicated by the graduations at only one temperature. For any other temperature than that at which the rod is standard a correction must be applied to reduce observed values to true values, the observed values being in terms of the unit of graduation of the rod, which is not an exact unit of length except at one temperature. The temperature correction therefore depends upon the coefficient of linear expansion of the rods, the temperature at the time of observation and the difference of elevation.

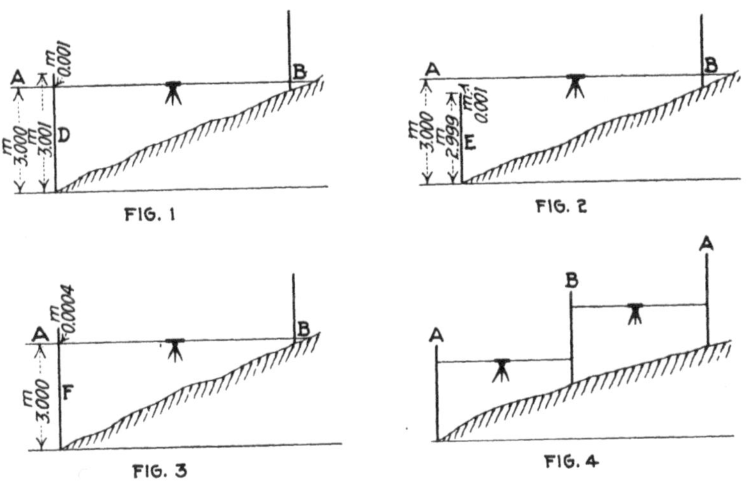


No. 10.—Rod Corrections.

Fig. 3, Illustration No. 10, illustrates this principle. Let us suppose that the rod F has an absolute length of exactly 3 meters at 0° C., and that the line of sight is 3 meters above the foot of the rod as in Figs. 1 and 2. At 32° C. the rod is 0.4 millimeter longer, and the rod reading at that temperature is therefore 0.4 millimeter too short. The temperature correction is then 0.4 millimeter, and the sign of the correction will be the same as the sign of the measured difference of elevation, unless the temperature is below the Centigrade zero.

The above reasoning is equally true if feet and the Fahrenheit thermometer are used in place of the meter and the Centigrade thermometer.

Where the leveling is done over a section which is practically level the total correction for temperature of the rod is insignificant or even zero. Where the section changes much in elevation, however, the effect is a vertical base line measured in very short sections, and the rod lengths and coefficients of expansion must be known just as the lengths and coefficients of the tape lines on horizontal base measurements are known.

The temperature correction is computed from the following formula:

$$\text{Correction in millimeters} = \Delta E \times T \times 0.004$$

where $\Delta$ E is the difference of elevation in meters and T is the temperature of the rod at the time of observation in degrees Centigrade.

In this work the temperature correction for a single section was sometimes as much as 5.8 millimeters, or 0.019 foot.

### The Index Correction.

In the index correction, account is taken of the fact that the zero of graduation and the foot of the rod are not exactly coincident. At times the index error is so small as to be practically zero, and is not applied except in the case where a reading is taken on a bench mark without the use of the rod. When the rods are in the field for a year or more the index error becomes appreciable on account of the wear on the bottom, and the correction must be entered for each section on the computation form.

The effect of the correction varies with the different ways in which the rods are used. For example, if two rods A and B, having index errors of $a$ and $b$ respectively ($a$ being greater than $b$) were used and the level section started from A as a backsight and ended on B as a foresight, the correction for this section would be $(a-b)$.

As $a$ and $b$ are always negative they would have the effect, in this case, of decreasing the difference of elevation if the readings on rod A were the greater and increasing it if the readings on rod B were the greater. If the level section started from B as a backsight and ended on A as a foresight, the correction would have the opposite effect. This can be seen from a study of Fig. 4 on page 33.

If the index correction for both rods is the same there is no correction for the section. Also, if the number of instrument stations be even and all readings are taken on the rods, there is no correction for the section. If the number of instrument stations be odd, only the last station need be considered and the amount of the index correction is the difference of the index errors of the two rods. If only one rod be used as in ordinary leveling, there is no index correction.

The average wear per year on the bottom of the leveling rods was 0.3 millimeter. This wear was detected by measuring the first decimeter of each rod by means of a finely graduated caliper square. As the first decimeter graduation is marked by a fine line on the silver plug inserted in the face of the rod, this measurement could be made very accurately and was repeated every two weeks.

## COMPUTATION OF PRECISE LEVELS.

The form on which the computations are made is shown on page 35.

The computation is discontinuous, showing results from different sections to illustrate the method of applying the corrections. The forward line from B. M. 3 to B. M. 4 on this form is that for which the record is given on page 31.

The fourth column on the left-hand page contains the sum of the rod intervals or stadia distances between bench marks. The fifth column contains the distances between the bench marks in kilometers and is derived from the fourth column by knowing that an interval of 300 millimeters subtended on the rod corresponds to 100 meters along the line, regardless of the lengths of the separate sights. The sixth column contains the difference between the sum of the total backsight intervals and

**Left-hand page.**

BOARD OF ESTIMATE & APPORTIONMENT
CITY OF NEW YORK.

## COMPUTATION OF PRECISE LEVELS.

LINE: *So. Ferry to U. S. Ship Canal.* Borough: *Manhattan*

| B. M'S. | Forward or Backward | No. of Stations | Sum of Rod Intervals | Distance in Kilometers | Rod Intervals Σ B–Σ F | Mean Rod Readings. Σ B | Σ F | Approximate Diff. of Elev. | Mean Temp. of Rods |
|---|---|---|---|---|---|---|---|---|---|
| | | | *mm.* | | *mm.* | *m.* | *m* | *m.* | °*C* |
| 1—2 | F | 12 | 2518 | 0.856 | — 6 | 18.4590 | 15.4065 | + 3.0528 | 17 |
| | B | 13 | 2618 | | —14 | 17.2293 | 20.2824 | — 3.0531 | 15 |
| 2—3 | F | 14 | 2098 | 0.714 | — 5 | 23.8577 | 17.7506 | + 6.1071 | 11 |
| | B | 14 | 2186 | | + 4 | 17.7193 | 23.8261 | — 6.1068 | 13 |
| 3—4 | F | 7 | 1440 | 0.478 | +14 | 11.8189 | 11.1353 | + 0.6836 | 9 |
| | B | 7 | 1427 | | + 7 | 11.1766 | 11.8589 | — 0.6823 | 14 |
| 41—42 | F | 16 | 2537 | 0.847 | +15 | 7.9620 | 42.5465 | —34.5845 | 23 |
| | B | 14 | 2544 | | + 4 | 42.2826 | 7.6979 | +34.5847 | 21 |
| 42—42*c* | F | 25 | 5340 | 1.780 | +10 | 37.2040 | 47.5256 | —10.3216 | 24 |
| | B | 27 | 5343 | | — 1 | 47.7688 | 37.4470 | +10.3218 | 23 |
| G—W | F | 57 | 9649 | 3.179 | +31 | 63.9932 | 121.8646 | —57.8714 | 21 |
| | B | 55 | 9427 | | +15 | 117.0606 | 59.1900 | +57.8706 | 25 |
| 110—111 | F | 11 | 2409 | 0.836 | — 1 | 16.6343 | 17.3765 | — 0.7422 | 29 |
| | B | 7 | 2401 | | +16 | 10.8159 | 10.0746 | + 0.7413 | 32 |
| 111—111*a* | F | 1 | 78 | 0.030 | +10 | B 1.7207 | RULE 0.3205 | + 1.4002 | 31 |
| | B | 1 | 103 | | —13 | RULE 0.1540 | B 1.5537 | — 1.3997 | 31 |

Abbreviations. S=sunshine. C=cloudy. S & C=alternate sunshine and shade.
Abbreviations, strength of wind: S=strong. M=Moderate. C=Calm.

**Right-hand page.**

## COMPUTATION OF PRECISE LEVELS.

*New York City, N. Y.* Observer: *F.W.K.* Year: *1911.*

| Corrections. Index Corr. | Curv. and R'f'n | Level | L'gth of Rod | Temp. of Rod | Difference of Elevation. Each Line. | Mean. | Divergence B–F | Toward or from Sun | Sunshine or Cl'dy | Wind | Date and Hour. |
|---|---|---|---|---|---|---|---|---|---|---|---|
| *mm.* | *mm.* | *mm.* | *mm.* | *mm.* | *m.* | *m.* | *mm.* | | | | |
| 0.0 | 0.0 | 0.0 | +0.4 | +0.2 | + 3.0534 | + 3.0535 | +0.3 | L. | S. | ML. | 9/28– 1:30 |
| 0.0 | 0.0 | 0.0 | —0.4 | —0.2 | — 3.0537 | | | | C. | C. | 2:45 |
| 0.0 | 0.0 | 0.0 | +0.8 | +0.3 | + 6.1082 | + 6.1080 | —0.3 | | C. | C. | 12/10– 8:30 |
| 0.0 | 0.0 | 0.0 | —0.8 | —0.3 | — 6.1079 | | | | C. | C. | 12:00 |
| 0.0 | 0.0 | 0.0 | +0.1 | 0.0 | + 0.6837 | + 0.6830 | —1.3 | | C. | C. | 12/10– 9:30 |
| 0.0 | 0.0 | 0.0 | —0.1 | 0.0 | — 0.6824 | | | | C. | C. | 11:30 |
| 0.0 | 0.0 | 0.0 | —4.6 | —3.2 | —34.5923 | —34.5923 | +0.1 | F. | S.&C. | C. | 10/ 9– 2:45 |
| 0.0 | 0.0 | 0.0 | +4.6 | +2.9 | +34.5922 | | | Tr. | S.&C. | C. | 3:50 |
| 0.0 | 0.0 | 0.0 | —1.3 | —1.0 | —10.3239 | —10.3239 | 0.0 | R. | S. | TM. | 10/10– 9:45 |
| 0.0 | 0.0 | 0.0 | +1.3 | +0.9 | +10.3240 | | | T. | S. | C. | 1:00 |
| 0.0 | 0.0 | +0.2 | 0.0 | —4.6 | —57.8758 | —57.8761 | —0.6 | L. | S.&C. | C. | 6/13–10:45 |
| 0.0 | —0.1 | +0.1 | 0.0 | +5.8 | +57.8764 | | | TL. | S.&C. | MR. | 6/14–11:30 |
| —0.4 | 0.0 | 0.0 | 0.0 | —0.1 | — 0.7427 | — 0.7423 | +0.9 | R. | S. | C. | 1910 6/20–10:00 |
| +0.4 | 0.0 | 0.0 | 0.0 | +0.1 | + 0.7418 | | | T. | S. | C. | 11:30 |
| —0.8 | 0.0 | 0.0 | 0.0 | +0.1 | + 1.3995 | + 1.3992 | —0.5 | | C. | C. | 1910 6/20–10:00 |
| +0.8 | 0.0 | 0.0 | 0.0 | —0.1 | — 1.3990 | | | | C. | C. | 11:30 |

Abbreviations, direction of progress relatively to sun: T/F=within 45° of directly toward/from sun. $T_R$/$F_R$=toward/from sun, but at an angle of more than 45° with sun to right. $T_L$/$F_L$=ditto with sun to left. R/L=sun to right/left and nearly at right angles to line. The same abbreviations also apply to the direction of progress relative to the wind.

the sum of the total foresight intervals for each line, and in the case of the first line shows the foresight distance to be 2 meters longer than the backsight distance.

The first column on the right-hand page shows the application of the index correction. On the forward line from B. M. 111 to B. M. 111a the backsight was taken on the rod B, which has an index error of —0.8 millimeter, and the foresight was taken on a metric rule having no index error. The backsight is therefore too great by 0.8 millimeter and the difference of elevation will be too great by the same amount; the index correction is therefore —0.8 millimeter or nearly 0.003 foot.

The corrections for curvature and refraction shown in the second column on the right-hand page are those due to the slight differences of corresponding foresights and backsights, no correction being necessary when the corresponding sights are exactly equal.

In applying the curvature and refraction correction to the level results, only the difference of rod intervals of the individual set-ups is considered, not the difference of rod intervals of the entire section. The correction is usually inappreciable and seldom exceeds 0.1 millimeter under actual conditions. It is applied very quickly by the use of properly prepared tables and a rapid inspection of the record books.

It is important to note that this is in the main a correction for curvature, a quantity which is not uncertain, the uncertain refraction correction being on an average about one-eighth as great as that for curvature. Curvature and refraction errors exist in every line of sight, but are eliminated if the foresight and backsight distances are kept equal.

The level correction shown in the third column on the right-hand page is equal to the constant C (previously defined) times the value in the sixth column of the left-hand page. Its sign is fixed by the signs of the two factors. In the forward line from bench mark G to bench mark W the value of C was + 0.007 and gave a correction for the level of + 0.2 millimeter. This correction will very seldom exceed 0.3 millimeter under actual conditions.

The fourth column on the right gives the correction due to the excess of length of the rods at 0° C. For the first five completed lines the rods A and C were used, their mean length being 0.133 millimeter too long on each meter. It will be seen from an inspection of the corrections in this column that the amount of the correction will vary with the measured difference of elevation, provided the correction per meter is the same for all rods.

In the line from B. M. 41 to B. M. 42 the difference of elevation was over 34 meters, and as the mean length of the rods at 0° C. was 0.133 millimeter too long on each meter, the length correction for that section is 4.6 millimeters or 0.015 foot. This is no inconsiderable correction in precise leveling. Again, in the line from B. M. G. to B. M. W., the difference of elevation was over 57 meters and there is no length correction; that is because the mean length of the rods used on that particular section is exactly 3.0 meters at 0° C.

The next column gives the correction due to the expansion of the rods from 0° C. to the temperature of observation, computed with the known coefficient of expansion of the rods, namely, 0.000004 per degree Centigrade. Here also we see that the amount of the correction varies with the measured difference of elevation, the greatest correction being + 5.8 millimeters or 0.019 foot for the backward run of the line from B. M. G. to B. M. W. The sum of the quantities in the fourth and fifth columns in any line gives the correction due to the excess of length of the rods at the temperature of observation.

The sixth column gives the result for each line after the corrections have been applied, and the seventh column the mean difference of elevation for each completed line. The next column gives the divergence (B-F) between the backward and forward lines for each section.

**Left-hand page.** **Right-hand page.**

BOARD OF ESTIMATE AND APPORTIONMENT
CITY OF NEW YORK.

ABSTRACT OF SPIRIT LEVEL RESULTS.

Borough: *Manhattan.* Instrument: *Level No. 1.* Rods: *A. & C.* Observer: *F. W. Koop.* Computers: *F. W. K. & J. S.*

| Date. | From B.M. to B.M. | Distance in Kilometers | Difference of Elevation. For-ward Line. | Difference of Elevation. Back-ward Line. | Difference of Elevation. Mean. | Discrepancy. Partial. | Discrepancy. Total Accumulated | No. of B.M. | Distance From B.M. No. S. | Adjusted Difference of Elevation. | Elevation Above Mean Sea Level. | | Locality. |
|---|---|---|---|---|---|---|---|---|---|---|---|---|---|
| | | | *m.* | *m.* | *m.* | *mm.* | *mm.* | U.S. | *km.* | *m.* | *m.* | *feet* | |
| 1911 | | | | | | | —0.4 | H | 14.278 | | 2.4382 | 7.999 | On sea wall at Governors Island. |
| Sept. 25–26.. | H— 1 | 0.947 | — 0.2923 | + 0.2923 | — 0.2923 | 0.0 | —0.4 | 1 | 15.225 | — 0.2922 | 2.1460 | 7.041 | |
| Sept. 28..... | 1— 2 | 0.856 | + 3.0534 | — 3.0537 | + 3.0535 | +0.3 | —0.1 | 2 | 16.081 | + 3.0536 | 5.1996 | 17.059 | |
| Dec. 10..... | 2— 3 | 0.714 | + 6.1082 | — 6.1079 | + 6.1080 | —0.3 | —0.4 | 3 | 16.795 | + 6.1080 | 11.3076 | 37.098 | |
| Dec. 10..... | 3— 4 | 0.478 | + 0.6837 | — 0.6824 | + 0.6830 | —1.3 | —1.7 | 4 | 17.273 | + 0.6830 | 11.9906 | 39.339 | |
| Dec. 10..... | 4— 5 | 0.136 | + 1.6464 | — 1.6461 | + 1.6463 | —0.3 | —2.0 | 5 | 17.409 | + 1.6463 | 13.6369 | 44.740 | Center of ⊤, S.W. cor. of City Hall. |
| Dec. 10..... | 5— 9 | 0.546 | — 1.9950 | + 1.9956 | — 1.9953 | —0.6 | —2.6 | 9 | 17.955 | — 1.9953 | 11.6416 | 38.194 | |
| Dec. 10..... | 9—970 | 0.657 | — 4.8243 | + 4.8246 | — 4.8244 | +0.3 | —2.3 | 970 | 18.612 | — 4.8244 | 6.8172 | 22.366 | |
| Dec. 17..... | 970— 10 | 0.540 | + 6.9025 | — 6.9025 | + 6.9025 | 0.0 | —2.3 | 10 | 19.152 | + 6.9025 | 13.7197 | 45.012 | |
| Dec. 17..... | 10— 11 | 1.346 | + 0.2209 | — 0.2200 | + 0.2205 | —0.9 | —3.2 | 11 | 20.498 | + 0.2206 | 13.9403 | 45.736 | |
| Dec. 17..... | 11— 12 | 0.425 | — 0.5931 | + 0.5911 | — 0.5921 | +2.0 | —1.2 | 12 | 20.923 | — 0.5921 | 13.3482 | 43.793 | |
| Dec. 17..... | 12— 13 | 0.405 | — 0.0150 | + 0.0162 | — 0.0156 | —1.2 | —2.4 | 13 | 21.328 | — 0.0156 | 13.3326 | 43.742 | |
| Feb. 6, 1912.. | 13— 14 | 0.924 | + 2.2592 | — 2.2592 | + 2.2592 | 0.0 | —2.4 | 14 | 22.252 | + 2.2592 | 15.5918 | 51.154 | |
| Feb. 6, 1912.. | 14— 15 | 0.673 | + 7.1966 | — 7.1957 | + 7.1961 | —0.9 | —3.3 | 15 | 22.925 | + 7.1961 | 22.7879 | 74.763 | |
| 1911 | | | | | | | | | | | | | |
| July 12–24... | 15— 16 | 0.496 | — 4.8245 | + 4.8252 | — 4.8247 | +0.7 | —2.6 | 16 | 23.421 | — 4.8247 | 17.9632 | 58.934 | |
| July 12–24... | 16— 17 | 0.356 | — 3.0341 | + 3.0341 | — 3.0341 | 0.0 | —2.6 | 17 | 23.777 | — 3.0341 | 14.9291 | 48.980 | |
| July 12–24... | 17–T1886 | 1.195 | —10.4676 | +10.4673 | —10.4674 | +0.3 | —2.3 | T.1886 | 24.972 | —10.4674 | 4.4617 | 14.638 | |
| | | | | | | | —3.3 | 15 | 22.925 | | 22.7879 | 74.763 | |
| July 12–13... | 15— 18 | 0.701 | + 1.7191 | — 1.7184 | + 1.7187 | —0.7 | —4.0 | 18 | 23.626 | + 1.7187 | 24.5066 | 80.402 | Main entrance to St. Patricks Cath. |
| July 14...... | 18— 19 | 0.757 | — 7.7174 | + 7.7170 | — 7.7172 | —0.4 | —3.6 | 19 | 24.383 | — 7.7172 | 16.7894 | 55.083 | |
| July 14...... | 19— 20 | 0.910 | + 9.6176 | — 9.6178 | + 9.6177 | +0.2 | —3.4 | 20 | 25.293 | + 9.6177 | 26.4071 | 86.637 | |
| July 18...... | 20— 21 | 0.740 | + 0.9985 | — 0.9996 | + 0.9990 | +1.1 | —2.3 | 21 | 26.033 | + 0.9990 | 27.4061 | 89.915 | |
| July 18...... | 21— 22 | 0.817 | + 0.2759 | — 0.2763 | + 0.2761 | +0.4 | —1.9 | 22 | 26.850 | + 0.2762 | 27.6823 | 90.821 | |
| July 18...... | 22— 23 | 0.105 | — 2.1376 | + 2.1373 | — 2.1374 | +0.3 | —1.6 | 23 | 26.955 | — 2.1374 | 25.5449 | 83.809 | |
| July 19...... | 23— 24 | 1.667 | + 6.2467 | — 6.2473 | + 6.2470 | +0.6 | —1.0 | 24 | 28.622 | + 6.2471 | 31.7920 | 104.304 | |
| Oct. 3....... | 24— 25 | 1.437 | —20.9705 | +20.9714 | —20.9709 | —0.9 | —1.9 | 25 | 30.059 | —20.9708 | 10.8212 | 35.503 | |

The last four columns on this form are for use whenever special studies are to be made to determine if possible the sources of the principal errors of leveling. It should be noted that the times of the backward and forward runnings of any section, as indicated in the last column, have no fixed relation to each other. The two runnings are sometimes made on the same day, sometimes on different days.

The abstract of results shown on page 37 is the form actually used in collecting the results of the computation form shown on page 35. While the computation is discontinuous, showing results from different sections, the abstract is continuous, and is the form in which the final results appear.

The most interesting columns on the left-hand page are those showing the discrepancies. In the seventh column the discrepancy between the backward and forward running of each section is shown, and from these quantities the probable error for any section of the level line may be computed. In the next column the total accumulated discrepancy from the beginning of the line which in this case is the Coast Survey tide-staff at Fort Hamilton, is shown.

At the head of the column it is seen that the sum of the discrepancies from Fort Hamilton to Governors Island was —0.4 millimeter for a distance of more than 14 kilometers, or at the rate of —0.029 millimeter per kilometer. At the bottom of the form the computed rate of accumulation of discrepancy between the forward and backward lines was —1.9 millimeters on 30 kilometers, or only —0.06 millimeter per kilometer. This is a remarkably small error and proves conclusively that there is no systematic error in the leveling.

In a level line of high precision, the accumulated discrepancy should not be very far from zero, and a measure of the precision of the leveling is obtained by an inspection of these sums.

For a number of kilometers on certain portions of the level line, the minus sign will sometimes predominate in this column, and then change gradually to plus and vice versa. The third column of the right-hand page gives the adjusted difference of elevation for any section, from which the elevations are at once derived. This column is not shown on the Coast Survey form of abstract, but it is found to be very convenient and keeps all the figures before the computer. An examination of the quantities in this column will show that the largest correction applied in order to close the several circuits of which this line forms a part, was at the rate of 0.038 millimeter per kilometer, or at the rate of 0.0002 foot per mile. This correction being so small, it was at once disposed of by adding 0.1 millimeter to a few of the longer lines.

The fourth and fifth columns contain the resulting elevations in meters and in feet respectively. The last column is for brief descriptions of the bench marks, a few of which are shown.

## CORRECTION TABLES.

For convenience there are inserted here three tables which are useful in making the foregoing computations. These tables were prepared by the United States Coast and Geodetic Survey.

The table of total correction for curvature and refraction is for use in computing C, in making river crossings, and in general wherever the total correction is required. In computing this table the refraction was assumed to be equal to one-eighth the curvature.

The table for the differential correction for curvature and refraction is for use in deriving the corrections shown in the second column of the right-hand page of the computation indicated on page 35. The table was computed upon the assumption that the refraction is one-eighth of the curvature, and that the stadia interval for

*Total correction for curvature and refraction.*

| Distance. | Correction to rod reading. | Distance. | Correction to rod reading. |
|---|---|---|---|
| *m.* *m.* | *mm.* | *m.* | *mm.* |
| 0 to 27 | 0.0 | 160 | −1.8 |
| 28 to 47 | −0.1 | 170 | −2.1 |
| 48 to 60 | −0.2 | 180 | −2.3 |
| 61 to 72 | −0.3 | 190 | −2.6 |
| 73 to 81 | −0.4 | 200 | −2.8 |
| 82 to 90 | −0.5 | 210 | −3.0 |
| 91 to 98 | −0.6 | 220 | −3.3 |
| 99 to 105 | −0.7 | 230 | −3.7 |
| 106 to 112 | −0.8 | 240 | −4.0 |
| 113 to 118 | −0.9 | 250 | −4.3 |
| 119 to 124 | −1.0 | 260 | −4.7 |
| 125 to 130 | −1.1 | 270 | −5.0 |
| 131 to 136 | −1.2 | 280 | −5.4 |
| 137 to 141 | −1.3 | 290 | −5.8 |
| 142 to 146 | −1.4 | 300 | −6.2 |
| 147 to 150 | −1.5 | | |

*Differential correction for curvature and refraction.*

| Mean length of sight in rod interval in millimeters. | Difference of sights in rod interval in millimeters. | | | | | | | | | | | | | | | | | | | | | | | | | | | | |
|---|---|---|---|---|---|---|---|---|---|---|---|---|---|---|---|---|---|---|---|---|---|---|---|---|---|---|---|---|---|
| | 2 | 4 | 6 | 8 | 10 | 12 | 14 | 16 | 18 | 20 | 22 | 24 | 26 | 28 | 30 | 32 | 34 | 36 | 38 | 40 | 42 | 44 | 46 | 48 | 50 | 52 | 54 | 56 | 58 |
| 10 | .0 | .0 | .0 | .0 | .0 | | | | | | | | | | | | | | | | | | | | | | | | |
| 20 | .0 | .0 | .0 | .0 | .0 | .0 | .0 | .0 | .0 | .0 | | | | | | | | | | | | | | | | | | | |
| 30 | .0 | .0 | .0 | .0 | .0 | .0 | .0 | .0 | .0 | .0 | .0 | .0 | .0 | .0 | .0 | | | | | | | | | | | | | | |
| 40 | .0 | .0 | .0 | .0 | .0 | .0 | .0 | .0 | .0 | .0 | .0 | .0 | .0 | .0 | .0 | .0 | .0 | .0 | .0 | .0 | | | | | | | | | |
| 50 | .0 | .0 | .0 | .0 | .0 | .0 | .0 | .0 | .0 | .0 | .0 | .0 | .0 | .0 | .0 | .0 | .0 | .0 | .0 | .0 | .0 | .0 | .0 | .0 | .0 | | | | |
| 60 | .0 | .0 | .0 | .0 | .0 | .0 | .0 | .0 | .0 | .0 | .0 | .0 | .0 | .0 | .0 | .0 | .0 | .0 | .0 | .0 | .0 | .0 | .0 | .0 | .0 | .0 | .0 | .1 | .1 |
| 70 | .0 | .0 | .0 | .0 | .0 | .0 | .0 | .0 | .0 | .0 | .0 | .0 | .0 | .0 | .0 | .0 | .0 | .0 | .0 | .0 | .0 | .0 | .0 | .1 | .1 | .1 | .1 | .1 | .1 |
| 80 | .0 | .0 | .0 | .0 | .0 | .0 | .0 | .0 | .0 | .0 | .0 | .0 | .0 | .0 | .0 | .0 | .0 | .0 | .0 | .0 | .1 | .1 | .1 | .1 | .1 | .1 | .1 | .1 | .1 |
| 90 | .0 | .0 | .0 | .0 | .0 | .0 | .0 | .0 | .0 | .0 | .0 | .0 | .0 | .0 | .0 | .0 | .0 | .0 | .1 | .1 | .1 | .1 | .1 | .1 | .1 | .1 | .1 | .1 | .1 |
| 100 | .0 | .0 | .0 | .0 | .0 | .0 | .0 | .0 | .0 | .0 | .0 | .0 | .0 | .0 | .0 | .0 | .1 | .1 | .1 | .1 | .1 | .1 | .1 | .1 | .1 | .1 | .1 | .1 | .1 |
| 110 | .0 | .0 | .0 | .0 | .0 | .0 | .0 | .0 | .0 | .0 | .0 | .0 | .0 | .0 | .1 | .1 | .1 | .1 | .1 | .1 | .1 | .1 | .1 | .1 | .1 | .1 | .1 | .1 | .1 |
| 120 | .0 | .0 | .0 | .0 | .0 | .0 | .0 | .0 | .0 | .0 | .0 | .0 | .0 | .1 | .1 | .1 | .1 | .1 | .1 | .1 | .1 | .1 | .1 | .1 | .1 | .1 | .1 | .1 | .1 |
| 130 | .0 | .0 | .0 | .0 | .0 | .0 | .0 | .0 | .0 | .0 | .0 | .0 | .1 | .1 | .1 | .1 | .1 | .1 | .1 | .1 | .1 | .1 | .1 | .1 | .1 | .1 | .1 | .1 | .1 |
| 140 | .0 | .0 | .0 | .0 | .0 | .0 | .0 | .0 | .0 | .0 | .0 | .1 | .1 | .1 | .1 | .1 | .1 | .1 | .1 | .1 | .1 | .1 | .1 | .1 | .1 | .1 | .1 | .1 | .1 |
| 150 | .0 | .0 | .0 | .0 | .0 | .0 | .0 | .0 | .0 | .0 | .1 | .1 | .1 | .1 | .1 | .1 | .1 | .1 | .1 | .1 | .1 | .1 | .1 | .1 | .1 | .1 | .1 | .1 | .1 |
| 160 | .0 | .0 | .0 | .0 | .0 | .0 | .0 | .0 | .0 | .0 | .1 | .1 | .1 | .1 | .1 | .1 | .1 | .1 | .1 | .1 | .1 | .1 | .1 | .1 | .1 | .1 | .1 | .1 | .1 |
| 170 | .0 | .0 | .0 | .0 | .0 | .0 | .0 | .0 | .0 | .1 | .1 | .1 | .1 | .1 | .1 | .1 | .1 | .1 | .1 | .1 | .1 | .1 | .1 | .1 | .1 | .1 | .1 | .1 | .2 |
| 180 | .0 | .0 | .0 | .0 | .0 | .0 | .0 | .0 | .0 | .1 | .1 | .1 | .1 | .1 | .1 | .1 | .1 | .1 | .1 | .1 | .1 | .1 | .1 | .1 | .1 | .1 | .1 | .2 | .2 |
| 190 | .0 | .0 | .0 | .0 | .0 | .0 | .0 | .0 | .1 | .1 | .1 | .1 | .1 | .1 | .1 | .1 | .1 | .1 | .1 | .1 | .1 | .1 | .1 | .1 | .1 | .2 | .2 | .2 | .2 |
| 200 | .0 | .0 | .0 | .0 | .0 | .0 | .0 | .0 | .1 | .1 | .1 | .1 | .1 | .1 | .1 | .1 | .1 | .1 | .1 | .1 | .1 | .1 | .1 | .1 | .2 | .2 | .2 | .2 | .2 |
| 210 | .0 | .0 | .0 | .0 | .0 | .0 | .0 | .1 | .1 | .1 | .1 | .1 | .1 | .1 | .1 | .1 | .1 | .1 | .1 | .1 | .1 | .1 | .1 | .2 | .2 | .2 | .2 | .2 | .2 |
| 220 | .0 | .0 | .0 | .0 | .0 | .0 | .0 | .1 | .1 | .1 | .1 | .1 | .1 | .1 | .1 | .1 | .1 | .1 | .1 | .1 | .1 | .1 | .2 | .2 | .2 | .2 | .2 | .2 | .2 |
| 230 | .0 | .0 | .0 | .0 | .0 | .0 | .0 | .1 | .1 | .1 | .1 | .1 | .1 | .1 | .1 | .1 | .1 | .1 | .1 | .1 | .1 | .2 | .2 | .2 | .2 | .2 | .2 | .2 | .2 |
| 240 | .0 | .0 | .0 | .0 | .0 | .0 | .1 | .1 | .1 | .1 | .1 | .1 | .1 | .1 | .1 | .1 | .1 | .1 | .1 | .1 | .2 | .2 | .2 | .2 | .2 | .2 | .2 | .2 | .2 |
| 250 | .0 | .0 | .0 | .0 | .0 | .0 | .1 | .1 | .1 | .1 | .1 | .1 | .1 | .1 | .1 | .1 | .1 | .1 | .1 | .2 | .2 | .2 | .2 | .2 | .2 | .2 | .2 | .2 | .2 |
| 260 | .0 | .0 | .0 | .0 | .0 | .0 | .1 | .1 | .1 | .1 | .1 | .1 | .1 | .1 | .1 | .1 | .1 | .1 | .2 | .2 | .2 | .2 | .2 | .2 | .2 | .2 | .2 | .2 | .2 |
| 270 | .0 | .0 | .0 | .0 | .0 | .0 | .1 | .1 | .1 | .1 | .1 | .1 | .1 | .1 | .1 | .1 | .1 | .1 | .2 | .2 | .2 | .2 | .2 | .2 | .2 | .2 | .2 | .2 | .2 |
| 280 | .0 | .0 | .0 | .0 | .0 | .1 | .1 | .1 | .1 | .1 | .1 | .1 | .1 | .1 | .1 | .1 | .1 | .2 | .2 | .2 | .2 | .2 | .2 | .2 | .2 | .2 | .2 | .2 | .2 |
| 290 | .0 | .0 | .0 | .0 | .0 | .1 | .1 | .1 | .1 | .1 | .1 | .1 | .1 | .1 | .1 | .1 | .2 | .2 | .2 | .2 | .2 | .2 | .2 | .2 | .2 | .2 | .2 | .2 | .3 |
| 300 | .0 | .0 | .0 | .0 | .0 | .1 | .1 | .1 | .1 | .1 | .1 | .1 | .1 | .1 | .1 | .1 | .2 | .2 | .2 | .2 | .2 | .2 | .2 | .2 | .2 | .2 | .2 | .3 | .3 |
| 310 | .0 | .0 | .0 | .0 | .0 | .1 | .1 | .1 | .1 | .1 | .1 | .1 | .1 | .1 | .1 | .2 | .2 | .2 | .2 | .2 | .2 | .2 | .2 | .2 | .2 | .2 | .3 | .3 | .3 |
| 320 | .0 | .0 | .0 | .0 | .0 | .1 | .1 | .1 | .1 | .1 | .1 | .1 | .1 | .1 | .1 | .2 | .2 | .2 | .2 | .2 | .2 | .2 | .2 | .2 | .2 | .3 | .3 | .3 | .3 |
| 330 | .0 | .0 | .0 | .0 | .1 | .1 | .1 | .1 | .1 | .1 | .1 | .1 | .1 | .1 | .2 | .2 | .2 | .2 | .2 | .2 | .2 | .2 | .2 | .2 | .3 | .3 | .3 | .3 | .3 |
| 340 | .0 | .0 | .0 | .0 | .1 | .1 | .1 | .1 | .1 | .1 | .1 | .1 | .1 | .1 | .2 | .2 | .2 | .2 | .2 | .2 | .2 | .2 | .2 | .2 | .3 | .3 | .3 | .3 | .3 |
| 350 | .0 | .0 | .0 | .0 | .1 | .1 | .1 | .1 | .1 | .1 | .1 | .1 | .1 | .1 | .2 | .2 | .2 | .2 | .2 | .2 | .2 | .2 | .2 | .3 | .3 | .3 | .3 | .3 | .3 |
| 360 | .0 | .0 | .0 | .0 | .1 | .1 | .1 | .1 | .1 | .1 | .1 | .1 | .1 | .2 | .2 | .2 | .2 | .2 | .2 | .2 | .2 | .2 | .3 | .3 | .3 | .3 | .3 | .3 | .3 |
| 400 | .0 | .0 | .0 | .0 | .1 | .1 | .1 | .1 | .1 | .1 | .1 | .1 | .2 | .2 | .2 | .2 | .2 | .2 | .2 | .2 | .3 | .3 | .3 | .3 | .3 | .3 | .3 | .3 | .4 |
| 440 | .0 | .0 | .0 | .1 | .1 | .1 | .1 | .1 | .1 | .1 | .1 | .2 | .2 | .2 | .2 | .2 | .2 | .2 | .3 | .3 | .3 | .3 | .3 | .3 | .3 | .3 | .4 | .4 | .4 |
| 480 | .0 | .0 | .0 | .1 | .1 | .1 | .1 | .1 | .1 | .1 | .2 | .2 | .2 | .2 | .2 | .2 | .2 | .3 | .3 | .3 | .3 | .3 | .3 | .4 | .4 | .4 | .4 | .4 | .4 |
| 520 | .0 | .0 | .1 | .1 | .1 | .1 | .1 | .1 | .1 | .2 | .2 | .2 | .2 | .2 | .2 | .3 | .3 | .3 | .3 | .3 | .3 | .3 | .4 | .4 | .4 | .4 | .4 | .4 | .4 |

the instrument is such that the distance from the instrument to the rod in meters is one-third of the interval subtended on the rod in millimeters. An inspection of the table will show that it is sufficiently accurate for use even though the stadia interval differs from that stated by 10 per cent. or more.

The sign of this correction is positive when the foresight is the longer, that is, when the stadia interval subtends more divisions on the rod for the foresight than for the backsight.

The differential correction for curvature and refraction may also be taken from the following table, which is sometimes more convenient. It gives, for a given difference of rod intervals, the lower limiting values of the mean rod interval for which the correction is (to the nearest tenth of a millimeter) 0.1 millimeter, 0.2 millimeter, etc.

Thus, for a difference of rod intervals of 37 millimeters, there are given in the

*Differential correction for curvature and refraction.*

LIMITING VALUES OF THE MEAN ROD INTERVAL.

| Difference of rod intervals. | Correction. | | | | Difference of rod intervals. | Correction. | | | |
|---|---|---|---|---|---|---|---|---|---|
| | 0.1 mm. | 0.2 mm. | 0.3 mm. | 0.4 mm. | | 0.1 mm. | 0.2 mm. | 0.3 mm. | 0.4 mm. |
| *mm.* | | | | | *mm.* | | | | |
| 8 | 409.7 | ........ | ........ | ........ | 35 | 93.7 | 281.0 | ........ | ........ |
| 9 | 364.2 | ........ | ........ | ........ | 36 | 91.0 | 273.2 | 455.2 | ........ |
| 10 | 327.8 | ........ | ........ | ........ | 37 | 88.6 | 265.8 | 442.9 | ........ |
| 11 | 298.0 | ........ | ........ | ........ | 38 | 86.3 | 258.8 | 431.3 | ........ |
| 12 | 273.2 | ........ | ........ | ........ | 39 | 84.0 | 252.1 | 420.2 | ........ |
| 13 | 252.1 | ........ | ........ | ........ | 40 | 81.9 | 245.8 | 409.7 | ........ |
| 14 | 234.1 | ........ | ........ | ........ | 41 | 79.9 | 239.8 | 399.7 | ........ |
| 15 | 218.5 | ........ | ........ | ........ | 42 | 78.0 | 234.1 | 390.2 | ........ |
| 16 | 204.9 | ........ | ........ | ........ | 43 | 76.2 | 228.7 | 381.1 | ........ |
| 17 | 192.8 | ........ | ........ | ........ | 44 | 74.5 | 223.5 | 372.5 | ........ |
| 18 | 182.1 | ........ | ........ | ........ | 45 | 72.8 | 218.5 | 364.2 | ........ |
| 19 | 172.5 | ........ | ........ | ........ | 46 | 71.3 | 213.8 | 356.3 | ........ |
| 20 | 163.9 | ........ | ........ | ........ | 47 | 69.7 | 209.2 | 348.7 | ........ |
| 21 | 156.1 | ........ | ........ | ........ | 48 | 68.3 | 204.9 | 341.4 | ........ |
| 22 | 149.0 | 447.0 | ........ | ........ | 49 | 66.9 | 200.7 | 334.5 | ........ |
| 23 | 142.5 | 427.5 | ........ | ........ | 50 | 65.6 | 196.7 | 327.8 | 458.9 |
| 24 | 136.6 | 407.9 | ........ | ........ | 51 | 64.3 | 192.8 | 321.4 | 449.9 |
| 25 | 131.1 | 393.3 | ........ | ........ | 52 | 63.0 | 189.1 | 315.2 | 441.2 |
| 26 | 126.1 | 378.2 | ........ | ........ | 53 | 61.8 | 185.5 | 309.2 | 432.9 |
| 27 | 121.4 | 364.2 | ........ | ........ | 54 | 60.7 | 182.1 | 303.5 | 424.9 |
| 28 | 117.1 | 351.2 | ........ | ........ | 55 | 59.6 | 178.8 | 298.0 | 417.2 |
| 29 | 113.0 | 339.1 | ........ | ........ | 56 | 58.5 | 175.6 | 292.7 | 409.7 |
| 30 | 109.3 | 327.8 | ........ | ........ | 57 | 57.5 | 172.5 | 287.5 | 402.5 |
| 31 | 105.7 | 317.2 | ........ | ........ | 58 | 56.5 | 169.5 | 282.6 | 395.6 |
| 32 | 102.4 | 307.3 | ........ | ........ | 59 | 55.6 | 166.7 | 277.8 | 388.9 |
| 33 | 99.3 | 298.0 | ........ | ........ | 60 | 54.6 | 163.9 | 273.2 | 382.4 |
| 34 | 96.4 | 289.2 | ........ | ........ | | | | | |

table the numbers 88.6, 265.8, and 442.9, under the respective headings 0.1 millimeter, 0.2 millimeter, 0.3 millimeter. This means that up to 88.6 millimeters the correction is 0.0 millimeter to the nearest tenth, from that point up to 265.8 the correction is 0.1 millimeter to the nearest tenth, and between 265.8 and 442.9 it must be taken as 0.3 millimeter.

The table is computed on the same assumptions as the preceding table, and the rule for the sign of the correction is the same, namely, positive when the foresight is the longer of the two sights, negative when it is the shorter. Numbers over 460 are omitted from the table.

The table of temperature corrections is for use in deriving the values shown in the fifth column of the right-hand page of the computation indicated on page 35, the length of the rod at zero degrees Centigrade having been used in deriving the fourth column. The table is computed on the assumption that the coefficient of expansion of the rod is four parts in a million per degree Centigrade. The sign of the correction is always the same as the sign of the measured difference of elevation unless the temperature is below the Centigrade zero.

*Correction for temperature (in millimeters).*

| Temp. C. | Difference of elevation in meters. | | | | | | | | | | | | | |
|---|---|---|---|---|---|---|---|---|---|---|---|---|---|---|
| | 1 | 2 | 3 | 4 | 5 | 6 | 7 | 8 | 9 | 10 | 11 | 12 | 13 | 14 |
| 1 | .0 | .0 | .0 | .0 | .0 | .0 | .0 | 0 | .0 | .0 | .0 | .0 | .0 | .1 |
| 2 | .0 | .0 | .0 | .0 | .0 | .0 | .1 | .1 | .1 | .1 | .1 | .1 | .1 | .1 |
| 3 | .0 | .0 | .0 | .0 | .1 | .1 | .1 | .1 | .1 | .1 | .1 | .1 | .2 | .2 |
| 4 | .0 | .0 | .0 | .1 | .1 | .1 | .1 | .1 | .1 | .2 | .2 | .2 | .2 | .2 |
| 5 | .0 | .0 | .1 | .1 | .1 | .1 | .1 | .2 | .2 | .2 | .2 | .2 | .3 | .3 |
| 6 | .0 | .0 | .1 | .1 | .1 | .1 | .2 | .2 | .2 | .2 | .3 | .3 | .3 | .3 |
| 7 | .0 | .1 | .1 | .1 | .1 | .2 | .2 | .2 | .2 | .3 | .3 | .3 | .4 | .4 |
| 8 | .0 | .1 | .1 | .1 | .2 | .2 | .2 | .3 | .3 | .3 | .4 | .4 | .4 | .4 |
| 9 | .0 | .1 | .1 | .1 | .2 | .2 | .2 | .3 | .3 | .4 | .4 | .4 | .5 | .5 |
| 10 | .0 | .1 | .1 | .2 | .2 | .2 | .3 | .3 | .4 | .4 | .4 | .5 | .5 | .6 |
| 11 | .0 | .1 | .1 | .2 | .2 | 3 | .3 | .4 | .4 | .4 | .5 | .5 | .6 | .6 |
| 12 | .0 | .1 | .1 | .2 | .2 | .3 | .3 | .4 | .4 | .5 | .5 | .6 | .6 | .7 |
| 13 | .0 | .1 | .2 | .2 | .3 | .3 | .4 | .4 | .5 | .5 | .6 | .6 | .7 | .7 |
| 14 | .1 | .1 | .2 | .2 | .3 | .3 | .4 | .4 | .5 | .6 | .6 | .7 | .7 | .8 |
| 15 | .1 | .1 | .2 | .2 | .3 | .4 | .4 | .5 | .5 | .6 | .7 | .7 | .8 | .8 |
| 16 | .1 | .1 | .2 | .3 | .3 | .4 | .4 | .5 | .6 | .6 | .7 | .8 | .8 | .9 |
| 17 | .1 | .1 | .2 | .3 | .3 | .4 | .5 | .5 | .6 | .7 | .8 | .8 | .9 | .9 |
| 18 | .1 | .1 | .2 | .3 | .4 | .4 | .5 | .6 | .6 | .7 | .8 | .9 | .9 | 1.0 |
| 19 | .1 | .2 | .2 | .3 | .4 | .5 | .5 | .6 | .7 | .8 | .8 | .9 | 1.0 | 1.1 |
| 20 | .1 | .2 | .2 | .3 | .4 | .5 | .6 | .6 | .7 | .8 | .9 | 1.0 | 1.0 | 1.1 |
| 21 | .1 | .2 | .2 | .3 | .4 | .5 | .6 | .7 | .8 | .8 | .9 | 1.0 | 1.1 | 1.2 |
| 22 | .1 | .2 | .3 | .4 | .4 | .5 | .6 | .7 | .8 | .9 | 1.0 | 1.1 | 1.1 | 1.2 |
| 23 | .1 | .2 | .3 | .4 | .5 | .6 | .6 | .7 | .8 | .9 | 1.0 | 1.1 | 1.2 | 1.3 |
| 24 | .1 | .2 | .3 | .4 | .5 | .6 | .7 | .8 | .9 | 1.0 | 1.1 | 1.2 | 1.2 | 1.3 |
| 25 | .1 | .2 | .3 | .4 | .5 | .6 | .7 | .8 | .9 | 1.0 | 1.1 | 1.2 | 1.3 | 1.4 |
| 26 | .1 | .2 | .3 | .4 | .5 | .6 | .7 | .8 | .9 | 1.0 | 1.1 | 1.2 | 1.3 | 1.5 |
| 27 | .1 | .2 | .3 | .4 | .5 | .6 | .8 | .9 | 1.0 | 1.1 | 1.2 | 1.3 | 1.4 | 1.5 |
| 28 | .1 | .2 | .3 | .4 | .6 | .7 | .8 | .9 | 1.0 | 1.1 | 1.2 | 1.3 | 1.4 | 1.6 |
| 29 | .1 | .2 | .4 | .5 | .6 | .7 | .8 | .9 | 1.0 | 1.2 | 1.3 | 1.4 | 1.5 | 1.6 |
| 30 | .1 | .2 | .4 | .5 | .6 | .7 | .8 | 1.0 | 1.1 | 1.2 | 1.3 | 1.4 | 1.6 | 1.7 |
| 31 | .1 | .2 | .4 | .5 | .6 | .7 | .9 | 1.0 | 1.1 | 1.2 | 1.4 | 1.5 | 1.6 | 1.7 |
| 32 | .1 | .3 | .4 | .5 | .6 | .8 | .9 | 1.0 | 1.2 | 1.3 | 1.4 | 1.5 | 1.7 | 1.8 |
| 33 | .1 | .3 | .4 | .5 | .7 | .8 | .9 | 1.1 | 1.2 | 1.3 | 1.4 | 1.6 | 1.7 | 1.8 |
| 34 | .1 | .3 | .4 | .5 | .7 | .8 | 1.0 | 1.1 | 1.2 | 1.4 | 1.5 | 1.6 | 1.8 | 1.9 |
| 35 | .1 | .3 | .4 | .6 | .7 | .8 | 1.0 | 1.1 | 1.3 | 1.4 | 1.5 | 1.7 | 1.8 | 2.0 |
| 36 | .1 | .3 | .4 | .6 | .7 | .9 | 1.0 | 1.2 | 1.3 | 1.4 | 1.6 | 1.7 | 1.9 | 2.0 |
| 37 | .1 | .3 | .4 | .6 | .7 | .9 | 1.0 | 1.2 | 1.3 | 1.5 | 1.6 | 1.8 | 1.9 | 2.1 |
| 38 | .1 | .3 | .5 | .6 | .8 | .9 | 1.1 | 1.2 | 1.4 | 1.5 | 1.7 | 1.8 | 2.0 | 2.1 |
| 39 | .2 | .3 | .5 | .6 | .8 | .9 | 1.1 | 1.2 | 1.4 | 1.6 | 1.7 | 1.9 | 2.0 | 2.2 |
| 40 | .2 | .3 | .5 | .6 | .8 | 1.0 | 1.1 | 1.3 | 1.4 | 1.6 | 1.8 | 1.9 | 2.1 | 2.2 |
| 41 | .2 | .3 | .5 | .7 | .8 | 1.0 | 1.1 | 1.3 | 1.5 | 1.6 | 1.8 | 2.0 | 2.1 | 2.3 |
| 42 | .2 | .3 | .5 | .7 | .8 | 1.0 | 1.2 | 1.3 | 1.5 | 1.7 | 1.8 | 2.0 | 2.2 | 2.3 |
| 43 | .2 | .3 | .5 | .7 | .9 | 1.0 | 1.2 | 1.4 | 1.5 | 1.7 | 1.9 | 2.1 | 2.2 | 2.4 |
| 44 | .2 | .3 | .5 | .7 | .9 | 1.1 | 1.2 | 1.4 | 1.6 | 1.8 | 1.9 | 2.1 | 2.3 | 2.5 |
| 45 | .2 | .3 | .5 | .7 | .9 | 1.1 | 1.3 | 1.4 | 1.6 | 1.8 | 2.0 | 2.2 | 2.3 | 2.5 |

## RIVER CROSSINGS.

All river crossings exceeding 500 feet in width were made by the simultaneous reciprocal method with uniformly good results. There were ten river crossings all told, eight of them ranging in width from 790 feet at Elizabethport, on the Kill Van Kull, to 4,400 feet at The Narrows. The two remaining crossings were those at Newtown Creek and at the Harlem River; these streams being less than 500 feet in width at the points crossed, were leveled over directly with one instrument, repeating the operation several times at each side. In such cases the individual results for difference of elevation between the bench marks located on opposite shores were well within the limit of error.

Before any river crossing was made each level was tested and put in very nearly perfect adjustment by means of the method previously described for the determination of C.

On account of the variability of refraction no true correction can be computed and applied to the rod readings of any one instrument when observations are taken over a wide body of water. If, therefore, it is desired to obtain the difference of elevation between two points on opposite shores with precision, the method of simultaneous reciprocal observations will have to be employed, as it eliminates all the effects of curvature and refraction changes, and no corrections are necessary. A description of the method as employed on this work is as follows:

### Field Work.

Two levels and two rods were used, one of the levels being set up on each side of the river about 20 feet distant from the B. M. or T. P. A rod reading was then taken by each observer on the B. M. or T. P. nearest his instrument; then, at a given signal the observations were commenced and the targets were set on each of the distant rods at nearly the same instant, and the reading recorded. Simultaneous observations were therefore made in opposite directions by each observer.

This operation was repeated as a rule ten times, allowing about five minutes between each reading so as to permit of a change in the atmospheric conditions. The near rod was read only at the beginning and at the end of the program of observation. The observers then exchanged stations, each man taking his instrument with him; the near rods were again read, and at a prearranged signal the observations on the distant rods were repeated in the opposite direction exactly as before. The arithmetical mean of each set of readings was taken as the final result.

When employing this method, it is very essential that both the instruments used should be very nearly alike in the magnifying power of their telescopes and in the sensitiveness of their spirit levels. On work of this kind the very sensitive bubble in the level vial of the geodetic level reveals the true qualities of the instrument, and exceptionally close readings can be made. It is also essential that the two targets be set as nearly simultaneously as possible in order that the same atmospheric conditions may affect both sights.

The diagram below (illustration No. 11) illustrates a simultaneous reciprocal level crossing of the East River from Astoria to Carl Schurz Park, and shows better than can be described the methods employed and the reasons therefor. The lines of sight as shown in the illustration, are, of course, exaggerated somewhat.

The heavy curved line is a line of true level, being parallel to the surface of still water and therefore at all points equidistant from the center of the earth. If the lines of sight through each instrument were lines of true level, they would take the position of the broken lines in the illustration; not being lines of true level they are shown by the use of full lines as they really are, normal to the radius of the earth through the center of each instrument. In this case, we desire to obtain the difference of elevation between B. M. 149B and B. M. 149C, the distance between the B. M.'s being 566 meters or nearly 1,937 feet.

When making the river crossing, it is best to consider the observations made from both sides of the river as being made in the forward direction as indicated by the arrow; that is, to consider the elevation in both cases as being carried from B. M. 149B to B. M. 149C, so as to give two forward runs, instead of one forward and one backward run as in ordinary leveling. This simplifies the computation, gives a better understanding of the work and causes less confusion in the signs.

On this crossing level No. 1 is set up near B. M. 149C, and level No. 2 is set up near B. M. 149B, as shown in the diagram. A backsight is taken on B. M. 149B with level No. 2 and then a foresight is taken on B. M. 149C with the same instrument.

The resulting difference of elevation between the two B. M.'s being uncorrected for curvature and refraction will be in error by the amount ab, which is the amount of the correction necessary to bring the line of sight down to the line of true level.

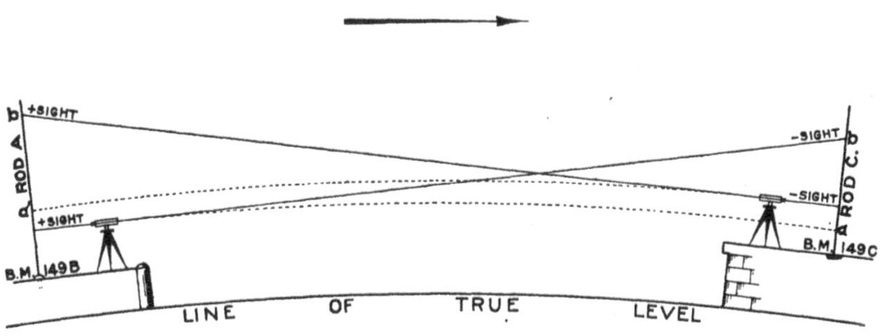


No. 11.—Method of Reciprocal Leveling for River Crossing.

In other words, the elevation of B. M. 149C as derived from B. M. 149B with level No. 2, will be too low by the amount ab. For the same reason, the difference of elevation between B. M. 149B and B. M. 149C as derived from observations with level No. 1 on the west side of the river, will be in error by the amount a'b', which is the sum of the curvature and refraction corrections for the line of sight through level No. 1, and the elevation of B. M. 149C will be too high by the amount a'b'.

As the observations taken with both instruments were simultaneous, the same atmospheric conditions affected both lines of sight, i.e., ab = a'b', and the actual difference of elevation is the mean of the two determinations. The operation is therefore a very simple one, and the reduction is similar to the application of a constant quantity to both backsight and foresight, which does not affect the true difference of elevation.

After a set of ten readings has been taken with each instrument at each station, the mean of the last two determinations of the difference of elevation is combined with the mean of the first two determinations, and the final mean is the accepted difference of elevation between the two bench marks.

On all of the river crossings the line of sight had an elevation of at least 12 feet above the surface of the water and was frequently much higher. The disturbance of the atmosphere is very great near the surface of land or water and decreases rapidly with the elevation of the line above the surface.

**Left-hand page.** | **Right-hand page.**

BOARD OF ESTIMATE AND APPORTIONMENT
CITY OF NEW YORK.

## RECIPROCAL LEVELS.

EAST RIVER CROSSING.

From B.M.149B. in Queens toB.M.149C. in Manhattan. Distance = 566 meters.

Date: *Dec. 20, 1911.* Rods, *A. & C.* Temp. 5°.0C. Sun: *C.* Wind: *C.* Time: *10 A.M. to 2 P.M.*

| Level No. 2. | | | | | | | | Level No. 1. | | | | | | | |
|---|---|---|---|---|---|---|---|---|---|---|---|---|---|---|---|
| Level on East Side of River. | | | | Level on West Side of River. | | | | Level on West Side of River. | | | | Level on East Side of River. | | | |
| + On 149B | — On 149C | v | $v^2$ | — On 149C | + On 149B | v | $v^2$ | — On 149C | + On 149B | v | $v^2$ | + On 149B | — On 149C | v | $v^2$ |
| *m.* | *m.* | *mm.* | *mm.* | *m.* | *m.* | *mm.* | *mm.* | *m.* | *m.* | *mm.* | *mm.* | *m.* | *m.* | *mm.* | *mm.* |
| 1.8450 | 0.993 | —2.6 | 6.76 | 1.5710 | 2.483 | —1.3 | 1.69 | 1.6910 | 2.617 | —0.1 | 0.01 | 1.9755 | 1.138 | —0.5 | 0.25 |
| | 0.987 | +3.4 | 11.56 | | 2.483 | —1.3 | 1.69 | | 2.613 | +3.9 | 15.21 | | 1.139 | —1.5 | 2.25 |
| | 0.991 | —0.6 | 0.36 | | 2.481 | +0.7 | 0.49 | | 2.618 | —1.1 | 1.21 | | 1.138 | —0.5 | 0.25 |
| | 0.989 | +1.4 | 1.96 | | 2.482 | —0.3 | 0.09 | | 2.617 | —0.1 | 0.01 | | 1.136 | +1.5 | 2.25 |
| | 0.985 | +5.4 | 29.16 | | 2.484 | —2.3 | 5.29 | | 2.617 | —0.1 | 0.01 | | 1.138 | —0.5 | 0.25 |
| | 0.986 | +4.4 | 19.36 | | 2.482 | —0.3 | 0.09 | | 2.615 | +1.9 | 3.61 | | 1.135 | +2.5 | 6.25 |
| | 0.989 | +1.4 | 1.96 | | 2.482 | —0.3 | 0.09 | | 2.619 | —2.1 | 4.41 | | 1.138 | —0.5 | 0.25 |
| | 0.993 | —2.6 | 6.76 | | 2.479 | +2.7 | 7.29 | | 2.617 | —0.1 | 0.01 | | 1.138 | —0.5 | 0.25 |
| | 0.996 | —5.6 | 31.36 | | 2.478 | +3.7 | 13.69 | | 2.617 | —0.1 | 0.01 | | 1.137 | +0.5 | 0.25 |
| 1.8450 | 0.995 | —4.6 | 21.16 | 1.5710 | 2.483 | —1.3 | 1.69 | 1.6910 | 2.619 | —2.1 | 4.41 | 1.9755 | 1.138 | —0.5 | 0.25 |
| M=1.8450 | M=0.9904 | | $\Sigma v^2$=130.40 | M=1.5710 | M=2.4817 | | $\Sigma v^2$=32.10 | M=1.6910 | M=2.6169 | | $\Sigma v^2$=28.90 | M=1.9755 | M=1.1375 | | $\Sigma v^2$=12.50 |
| | *mm.* $r = 2.57$ $r_o = 0.81$ | | | | *mm.* $r = 1.27$ $r_o = 0.40$ | | | | | *mm.* $r = 1.21$ $r_o = 0.38$ | | | | *mm.* $r = 0.80$ $r_o = 0.25$ | |

**Left-hand page.**

Date: *Dec. 21, 1911.* Rods: *A. & C.* Temp., *8° C.*

LEVEL No. 2.

| Level on West Side of River. | | | | Level on East Side of River. | | | |
|---|---|---|---|---|---|---|---|
| — On 149C | + On 149B | v | v² | + On 149B | — On 149C | v | v² |
| *m.* | *m.* | *mm.* | *mm.* | *m.* | *m.* | *mm.* | *mm.* |
| 1.5720 | 2.473 | +2.2 | 4.84 | 1.9050 | 1.032 | +2.8 | 7.84 |
| | 2.473 | +2.2 | 4.84 | | 1.038 | —3.2 | 10.24 |
| | 2.476 | —0.8 | 0.64 | | 1.035 | —0.2 | 0.04 |
| | 2.476 | —0.8 | 0.64 | | 1.033 | +1.8 | 3.24 |
| | 2.476 | —0.8 | 0.64 | | 1.033 | +1.8 | 3.24 |
| | 2.474 | +1.2 | 1.44 | | 1.035 | —0.2 | 0.04 |
| | 2.477 | —1.8 | 3.24 | | 1.035 | —0.2 | 0.04 |
| | 2.476 | —0.8 | 0.64 | | 1.037 | —2.2 | 4.84 |
| | 2.476 | —0.8 | 0.64 | | 1.035 | —0.2 | 0.04 |
| 1.5720 | 2.475 | +0.2 | 0.04 | 1.9046 | 1.035 | —0.2 | 0.04 |
| M = 1.5720 | M = 2.4752 | | Σv² = 17.60 | M = 1.9048 | M = 1.0348 | | Σv² = 29.60 |

*mm.* $r = 0.94$, $r_o = 0.30$ *mm.* $r = 1.22$, $r_o = 0.39$

$$r = 0.6745\sqrt{\frac{\Sigma v^2}{n-1}}$$

**Right-hand page.**

Sun: *C.* Wind: *C.* Time: *1 P.M. to 3 P.M.*

LEVEL No. 1.

| Level on East Side of River. | | | | Level on West Side of River. | | | |
|---|---|---|---|---|---|---|---|
| + On 149B | — On 149C | v | v² | — On 149C | + On 149B | v | v² |
| *m.* | *m.* | *mm.* | *mm.* | *m.* | *m.* | *mm.* | *mm.* |
| 2.0307 | 1.166 | —3.6 | 12.96 | 1.7083 | 2.611 | +1.1 | 1.21 |
| | 1.164 | —1.6 | 2.56 | | 2.611 | +1.1 | 1.21 |
| | 1.165 | —2.6 | 6.76 | | 2.614 | —1.9 | 3.61 |
| | 1.162 | +0.4 | 0.16 | | 2.610 | +2.1 | 4.41 |
| | 1.165 | —2.6 | 6.76 | | 2.612 | +0.1 | 0.01 |
| | 1.163 | —0.6 | 0.36 | | 2.612 | +0.1 | 0.01 |
| | 1.163 | —0.6 | 0.36 | | 2.615 | —2.9 | 8.41 |
| | 1.158 | +4.4 | 19.36 | | 2.610 | +2.1 | 4.41 |
| | 1.158 | +4.4 | 19.36 | | 2.614 | —1.9 | 3.61 |
| 2.0301 | 1.160 | +2.4 | 5.76 | 1.7083 | 2.612 | +0.1 | 0.01 |
| M = 2.0304 | M = 1.1624 | | Σv² = 74.40 | M = 1.7083 | M = 2.6121 | | Σv² = 26.90 |

*mm.* $r = 1.94$, $r_o = 0.61$ *mm.* $r = 1.17$, $r_o = 0.37$

$$r_o = \frac{r}{\sqrt{n}} = 0.6745\sqrt{\frac{\Sigma v^2}{n(n-1)}}$$

## Record Forms.

The form of record used for recording the notes of such a river crossing is shown on pages 44 and 45. The field book is prepared as shown, separate columns being provided for the readings with each instrument on each side of the river, and also for the residual errors and the squares of the residuals. It will be noted that all readings taken on B. M. 149B are shown as plus or backsight readings, no matter from which side of the river they are taken, and all readings on B. M. 149C are therefore minus. This is in accordance with the principle previously stated that both crossings be considered as proceeding in a forward direction.

The location of each instrument is noted, also the distance between the bench marks, the date, the time of day, the rods used, the mean temperature, and the conditions of sun and wind.

To avoid any possibility of error, each rod reading is connected with the point on which it was taken by the note at the head of the column, and the arithmetical mean is shown at the foot.

## Precision of Observations.

In order to obtain a measure of the precision of the observations, each rod reading is subtracted from the mean of its set and the algebraic differences are entered in the column headed v. These differences are the residuals from the mean and are called the residual errors. The algebraic sum of the residuals must always equal zero.

An inspection of the residuals will show how much each reading differs from the mean, and these differences must be limited. On river crossings from 600 to 800 meters in width, any reading differing more than 1 centimeter from the mean would be rejected. The column headed $v^2$ contains the squares of the residuals, and from their sum is obtained the probable error of a single observation and also the probable error of the mean. In the form shown above, the probable error of a single observation is shown equal to $r$, and the probable error of the mean is shown equal to $r_0$. This means that any one observation may be in error by the amount $\pm r$ and the arithmetical mean by the amount $\pm r_0$. The formulas employed in obtaining the probable errors are shown below the readings, in which $n$ is the number of observations taken. The values of $r$ and $r_0$ should not differ very much providing the conditions are the same for each set of readings. However, a flash of sunlight or a gust of wind will sometimes cause erratic readings on one side of the river, and so increase the probable error.

The method of reduction is illustrated in the form shown on page 47. But little explanation is needed in connection with this form. The third column on the left-hand page shows the direction in which the line progresses. The fourth column, the distance between the bench marks in kilometers. The second column on the right-hand page gives the correction due to the excess of length of the rods at 0°C., the mean length of the rods A and C being 0.133 millimeter too long on each meter. There is no temperature correction necessary in this case, and there would not be any until the temperature should rise to 15° C. or more. At 15° C., the correction applied would be + 0.1 millimeter.

It will be seen that the mean difference of elevation $+0^{m}.8824$ for both levels is the mean derived from all the observations made on December 20, and that the mean difference of elevation $+0^{m}.8863$ is the mean derived from the observations of December 21. The final mean difference of elevation $+0^{m}.8843$ is the mean of the determinations of both days.

When the difference of elevation between the two bench marks on opposite sides of a river is very small, it is especially necessary to be careful of the signs; for, even

**Left-hand page.** **Right-hand page.**

BOARD OF ESTIMATE AND APPORTIONMENT
CITY OF NEW YORK

## COMPUTATION OF RECIPROCAL LEVELS.

Line: *Astoria to Carl Schurz Park.* *New York City, N. Y.* Observers: *F.W.K. & J.S.* Year: *1911.*

| B.M'S. | Position of Instrument. | For-ward or Back-ward | Dis-tance in Kilo-meters | Mean Rod Readings. | | Approx-imate Difference of Elev. | Mean Temp. of Rods. | Corrections. | | | Difference of Elevation. | | Sun-shine or Cl'dy | Wind | Date and Hour. |
|---|---|---|---|---|---|---|---|---|---|---|---|---|---|---|---|
| | | | | | | | | Index Correc-tion | Length of Rod | Temp. of Rod | Each Line | Mean | | | |
| | | | | *m.* | *m.* | *m.* | °*C.* | *mm.* | *mm.* | *mm.* | *m.* | *m.* | | | |
| 149B–149C | Level No. 2, on E/S of river.... | F | 0.566 | 1.8450 | 0.9904 | +0.8546 | 5 | 0.0 | +0.1 | 0.0 | +0.8547 | +0.8827 | C. | C. | 12/20–10:00 |
| 149B–149C | Level No. 2, on W/S of river.... | F | | 2.4817 | 1.5710 | +0.9107 | 5 | 0.0 | +0.1 | 0.0 | +0.9108 | | C. | C. | 2:00 |
| 149B–149C | Level No. 1, on W/S of river.... | F | 0.566 | 2.6169 | 1.6910 | +0.9259 | 5 | 0.0 | +0.1 | 0.0 | +0.9260 | +0.8820 | C. | C. | 12/20–10:00 |
| 149B–149C | Level No. 1, on E/S of river.... | F | | 1.9755 | 1.1375 | +0.8380 | 5 | 0.0 | +0.1 | 0.0 | +0.8381 | | | | 2:00 |
| | | | | | | | | Mean for both levels = | | | | +0.8824 | | | |
| 149B–149C | Level No. 2, on W/S of river.... | F | 0.566 | 2.4752 | 1.5720 | +0.9032 | 8 | 0.0 | +0.1 | 0.0 | +0.9033 | +0.8867 | C. | C. | 12/21– 1:00 |
| 149B–149C | Level No. 2, on E/S of river.... | F | | 1.9048 | 1.0348 | +0.8700 | 8 | 0.0 | +0.1 | 0.0 | +0.8701 | | | | 3:00 |
| 149B–149C | Level No. 1, on E/S of river.... | F | 0.566 | 2.0304 | 1.1624 | +0.8680 | 8 | 0.0 | +0.1 | 0.0 | +0.8681 | +0.8860 | C. | C. | 12/21– 1:00 |
| 149B–149C | Level No. 1, on W/S of river.... | F | | 2.6121 | 1.7083 | +0.9038 | 8 | 0.0 | +0.1 | 0.0 | +0.9039 | | C. | C. | 3:00 |
| | | | | | | | | Mean for both levels = | | | | +0.8863 | | | |
| | | | | | | | | Final difference of elev. = | | | | +0.8843 | | | |

Abbreviations. S=sunshine. C=cloudy. S & C=alternate sunshine and shade.
Abbreviations, strength of wind: S=strong. M=moderate. C=calm.

Abbreviations, direction of progress relatively to sun: $\frac{T}{F}$=within 45° of directly $\frac{\text{toward}}{\text{from}}$ the sun. $\frac{T_R}{F_R}=\frac{\text{toward}}{\text{from}}$ sun, but at an angle of more than 45° with sun to right. $\frac{T_L}{F_L}$=ditto, with sun to left. $\frac{R}{L}$=sun to $\frac{\text{right}}{\text{left}}$ and nearly at right angles to line. The same abbreviations also apply to the direction of progress relative to the wind.

though both crossings are assumed to be made in the forward direction, one of the values may be plus and the other minus. In such a case, the true difference of elevation is the algebraic sum of the two differences of elevation divided by two. No separate abstract of results is necessary for this work, these being incorporated in the usual abstract covering that territory.

Figs. 1, 2 and 3 in illustration No. 12 show the types of targets used in the reciprocal leveling. These targets were made by the men employed on the work.

## Conclusions on the Accuracy of Reciprocal Leveling.

The best time of day for reciprocal leveling is between 10 a.m. and 3 p.m., as between these hours the vertical refraction is less variable than at other times of the day. On a wide crossing, observations should be taken on at least two days so as to average up the atmospheric conditions. The higher the line of sight above the water, the more accurate the results are apt to be.

It has been frequently stated that a river crossing with sights 1 kilometer long or longer would be much weaker than the remainder of a precise level line, even if the best of targets were used and reciprocal simultaneous sights were taken. It can be here positively stated, however, that the results of river crossings made on different days by the method above described do not as a rule vary beyond the limit of error fixed for land leveling, and that wherever these crossings form part of level circuits large or small that portion of the circuit has proven to be as strong as the remainder. In fact, some of the best circuit closures are those representing both large and small circuits containing one or more river crossings.

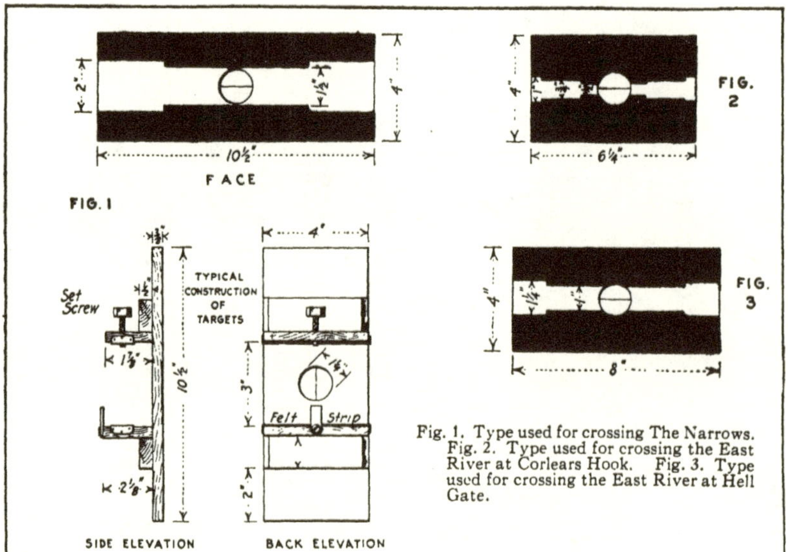


Fig. 1. Type used for crossing The Narrows. Fig. 2. Type used for crossing the East River at Corlears Hook. Fig. 3. Type used for crossing the East River at Hell Gate.

No. 12.—Types of Targets Used in Reciprocal Leveling.

A number of engineers have expressed the opinion that river crossings would be more accurate if the method of trigonometric leveling were employed. In no case can this work attain a degree of precision comparable with that done by spirit-leveling.

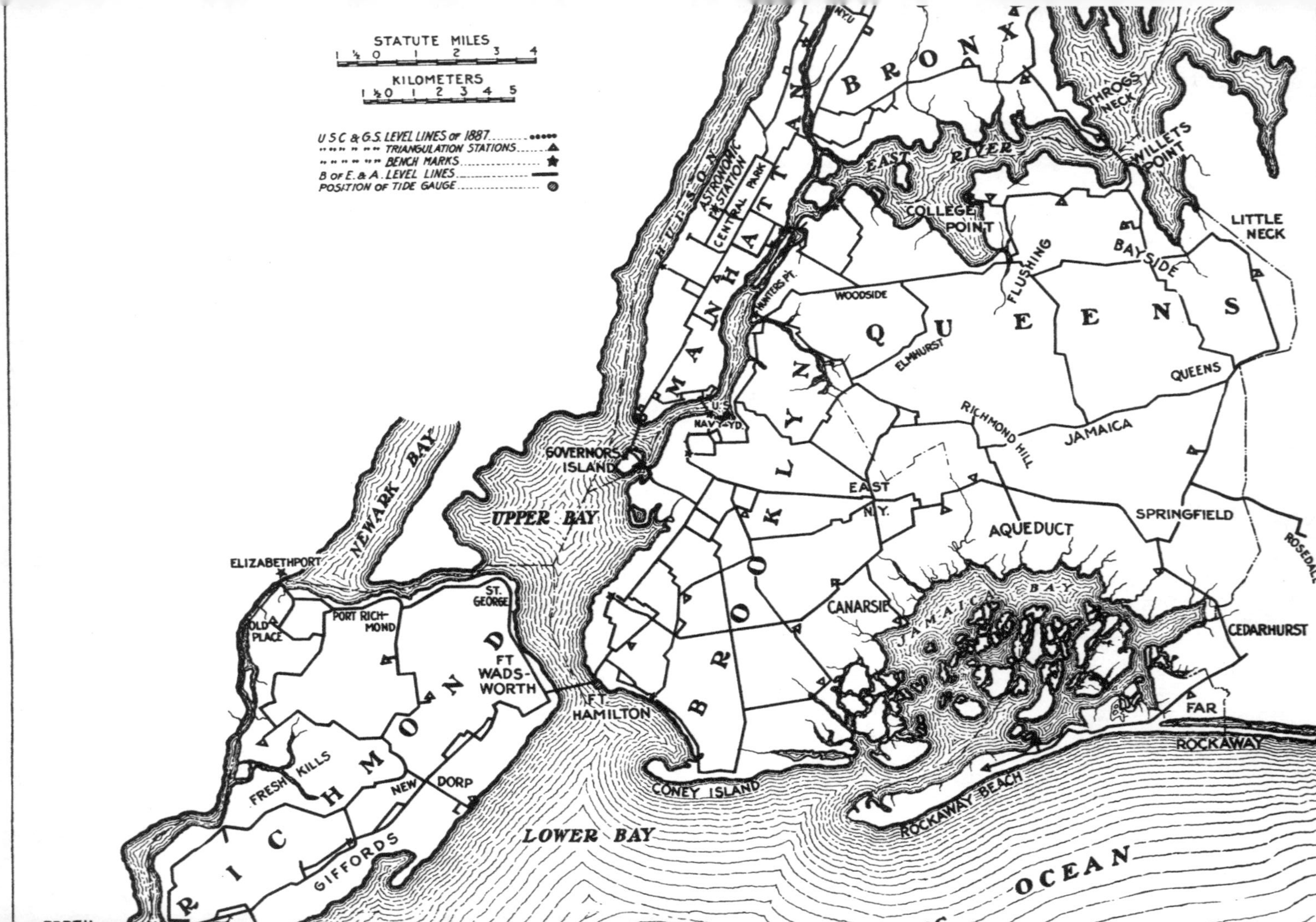

STATUTE MILES
1 ½ 0 1 2 3 4
KILOMETERS
1 ½ 0 1 2 3 4 5
U S C & G.S. LEVEL LINES of 1887
TRIANGULATION STATIONS
BENCH MARKS
B of E. & A. LEVEL LINES
POSITION OF TIDE GAUGE
BRONX
THROGS NECK
WILLETS POINT
EAST RIVER
COLLEGE POINT
LITTLE NECK
BAYSIDE
FLUSHING
HUDSON
ASTRONOMIC STATION
CENTRAL PARK
MANHATTAN
N.Y.U.
HUNTERS PT.
WOODSIDE
QUEENS
ELMHURST
QUEENS
RICHMOND HILL
JAMAICA
U.S. NAVY YD.
GOVERNORS ISLAND
BROOKLYN
EAST N.Y.
SPRINGFIELD
AQUEDUCT
ROSEDALE
NEWARK BAY
UPPER BAY
ELIZABETHPORT
ST. GEORGE
PORT RICHMOND
OLD PLACE
FT WADSWORTH
FT HAMILTON
CANARSIE
JAMAICA BAY
CEDARHURST
FAR ROCKAWAY
RICHMOND
FRESH KILLS
NEW DORP
GIFFORDS
CONEY ISLAND
ROCKAWAY BEACH
LOWER BAY
OCEAN

The Coast Survey has done a large amount of trigonometric leveling in connection with its triangulation work, and in order to prevent an accumulation of errors in the elevations thus determined, connection is made at various points with precise level bench marks, and the trigonometric leveling is adjusted to fit the precise leveling between these points. Vertical angles for determining differences of elevation are usually very small and a large probable error may occur in a computed elevation, even when the probable error of the vertical angle is not large.

## THE PRECISE LEVEL NET.

Illustration No. 13 shows the Board of Estimate and Apportionment precise level net. This system extends from Perth Amboy, N. J., to Dobbs Ferry, N. Y., encircling and traversing each of the boroughs en route, as shown by the full heavy lines on the map.

In 1881 the U. S. Coast Survey transcontinental line of spirit-levels was commenced at the Sandy Hook tide-gage and continued westward. The elevations of this line were based upon continuous tidal observations made with a self-registering tide-gage during the six years 1876-1881 inclusive. The probable error of the mean sea level determination resulting from these tidal observations is $\pm$ 9.1 millimeters.

In 1886-87, the U. S. Coast Survey ran a precise level line from Sandy Hook, N. J., to Dobbs Ferry, N. Y., and the elevations along this line were based upon the same tidal observations at Sandy Hook as were used in connection with the transcontinental line. At the Coast Survey bench mark F, on Raritan Bay, a difference of 32.3 millimeters was developed between the two level lines, the elevation given by the line of 1886-87 being the lower. After an exhaustive examination of the tidal observations taken at twenty-two points in New York Harbor in addition to those made at Sandy Hook, and with which the line to Dobbs Ferry had been connected, it was decided to use the elevation given by the 1881 line for bench mark F at Raritan Bay. To all other elevations along the Sandy Hook-Dobbs Ferry line there was applied a correction of + 32.3 millimeters in order to make the elevation of bench mark F, as determined in 1886-87, agree with the elevation fixed for this point in connection with the line run in 1881. All elevations subsequently referred to in this report as having been established by the Coast Survey level line of 1886-87 are therefore understood to mean the corrected elevations. The Coast Survey precise level line of 1886-87, from Sandy Hook to Perth Amboy is shown as a dotted line on the map, and the few bench marks that are left, from Perth Amboy to Dobbs Ferry, are shown as heavy stars.

When the running of the precise level lines in New York City was commenced by the Board of Estimate it was decided to determine whether or not the old bench marks established by the Coast Survey in 1886-87 had changed in elevation. It was also decided to ascertain whether mean sea level at Sandy Hook was at the same elevation as mean sea level at Fort Hamilton, as determined by the long series of tidal observations taken there by the Coast Survey.

While it had been intended in the beginning to base the elevations resulting from the Board of Estimate leveling upon the elevations of the existing Coast Survey bench marks, a number of discrepancies were developed as a result of the first season's leveling and it was then decided not to fix any elevations until all the old Coast Survey bench marks had been connected and the results examined.

The Board of Estimate system of precise level lines begins at a bench mark in Perth Amboy, N. J., the elevation of which, above mean sea level at Sandy Hook, had been established in 1887 by the Coast Survey. This bench mark had been connected with the transcontinental line of spirit-levels and is also a bench mark of the State Geological Survey.

Starting from this bench mark the level line crossed the Arthur Kill to Tottenville, and proceeding along the easterly side of Richmond Borough to Fort Wadsworth, crossed the Narrows to Fort Hamilton and connected with the Coast Survey tide-staff. When the resulting difference of elevation was applied to the bench mark at Perth Amboy, a discrepancy of 4.0 millimeters was developed. As the probable error of the determination of mean sea level at Sandy Hook is $\pm$ 9.1 millimeters, the discrepancy of 4.0 millimeters developed at Fort Hamilton tide-staff is well within the probable error, and it can reasonably be assumed that mean sea level is the same at both places. Also, this difference of 4.0 millimeters up to Fort Hamilton tide-staff corresponds to the allowable error on one kilometer of leveling and when distributed along the level line from Perth Amboy is only 0.14 millimeter per kilometer. While the difference of 4.0 millimeters could therefore be safely assumed to be due to errors in the leveling, at the same time on the other hand, it might with even greater likelihood be attributed to error in the mean sea level determination at Sandy Hook. In view of this evidence, therefore, the conclusion must be reached that mean sea level at Fort Hamilton is at the same elevation as mean sea level at Sandy Hook.

Accepting the Coast Survey elevation of 1.8410 meters for the top of the fixed tide-staff at Forth Hamilton, a determination based upon 19 years of continuous tidal observations, and applying to it the measured difference of elevation up to primary bench mark J at Bay Ridge, we obtain for J an elevation of 13.2995 meters, which is only 0.5 millimeter or less than 0.002 foot below the elevation established for it by the Coast Survey precise level line of 1887. Again, by applying to the Coast Survey elevation of the tide-staff the measured difference of elevation up to primary bench mark K at Bath Beach, we obtain for K an elevation of 8.3277 meters, which is only 1.2 millimeters or about 0.004 foot below the Coast Survey elevation established for it in 1887. It may be of interest to note in this connection that had the mean of sixteen (instead of nineteen) years of tidal observations been used as a basis of comparison, the elevations of bench marks J and K would have been slightly greater, instead of slightly less, than those established in 1887. These additional facts also prove that mean sea level at Sandy Hook and at Fort Hamilton are at the same elevation.

Having reached this conclusion, the Board of Estimate level lines in the Borough of Richmond were adjusted to a datum between the initial bench mark at Perth Amboy and mean sea level at Fort Hamilton, as determined from 19 years of continuous tidal observations by the United States Coast and Geodetic Survey. In this adjustment, the difference developed at Coast Survey bench mark N, at Giffords, was so small that the elevation established for it in 1887 was held.

The large circuit in the Borough of Richmond embraces almost the entire borough and is divided into five smaller circuits, one of the dividing lines passing through the highest land in the city. The largest circuit has a perimeter of 67 kilometers or 42 miles, and, before the rod length and rod temperature corrections were applied, failed to close by 6 millimeters or 0.02 foot. The peculiar loop in the westerly part of the borough may need some explanation.

The reason for the loop is that the Fresh Kills lie just west of it and although Bridge Avenue runs across the Kills, it is a narrow road with a tar-like pavement and affords no sites whatever for permanent bench marks. As a good route for the level line ran through New Springville and Richmond and the country was liberally supplied with sites for good bench marks, the connection between Linoleumville and Fresh Kills was made by way of these villages.

It will be seen from the map that the level lines extend easterly to Little Neck, Floral Park, Springfield, Cedarhurst, Far Rockaway, and Belle Harbor at Rockaway Beach. At Seaside, the level line connects with the Department of Docks and Ferries tide-gage in Jamaica Bay. On the north the level lines traverse the cities of Yonkers and Mount Vernon, passing through such country as afforded the best sites for permanent bench marks.

The small heavy triangles shown on the map indicate the Coast Survey triangulation stations for which elevations were established by the Board of Estimate leveling. The peculiar convention just west of Central Park indicates the position of a Coast Survey astronomic station in the grounds of the American Museum of Natural History. Its elevation was also determined.

## Changed Elevations of Old Bench Marks.

Many important and interesting facts were developed as a result of the Board of Estimate leveling. The most noteworthy of these is the fact that many of the Coast Survey bench marks established in 1886-87 have apparently changed materially in elevation. These changes have occurred no doubt on account of the settlement of the buildings on which the bench marks were placed, and not, in the opinion of the writer, on account of coastal subsidence.

Instead of holding these bench marks and thus vitiating the high precision of the Board of Estimate leveling, new elevations were established for most of them. The old and new elevations are shown in the table on page 52, and the amount that the bench marks have settled is shown in the last column. A study of these differences will prove how unwise it would have been to hold the old elevations established by the Coast Survey.

While the establishment of new elevations for the Coast Survey bench marks may result in some temporary confusion in the bureaus whose level systems were adjusted to them, it is believed there is a greater desire for the correct elevations, and that new work will rapidly adjust itself to the new figures.

As the assignment of new elevations to these bench marks will no doubt occasion some comment it is proposed briefly to explain the reason for such action.

It has long been a matter of common knowledge that Coast Survey bench marks E at Corlears Hook and F at Hudson Street, Brooklyn, have not maintained the elevations established for them in 1887. It is moreover to be noted that although the elevations of some other points established at that time were likewise suspected to have undergone changes, the absence of a general system of precise levels connecting them all made it impossible to arrive at any definite conclusions in the matter. It is natural to suppose that a heavy building or other structure at or near the waterfront will settle more or less in the course of time, particularly if not well founded. Signs of such settlement are very frequent. If, therefore, the bench marks established on such structures several decades ago are examined and new determinations of their elevations made, it is reasonable to expect evidence of some discrepancies.

An examination of the Coast Survey bench marks affected shows the following conditions: Considering first the tidal bench mark at Elizabethport, N. J., where a discrepancy of 0.0403 foot was developed, it is seen that this bench is located on an old foundry building erected on filled-in ground, and only about 100 feet from the waterfront. A settlement of the character noted might well be anticipated under such circumstances after the lapse of nearly a quarter of a century.

Bench mark H at Governors Island is at the northeast corner of a large stone on top of the sea wall and it is perfectly safe to assume that, in such a structure, a settlement of the amount indicated has occurred since 1887.

At bench mark I, a much larger discrepancy is found and the reasons therefor are very plain. Heavy trucking takes place over the sill on which the bench mark is located, and the wall adjacent to the bench on the south has been repaired several times.

Bench mark C at Brooklyn is located on the Kings County Hall of Records, a building which was practically reconstructed about 12 years ago. On this account the bench has always been viewed with suspicion. Subsequently, the Interborough Subway

ELEVATIONS OF U.S.C. & G.S. BENCH MARKS

IN NEW YORK CITY AND VICINITY.

| Place. | Designation of B.M. | Elevation U. S. C. & G. S. | | Elevation B. of E. & A. | B. of E. & A. Minus U.S.C. & G.S |
|---|---|---|---|---|---|
| | | *Meters* | *Feet* | *Feet* | *Feet* |
| At Perth Amboy, N. J | State Geological Survey | 18.5763 | 60.9458 | 60.9458 | 0.0 |
| Near Gifford, N. Y | N. | 21.6607 | 71.0651 | 71.0651 | 0.0 |
| At Elizabethport, N. J | Elizabethport Tidal | 3.7910 | 12.4376 | 12.3973 | —0.0403 |
| At Bath Beach, N. Y | K. | 8.3289 | 27.3257 | 27.3257 | 0.0 |
| At Bay Ridge, N. Y | J. | 13.3000 | 43.6351 | 43.6351 | 0.0 |
| At Governors Island, N. Y | H. | 2.4439 | 8.0180 | 7.9993 | —0.0187 |
| At Governors Island, N. Y | I. | 2.6313 | 8.6328 | 8.5394 | —0.0934 |
| At Brooklyn, N. Y | C. | 17.0737 | 56.0160 | 55.9694 | —0.0466 |
| At Brooklyn, N. Y | F. | 3.1708 | 10.4029 | 10.1893 | —0.2136 |
| At New York, N. Y | E. | 4.0819 | 13.3920 | 11.6289 | —1.7631 |
| At Hunters Point, N. Y | B. | 2.4207 | 7.9419 | 7.8815 | —0.0604 |
| At Pot Cove, N. Y | No. 3 | 5.1032 | 16.7428 | 16.6883 | —0.0545 |
| At Polhemus Dock, N. Y | No. 4a | 2.0778 | 6.8169 | 6.7621 | —0.0548 |
| At New York, N. Y | Tidal, Foot of W. 42nd Street | 4.5927 | 15.0679 | 14.6381 | —0.4298 |
| At Dobbs Ferry, N. Y | U. | 4.5375 | 14.8868 | 14.6978 | —0.1890 |
| At Dobbs Ferry, N. Y | V. | 2.9357 | 9.6315 | 9.4425 | —0.1890 |
| At Dobbs Ferry, N. Y | W. | 2.8174 | 9.2434 | 9.0544 | —0.1890 |
| At College Point, N. Y | Tidal Station No. 68 | 2.9225 | 9.5882 | 9.5882 | 0.0 |
| At New York, N. Y | No. 5 | 2.2991 | 7.5430 | 7.5430 | 0.0 |

was built passing directly alongside the building on Fulton Street, and large cracks are now apparent only a few feet south and west of the bench mark. It is not to be wondered at, therefore, that the bench mark has settled 0.0466 foot.

Similar conditions prevail at many points along the line of the subway. Public Service Commission bench mark 18b, at the southwest corner of the Brooklyn Borough Hall, is now 1.603 feet below Public Service Commission bench mark 18a, on the old Dime Savings Bank almost opposite. Before subway construction began the difference of elevation between the two bench marks was 1.562 feet.

Bench mark F, at the foot of Hudson Street, Brooklyn, is on a dilapidated stone building on filled-in ground directly at the waterfront. This building has long been used for the storage of coal and the fact of its having settled is well known.

Bench mark E at Corlears Hook is located on a brick building still in use as a warehouse. As the discrepancy developed here is very large some definite reason therefore was sought.

The bench mark corresponds with its description and hence there is no question as to its identity. The building containing the bench mark is a heavy structure founded on piling and situated directly on the waterfront. During recent years repairs have been made to this building. A structure of this character used for the storage of hundreds of tons of merchandise and depending on friction for its stability does not afford a suitable site for a permanent bench mark.

Bench mark B at Hunters Point is on a brick building founded on filled-in ground less than 100 feet from the East River. Moreover, the East River tunnels of the Pennsylvania Railroad pass very close to the building and during construction work signs of settlement were very apparent throughout the entire locality. Under such conditions, therefore, it would be impossible for bench mark B to have maintained its original elevation.

Bench mark No. 3 between Pot Cove and Polhemus, is located on an old stone step at the waterfront.

Bench mark No. 4a is at the corner of an old stone dock and is now covered with several courses of masonry.

The tidal bench mark at the foot of West 42nd Street is on a brick building founded on filled-in ground about 100 feet from the Hudson River. This building is frequently surrounded by thousands of tons of coal and a material settlement under such conditions is to be expected.

Bench mark V, at Dobbs Ferry, is on a brick building erected on filled-in ground only a few feet distant from the Hudson River, and the wall containing the bench mark has apparently been repaired several times since the bench mark was established. Large cracks in the south side of the building and just west of the bench mark show that considerable settlement has taken place. As bench marks U and W were subsequently established from bench mark V after the latter had already settled, they are subject to the same correction as bench mark V. The measured difference of elevation between bench marks U, V and W differed so little from those determined by the Coast Survey that they were accepted as being correct. The fact of bench mark V having changed in elevation is also proven by the difference of elevation obtained as a result of the leveling done by the Board of Water Supply in 1907.

The unadjusted difference of elevation resulting from the Board of Estimate leveling between Coast Survey bench mark No. 12 at Fort Hamilton and Coast Survey bench mark W at Dobbs Ferry, is —30.143 feet. The unadjusted difference of elevation resulting from the Board of Water Supply leveling between the same bench marks is —30.130 feet. As Coast Survey bench marks U and W were established from bench mark V, and the Coast Survey differences of elevation between them were checked, it is seen that the Board of Water Supply leveling checks the discrepancies developed at Dobbs Ferry.

Another important fact developed is that the old T cut bench mark at the southwest corner of the New York City Hall is almost 3 inches lower than its old elevation indicates. When this bench mark was established in 1840 its elevation was fixed at 24.413 ft. above the old monument in the grounds of Bellevue Hospital. It may be argued that the monument has settled; this is disproved by the fact that the monument still maintains its exact relation to the original line of bench marks established along 26th Street in 1880. The monument could not therefore have settled unless all of 26th Street had settled also and by the same amount.

## BOARD OF ESTIMATE AND APPORTIONMENT DATUM PLANE CHART.

One of the results of the Board of Estimate leveling is a new Datum Plane Chart, shown in illustration No. 14. This chart shows the relation between the departmental datum planes at or near their origin and mean sea level at Sandy Hook, N. J.

When the origin and elevation of the initial bench mark of a system of levels were available, and there was no evidence to show that any settlement had taken place, the difference between the planes was at once obtained by a comparison with the plane of mean sea level as developed over the city by the Board of Estimate leveling. When no origin was available, the mean of a large number of determinations on all sides of the origin and in its immediate vicinity, was taken.

All the figures on the chart above referred to differ from those furnished by the Datum Plane Committee of the Municipal Engineers in 1905, because the Board of Estimate chart is based on a direct comparison between the various datums and one extensive level system over the entire city.

Many of the figures shown on this chart are already in use by a number of the city departments. As only a small portion of the Board of Water Supply elevations are referred to mean tide at Fort Hamilton, instead of to mean sea level, the datum here assumed for that department is the same as that for the Board of Estimate. It is readily seen, however, that the elevations established for any common point in New York City by these two departments will not be identical for the reason above noted as well as on account of the fact that the Board of Water Supply accepted as correct the old elevation of Coast Survey bench marks C and B, and adjusted their results to them.

As a result of the Board of Estimate leveling the elevation of the Public Works Datum above mean sea level at Sandy Hook has been shown to be 2.75 feet instead of 2.67 feet as at present used. The difference of 2.75 feet is very constant in the neighborhood of 26th Street, but varies somewhat a few miles away. However, the place to determine the difference between the datums is at the origin, and at Bellevue Hospital the difference is 2.75 feet.

## BENCH MARKS.

Since a bench mark is designed to serve as a lasting record of the results of the leveling in a particular locality, the importance of insuring its permanency and freedom from any interference likely to change its position will be at once apparent. For this reason, in the selection and preparation of the bench marks to be established along the primary level lines it was decided to adopt the type used in large cities by the various government surveys, a type which experience has shown to be well adapted to meet the requirements above noted.

The principal types of bench marks established on the Board of Estimate leveling are shown in illustrations Nos. 15 to 17. The use of several different types of bench marks was made necessary by reason of the varying conditions in different parts of

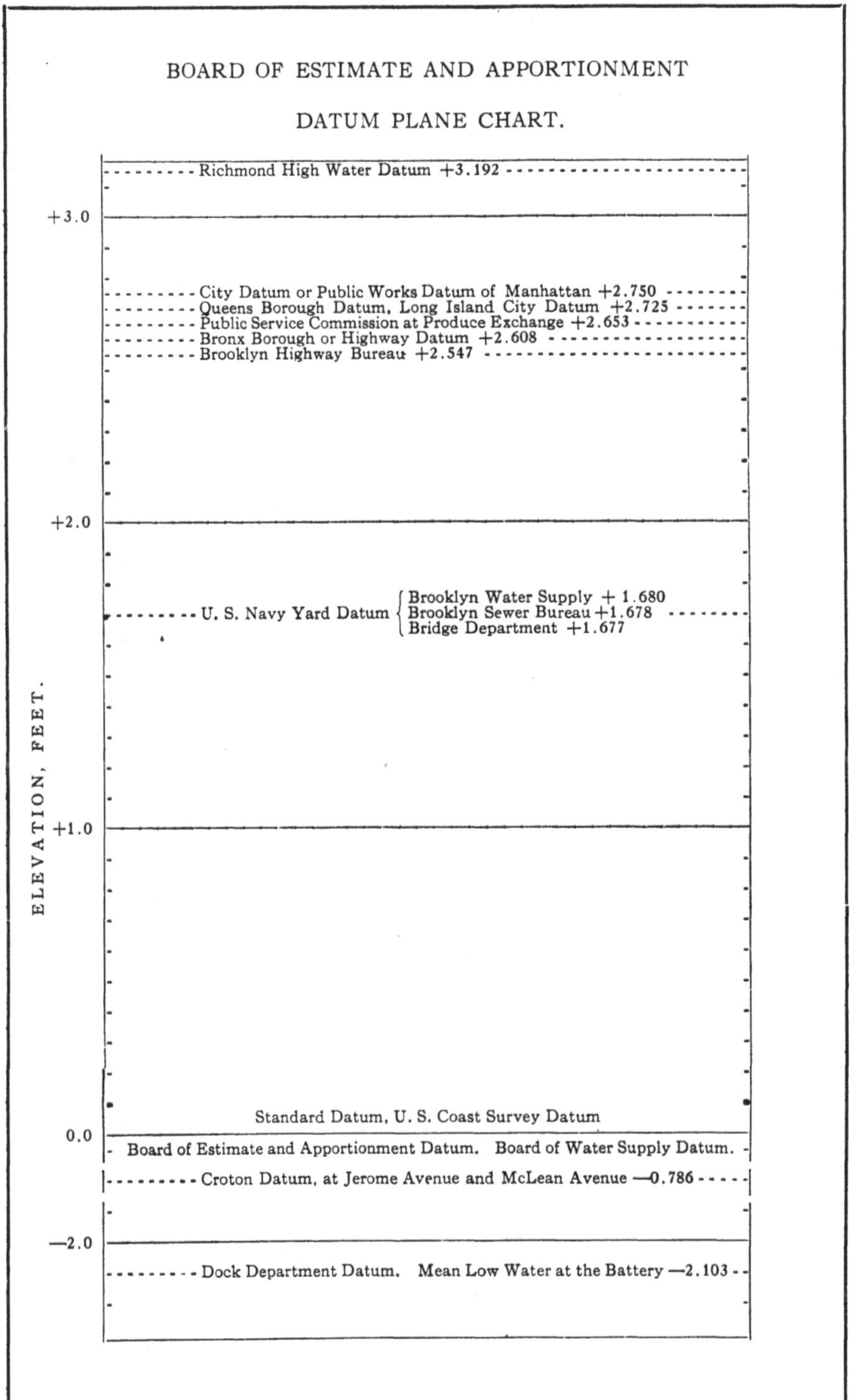


No. 14.—The New Datum Plane Chart.

the city, the guiding intent being to leave no long stretches devoid of bench marks, even though no ideal locations for placing them might be available.

A detailed description of these, as well as other types of bench marks established during the course of the work, is given in the notes on pages 89 to 91.

The types of bench marks used on primary level lines in large cities may be grouped under four general headings, viz: Marks on permanent structures, set-bolt bench marks, ground monuments and projecting bench marks.

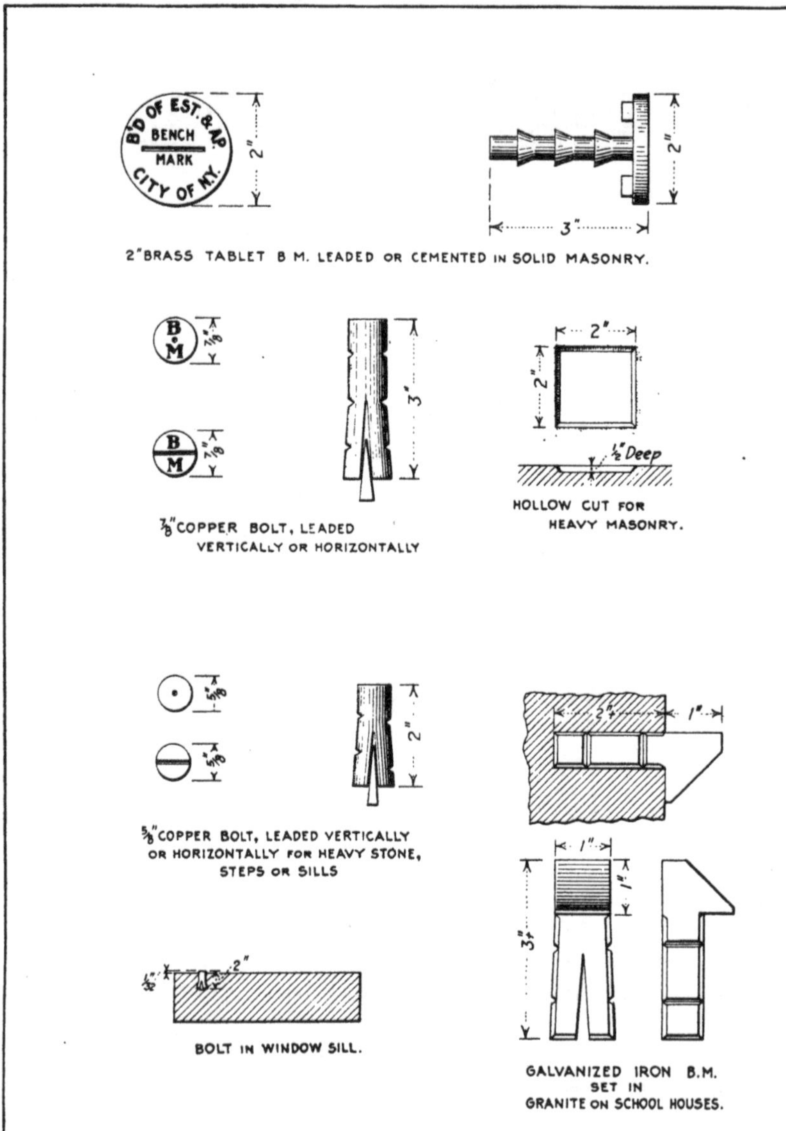


No. 15.—Standard Bench Marks. Set-bolt Types.

Observations extending over a long period of years have demonstrated that marks on massive structures are permanent and reasonably easy to locate. When the mark takes the form of an ordinary cut it is inconspicuous and for this reason unlikely to suffer from wilful interference. Such marks are usually placed on public buildings, bridge abutments, foundations, etc. It is, of course, understood that such a mark would be placed only on a structure which had been erected for a sufficient length of time to have allowed any possible settlement to have taken place.

When it was necessary to establish a bench mark of this type the mark was usually

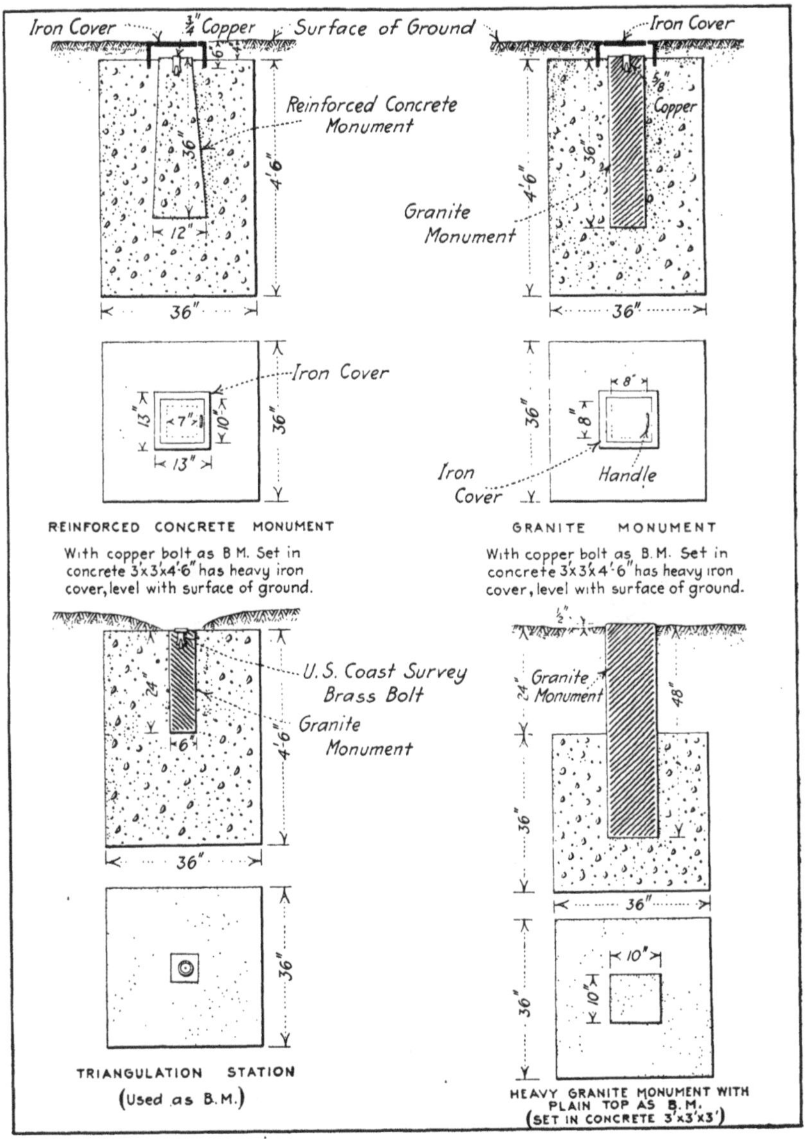


No. 16.—Standard Bench Marks. Ground Monument Types.

made in the form of a large T or a + of uniform width with V-shaped indentations at least ¼-inch deep.

An advantage of this type of bench mark is the ease and rapidity with which it can be prepared and the absence of any probability that it will be defaced or tampered with. In general, this type more nearly meets all the requirements of an ideal bench mark than does any other type.

The set-bolt type of bench mark has been extensively used on the national surveys. Marks of this character are permanent and are very readily located. If desired, they

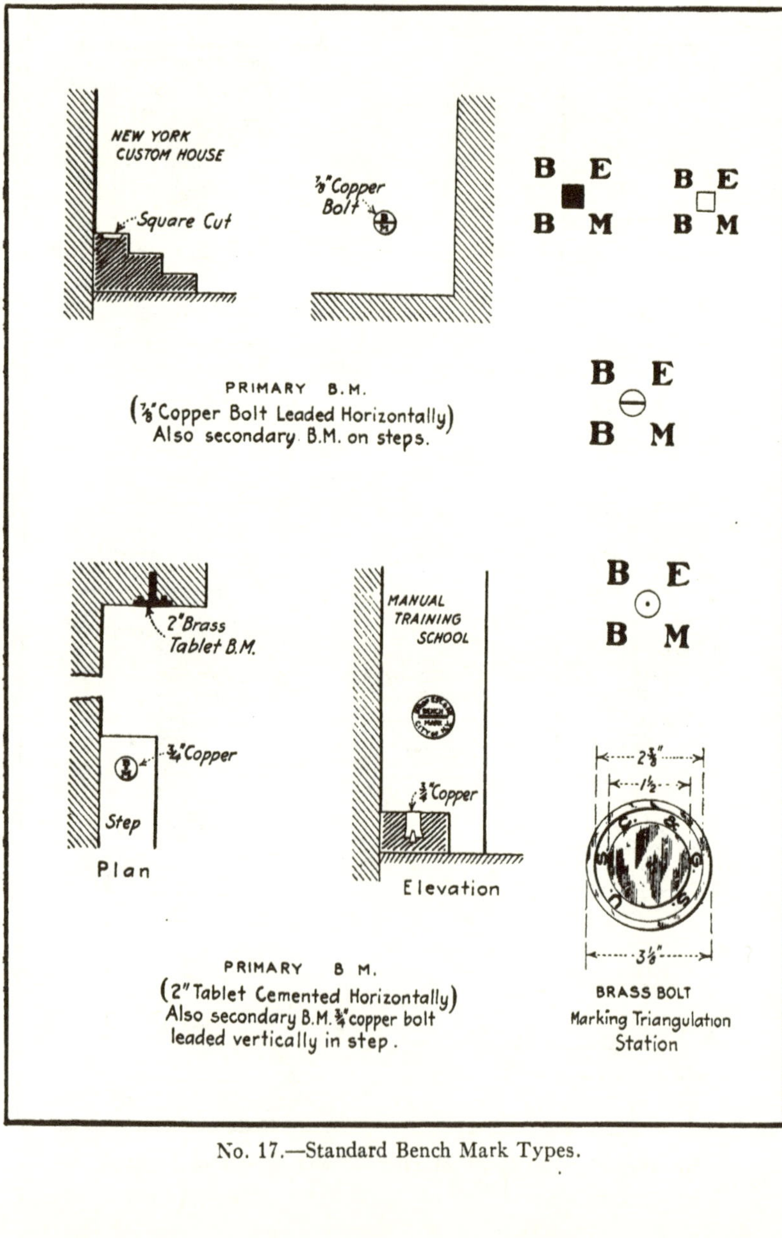


No. 17.—Standard Bench Mark Types.

can, moreover, be lettered and numbered in a manner which at once identifies them and establishes their origin. For primary bench marks this type consists mainly of brass tablets suitably inscribed and set in the masonry of substantial public buildings. This type of bench mark is very permanent, easy of access, and readily identified.

While it is at times difficult to take a reading on bench marks of the two types above described, this disadvantage is more than offset by their permanence and freedom from normal causes of interference.

Illustration No. 18 shows the method used on the Board of Estimate leveling for taking rod readings on bolts set horizontally into masonry. When the bolt was set near the ground and the wall containing it was surrounded by earth, a stake was driven until the hand level could be set upon it with the steel flange inserted in the center of the bolt. Care was always taken to ascertain that the bubble was in the center of the level vial. (See Fig. 1.)

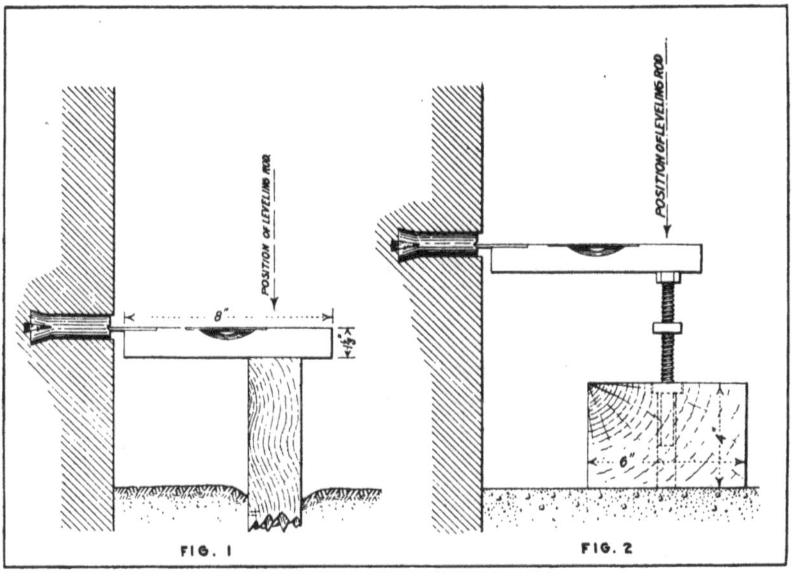


No. 18.—Method Used for Taking Rod Readings on Bolts Set Horizontally into Masonry.

If a stone or concrete walk was adjacent to the bench mark, the device shown in Fig. 2 was employed. This consists of a rectangular block of wood with a nut inserted to hold the screw upon which the level rests. The screw is raised or lowered until the bubble is in the center of the level, and it is then secured by clamping the loose nut against the block. Whenever such marks were established a secondary bench was usually placed on a step or sill nearby, as being more convenient to hold upon. It will be readily apparent that such a secondary mark would be much less permanent than the primary mark, being subject to removal or displacement at any time.

Experience has shown ground monuments set in concrete to be permanent, providing that the grade remains unchanged. This type possesses, also, the further advantage of permitting a direct reading of the rod to be taken from the bench itself. In using ground monuments some inconvenience has been encountered in winter by reason of the presence of ice between the metal cover and the top of the monument. This consideration led the writer to devise a variant of the ground monument type of bench mark, in which the top of the monument, this consisting of a massive granite block, is set slightly above the surface. No provision is made for inserting a bolt, the top of the monument being accurately set to serve as the bench.

A ground monument of this modified type was set in Poe Park, Borough of The Bronx, and is shown in illustration No. 16.

The fourth general type of bench mark used on primary level lines, namely the projecting bench mark, has been found unsatisfactory in practice for the reason that it is frequently removed, may become bent and is in general, by reason of its exposed position, subject to mishaps calculated to destroy its accuracy.

Some of the engineers interested in precise leveling in the city, however, expressed a desire for the installation of projecting bench marks for the reason that this type permits of an easy reading on the rod. To meet their views, the writer designed a special type of projecting mark, adapted to overcome, in as great a degree as possible, the objections above noted. This bench mark, which is shown in illustration No. 15, is of very substantial construction and was set on schoolhouses, where, by reason of the protection afforded by a fence, the danger of displacement was reduced to a minimum.

During the progress of the leveling in the remoter and less densely settled portions of the city it was frequently found necessary to make detours in order to establish bench marks on the most permanent structures available. Under such a procedure the benches affected are somewhat more difficult to reach, but it was held that ease of access should yield precedence to the more important considerations of permanency and freedom from interference.

The small loops and spurs appearing upon the map (illustration No. 13), are to be interpreted in the light of the above discussion.

In sections of the city where new development is proceeding at a rapid rate, notably in the outlying districts, it is frequently impracticable, even with the utmost care, to establish bench marks in a manner affording any prospect of permanence. The opening up of new thoroughfares, the widening of old roads and frequent regrading and repaving result in destroying benches on city monuments set at or near the natural surface, while marks upon buildings often lose their value by reason of changes made in the elevations of the structures to conform with the new street surface.

As an illustration of the condition above noted, may be mentioned the case of the monument on the northeast corner of Jackson Avenue and Junction Avenue, Borough of Queens. The original elevation established for this monument by the Board of Estimate leveling was 21.6598 meters, while subsequent to regrading the elevation was found to be 22.0772 meters. Were this monument to be used as an origin of levels by any one unfamiliar with the fact that regrading had been carried out, an error of 0.4174 meter or 1.369 feet, would be introduced into the work.

Arbitrary changes in the elevation of city monuments in cases where no regrading has occurred, occasionally form a basis for possible error. In the case of the monument on the northwest corner of Third Avenue and Thirteenth Street, College Point, a change in elevation of 0.0849 meter or 0.279 foot was found to have resulted from the raising of the monument from a position below the surface to one flush with the sidewalk. Occurrences of this kind emphasize the necessity of an accurate description for every bench mark, in order that the engineer having occasion to use such a mark may assure himself that the bench still occupies the same position as at the time when the original determination of its elevation was made.

In planning the marking of the Board of Estimate level lines in so far as the marks are located on buildings, etc., care was taken to place bench marks on the foundations or base courses of massive structures whenever possible, and benches were not established on copings except when provision could be made for tieing them into more permanent points in the vicinity. Bench marks on bridges were either placed as low as possible, or were located on the bridge seat.

In consequence of the action of successive frosts in the joints of masonry, it was deemed wise, whenever it was necessary to place benches in the immediate vicinity of the waterfront, to select for their location the inshore portions of the structure or wall concerned, at points as remote as possible from the water.

Progressive lifting of ashlar masonry is no doubt caused by the action of frost on water in the joints of such masonry. The increase in elevation due to this cause is, in some instances, quite appreciable and experience indicates that, upon the melting of the ice, fine sand may sift into the microscopic cracks formed by the freezing of the water, thus preventing the structure from settling back into its original position.

The considerations noted show a natural rock ledge to be the ideal location for a bench mark, and in the rather infrequent instances when such locations were available the opportunity to place benches on such outcroppings was always utilized.

Practically all of the bench marks established on this work were prepared before the level lines were run.

## Supports for Rods and Instrument.

Rod supports consisted mostly of cement sidewalks, large flagstones, stone steps and copings. The use of foot plates for rod supports on the sidewalks in the built-up portions of the city was found to be impracticable, as the danger of lateral motion was very great.

When the leveling was being carried on through the city streets the rodman was instructed to select for the turning point a smooth portion of the sidewalk. After selecting such a location the rodman would mark with a piece of red or yellow kiel a cross (+) about 2 inches long, and place his rod carefully upon the intersection. The use of such a mark insures the rod being held upon the same point for both back-sight and foresight readings. The instrument was also set upon the sidewalk and, in the case of flagstone pavement, care was taken to set the tripod legs upon stones other than those upon which the observer stood.

A turning point defined as above described gives very satisfactory results in large cities, where cement sidewalks are common, and has the advantage of requiring no preparation other than the marking. When the sidewalks were in bad condition the turning points were taken on smooth portions of stone steps or copings, and occasionally on street car tracks. The use of such turning points increased both the speed and accuracy of the leveling.

Even on steep grades a cross marked on a sidewalk or brick pavement makes an excellent turning point. For example: When leveling along East Sixth Street, Mount Vernon, between B. M. 1151 and B. M. 1153, a distance of only 1 kilometer, the difference of elevation was found to be + 25.3391 meters. On some parts of this street the grade was found to be as much as 6.8%. The pavement between the car tracks on this street was of brick and on the forward running of the line the foot plates were used as turning points. On the backward run the turning points were kiel marks on the brick pavement. The divergence (B — F) was + 1.7 millimeters. Similar comparisons made elsewhere showed that a point marked upon a sidewalk or a brick or stone pavement was just as satisfactory a turning point as the foot plate.

The very small values for the total accumulated discrepancies on the Board of Estimate level lines, together with the excellent circuit closures, prove conclusively that there was no systematic error in the leveling and that the use of turning points of the character above considered prevented any rising or settling of the rod. On many large surveys this has been a most prolific source of error.

When striving for good results in precise leveling, it is, of course, necessary to place as little dependence as possible on the rodman's judgment of what constitutes a good turning point. However, after a few weeks of careful training in the field almost any rodman is capable of selecting a good turning point, and if such be not the case, he does not belong in a precise level party.

In suburban districts the turning points consisted of foot plates and foot pins, and of marks on boulders, on street car tracks and on railroad rails.

When the leveling was done along railroad lines the top of the rail was used for turning points almost exclusively. This has been the general practice on Coast Survey precise leveling for many years and is conducive to both accuracy and economy in the work. Before taking the rod reading a cross (+) was marked on the top of the rail. The exact spot on the rail which was used for the support of the rod was the intersection of the right angle lines of the cross. This mark was recoverable, even after a train had passed over it. It may be surprising, but it is nevertheless a fact, that a railroad rail will return to exactly its former elevation almost immediately after a train has passed over it. This was proven many times during the progress of the work. Exceptional cases may occur when the rail fails to promptly resume its former position, but the results of the leveling indicate that no such instance occurred during the work. When leveling along the railroad lines the instrument was very often set up between the rails with two of the tripod legs resting on the flange at the inner base of the rail, and the third leg resting on one of the ties. In this position it was possible for the observer to stand on a support separate from that on which the instrument rested, thus insuring the stability of the instrument. Such a procedure was particularly satisfactory when the leveling was done along an embankment.

When the leveling was to be done over long stretches of asphalt pavement in the suburban districts, where no sidewalks were available, such work was reserved for cold weather. On account of the heavy automobile traffic over such roads at all times, the surface of the road becomes almost like glass. The Eastern Boulevard in The Bronx is an example. As the use of foot plates or foot pins was impracticable under such circumstances, short flat-headed nails, known as plasterers' nails, were driven full length into the asphalt and the rod readings were taken upon them. These nails, having a large head, made excellent turning points. As the temperature at the time of the leveling was usually below 0° C. the instrument was set upon the asphalt without any sign of settlement whatever.

On dirt and macadam roads foot plates were used almost exclusively.

## Settling of Rods and Instrument.

If a rod settles after a foresight is taken upon it and before the backsight is taken, the computed elevations beyond this point of the line will be too great by the amount of the settling, for in the computation the same elevation is ascribed to the rod after the settling actually took place as before.

In accordance with paragraph 9 of the General Instructions for Precise Leveling, "the backsight is to be taken before the foresight at stations of odd numbers, and at even stations the foresight is to be taken before the backsight." Compliance with this paragraph insures that any constant or nearly constant tendency on the part of the instrument to rise or to settle during the process of the observation will have no effect upon the computed elevations. In the ordinary method of leveling, with each backsight taken before the corresponding foresight, if the rods or instrument settle by a microscopic amount between each backsight and the corresponding foresight, the sum of such settlements enters to make the computed elevation at the end of the line too great.

The instructions noted in the preceding paragraph are particularly applicable to work in the open country, where the ground leveled over affords few suitable supports.

for rod or instrument. Within the built-up sections of the city literal compliance with the requirements of paragraph 9 was not deemed essential by reason of the excellent support afforded by stone or concrete sidewalks.

### Winter Work.

Experience on the Board of Estimate leveling demonstrates the fact that little or no precise leveling can be accomplished during a stormy winter. Every engineer knows that an instrument will "creep" when set up on ice or on frosty ground. Under such circumstances the instrument gradually settles and in the case of the level, introduces systematic error into the work. While some precise leveling can be done along cement sidewalks that have been cleared of snow, the results in general cannot be depended upon. During the winter months long periods of clear weather may occur during which the atmospheric conditions are more favorable than at other times. The supports for the rod and instrument, however, are not so good, on account of frost in the ground. A comparison of the results of the leveling done on certain sections at different seasons of the year, showed in general that the leveling done during the winter months had a degree of precision no greater than that of ordinary leveling.

During the winter months most of the time was employed in making the several river crossings, preparing permanent bench marks, making reconnaissances, etc.

### So-called Precise Leveling.

A great deal of work has been done in New York City and vicinity and is still being done by various organizations, under the name of "precise leveling." Just why this title is given to the work referred to is difficult of comprehension as there is an entire absence of those elements which constitute precise or geodetic leveling. In many instances it was found that this so-called precise leveling was done with antiquated wye levels having a 30 to 60 second bubble, and that no record was kept of the relation between the axis of the level vial and the line of collimation. In some instances the testing of the level used was absolutely unheard of and no attempt was made to shade the instrument from the direct rays of the sun. The methods of observation also differed radically. Only one rod was used, and long intervals were permitted to elapse between successive readings of the backsights and foresights, thereby affording ample time for the introduction of errors due to varying refraction. It was not considered necessary even to equalize backsights and foresights or to plumb the rod with a spirit-level or plumb line. No attempt was made to eliminate systematic errors, and in the great majority of instances the rods had never been standardized so that their actual length at any given temperature was unknown. No record was kept of wear on the bottom of the rod and no corrections for varying temperature were made. Applied to work of this character, the term "precise leveling" is simply a misnomer.

## COMPUTATION AND ADJUSTMENT.

### Circuit Closures.

As the closing errors of the various circuits forming the level net afford a measure of the accuracy of the work, an inspection of the principal circuits on the Board of Estimate leveling may prove of interest.

On such an extensive level system it is, of course, almost impossible to consider each individual circuit. For example, in the Boroughs of Brooklyn and Queens, it is possible, by making various combinations, to obtain a large number of different circuits, and, in a similar manner, by taking into consideration all the circuits in Manhattan,

The Bronx, Queens and Brooklyn, a very great number of such combinations could be formed.

The larger and more important circuits, upon the adjustment of which the greater number of elevations depend, are shown in the table on pages 65 and 66. The probable error of the closure of each circuit, as shown in the last column of the table, was computed in the following manner. After determining the probable error of the complete leveling per kilometer for each circuit, the mean of these probable errors was obtained, and was found to be ± 0.38 millimeter. The product of this quanity into the square root of the distance leveled gives the figures shown in the last column. The small closing errors show that systematic errors have been largely eliminated.

If the errors of leveling were all accidental, the probable error of closure of any circuit should be the square root of the circumference of the circuit in kilometers times the probable error of leveling for a single kilometer. Experience shows, however, that the probable errors computed from circuit closures are larger than those computed from discrepancies on short sections of the separate lines.

The real test of the accuracy is the most severe test, namely, the corrections which are found necessary when the lines are adjusted as part of a net, these being the corrections which must be made in order to close all circuits.

## Adjustment.

The principal level nets were adjusted by the method of least squares and, as the leveling was done with the one type of instrument by a standardized method, and under uniform conditions, the computations were very much simplified.

In the adjustment of the level nets some of the corrections applied were so small that it would not have made any material difference if some scientific method other than that of least squares had been followed. For example, in the determination of the elevation of bench mark No. 342 at East New York according to the method of least squares, the adjusted elevation is 21.0458 meters. By taking the arithmetical mean of the three elevations determined for this bench by leveling from Coast Survey bench marks J, C and B, the resulting elevation is 21.0460 meters. The difference in the results by the two methods is seen to amount to only 0.2 millimeter.

When several values were obtained for the elevation of a point (called a junction point), common to two or more circuits and the close agreement of these values indicated that the adjustment by the method of least squares would give a value but slightly different from that obtained by taking the arithmetical mean, the latter was always used.

To have applied the method of least squares to all the figures in the level net would require more computing than would be justified by the very slight gain in accuracy. When only one side of a circuit was affected by the adjustment, the closing error was distributed uniformly around the remainder of the circuit.

The amount of time and labor required to solve the many normal equations on such an extensive level net can hardly be realized. The aids in the arithmetical work of computing these equations are a table of squares, a table of products (Crelle's), a table of reciprocals, and a table of logarithms. A machine for performing multiplication and division is also of great use in computations of this kind. As a matter of fact, however, such aids as the latter were not available on this work.

The computer is not bound by any hard and fast rule in making the computations and when the circuit closures indicate that the corrections to be applied are sure to be very small, he is at liberty to follow such methods of adjustment as will simplify the work and lighten his labors.

Contrary to general opinion, good results cannot be derived from poor observations even if the method of least squares is applied in their adjustment. This method is

## CLOSING ERRORS OF CIRCUITS.

| No. | Circuit. | Closing Error* | Circumference of Circuit | Closure per Kilometer | Probable Error of Closing. |
|---|---|---|---|---|---|
| | | mm. | km. | mm. | mm. |
| 1 | Richmond: Howland Hook—Port Richmond—St. George—Fort Wadsworth—New Dorp—Giffords—Huguenot—Richmond Valley—Kreischerville—Rossville—Greenridge—Richmond—New Springville—Linoleumville—Old Place—Howland Hook | —2.8 | 67 | —0.042 | ±3.11 |
| 2 | Richmond: Arlington—Port Richmond—Castleton Corners—New Springville—Linoleumville—Summerville—Arlington | —3.6 | 40 | —0.09 | ±2.40 |
| 3 | Richmond: Port Richmond—New Brighton—St. George—Stapleton—Fort Wadsworth—New Dorp—Castleton Corners—Port Richmond | —6.6 | 29 | —0.23 | ±2.05 |
| 4 | Bay Ridge—Governors Island—Manhattan—Yonkers—Mount Vernon—Throgs Neck—Willets Point—College Point—Flushing—Hunters Point—Borough Hall—Bay Ridge | —3.8 | 118 | —0.03 | ±4.13 |
| 5 | 86th St. and 5th Ave.—Yonkers—Mount Vernon—Throgs Neck—Willets Point—College Point—Flushing—Woodside—Astoria—East 86th St.—86th St. and 5th Ave | —2.7 | 66 | —0.04 | ±3.09 |
| 6 | 110th St. and 5th Ave.—Yonkers—Mount Vernon—Schuylerville—Unionport—Westchester Ave.—Mott Haven—110th St. and 5th Ave | +1.4 | 33.2 | +0.04 | ±2.19 |
| 7 | 207th St. and Broadway—Yonkers—Mount Vernon—Pelham Bay Park—Fordham—207th St. and Broadway | —4.8 | 33.8 | —0.14 | ±2.21 |

CLOSING ERRORS OF CIRCUITS—Continued.

| No. | Circuit. | Closing Error* | Circumference of Circuit | Closure per Kilometer | Probable Error of Closing. |
|---|---|---|---|---|---|
| | | mm. | km. | mm. | mm. |
| 8 | Brooklyn: Bay Ridge—Borough Hall—East New York—New Lots—Flatbush—Parkville—Bay Ridge | +1.0 | 31.8 | +0.03 | ±2.14 |
| 9 | Queens: Hunters Point—Flushing—Little Neck—Floral Park—Springfield—Woodhaven—East New York—Maspeth—Hunters Point | +4.9 | 60 | +0.08 | ±2.94 |
| 10 | Queens: Flushing—Bayside—Little Neck—Floral Park—Queens—Hollis—Flushing | +4.5 | 33 | +0.14 | ±2.18 |
| 11 | Queens: Flushing—College Point—Whitestone—Willets Point—Bayside—Flushing | +0.1 | 17.6 | +0.01 | ±1.59 |
| 12 | Central Park: 59th Street—Central Park West—110th Street—5th Ave.—59th Street | +2.5 | 10.6 | +0.24 | ±1.24 |

* A + closing error means that an elevation carried around the circuit in a clockwise direction is too high.

one of computation and not of observation and serves merely to aid the computer in securing in the computed results the highest grade of accuracy possible from a given series of observations but cannot increase the accuracy of observations already taken.

The largest correction arising from the necessity of closing all circuits, the most severe test of accuracy which can be applied, was 0.6 millimeter per kilometer. Corrections as large as this were the exceptions, however, and arose only in connection with the smaller circuits. On more than one-half of the work the average correction applied in order to close the several circuits was less than 0.2 millimeter per kilometer.

## Precision of the Results.

In accordance with paragraph 5 of the General Instructions for Precise Leveling the discrepancy between the forward and backward measures on any section must not be more than $4^{mm}.0\sqrt{K}$, in which K is the distance leveled between adjacent bench marks in kilometers. This maximum discrepancy amounts to 5 millimeters or 0.017 foot on a section one mile long. Such a limitation at once establishes a measure of precision for the work and on the Board of Estimate leveling there was no difficulty encountered in keeping well within the limit set. In fact the actual discrepancy between the forward and backward measures of the majority of the sections on this work was less than $2^{mm}.0\sqrt{K}$. It was a rare occurrence when the limit of error was exceeded and this accounts for the very small percentage of re-running. The accuracy attained was due to the methods used and precautions taken in making the observations without materially delaying the work for that purpose.

A limiting error of $0^{ft}.02\sqrt{\text{dist. in miles}}$ has long been considered very low and has been the limit adopted on precise leveling by various organizations throughout the United States. On the Mississippi River Survey the limit of discrepancy between duplicate lines was $5^{mm}.0\sqrt{K}$ or $0^{ft}.021\sqrt{\text{dist. in miles}}$. On the U. S. Lake Survey the limit of error was $10^{mm}.0\sqrt{K}$ or about $0^{ft}.041\sqrt{\text{dist. in miles}}$.

On the national surveys the territory to be covered with precise level nets is so great and the necessity of completing a large number of miles per month is so apparent, that the limiting error of $4^{mm}.0\sqrt{K}$ is to be considered an economic one. Also, on account of much of this work being done in the open country where foot plates and foot pins are very frequently used, the possibilities of error arising from unstable supports for both rod and instrument are very great. It is believed that the stable supports for both rods and instrument on the Board of Estimate leveling account very largely for the small divergence (B — F) between the two runnings of each line.

Experience on other large surveys indicates that a large steady rate of divergence is in general due to a systematic rising or settling of rod supports, and a rapidly accumulating discrepancy with a minus sign is very good evidence of the settling above noted.

The total accumulated discrepancy between the backward and forward runnings of the principal level lines together with the discrepancy in millimeters per kilometer are shown in the table on page 68.

The discrepancy in millimeters per kilometer was obtained by dividing the total discrepancy on the main line by the length of the main line.

The probable error of the mean result for a section was computed in millimeters by the formula

$$r'' = 0.674\sqrt{\frac{\Sigma d^2}{4s}}$$

in which $d$ is the discrepancy between the forward and backward leveling over a section and $s$ is the number of sections. The probable error for 1 kilometer, $r_1$, was

VALUES OF DISCREPANCY (B–F) IN PRINCIPAL LINES.

| Bench Mark to Bench Mark. | Line. | Distance in Kilometers | B–F | B–F Per Kilometer. |
|---|---|---|---|---|
| | | | mm. | mm. |
| 8– 117 | Tottenville — St. George — Howland Hook.......................... | 38 | +2.0 | +0.05 |
| 180– 737 | Fort Hamilton—Hunters Point—Willets Point........................ | 45 | —3.5 | —0.08 |
| 646– 737 | Hunters Point—Glendale—East New York — Springfield — Queens — Bayside—Willets Point............ | 44 | +4.2 | +0.10 |
| 258– 342 | Bay Ridge — New Lots — East New York.......................... | 15 | +1.0 | —0.07 |
| 742–1185 | Battery — Spuyten Duyvil — Yonkers —Dobbs Ferry.................. | 53 | —3.2 | —0.06 |
| 1163– 999 | Yonkers — Mount Vernon — Pelham —Throgs Neck.................. | 22 | +3.0 | +0.14 |

derived by assuming that the average length of a section is to 1 kilometer as $(r'')^2$ is to $r_1^2$.

### Condition of Atmosphere and Length of Sight.

On the Board of Estimate leveling the work was continued throughout the day during all seasons of the year, with the exception of stormy periods in winter. In very warm weather the atmospheric conditions were found to be sensibly as good as at other times, unless the leveling were done over sandy soil or cinders. Adverse conditions likewise arose when the wind blowing across the line of sight had swept across sandy or cindery areas.

Excessive radiation, it was found, was more troublesome in the open country than in the built-up portions of the city. Whenever there was moisture in the air the rod appeared remarkably distinct and steady.

When the leveling was done along lines which ran toward or away from the sun, there was usually more chance of error due to accidental misreadings of the rod, by reason of the fact that the sun shone brightly on that rod which faced it, making the readings uncertain. When leveling toward or away from the sun near sunset, the rod nearest the sun would become hazy and dark and the readings on this rod would be very uncertain. Under the conditions above noted the length of sight was reduced and the rate of progress was correspondingly slower.

When leveling on steep grades the errors due to the varying refraction were avoided by never letting the line of sight come nearer than 1 foot to the ground.

The proper length of sight depends upon the distance at which the rod appears distinct and steady, upon the variations between the upper and lower thread intervals and upon the degree of precision required. On this work the average length of sight throughout the city was 61 meters or about 200 feet. In the suburbs, sights of 300 feet or more were taken whenever the ground was not too hilly. As a rule, however, the length of sight did not exceed 300 feet, as it was found that the danger of erroneous rod readings increased when this length was exceeded. The likelihood of such error could no doubt have been diminished by a greater magnifying power of the telescope. The opportunities for taking such long sights, however, were not very frequent throughout the city. Short sights were usually the rule and, while the rate of progress was necessarily slower, this was unavoidable in such a large city.

## RATE OF PROGRESS.

The average rate of progress throughout the city in a field day of 6 working hours was 5 kilometers or about 3.1 miles of single level line and in addition thereto one move between sections was usually made. A run of 6 kilometers per day was considered good.

When the bench marks were close together, in some instances three or more to 1 kilometer, or somewhat difficult to take readings on, the rate of progress was often much less; whereas, in the suburban districts, with bench marks 1 kilometer or more apart, 8 kilometers or 5 miles of single level line were very often covered in a field day of 6 working hours without any difficulty.

The question of rate of progress may also be approached from the viewpoint of the number of stations occupied in a given length of time. The records of the Board of Estimate leveling show that the average elapsed time between arrival at two consecutive instrument stations amounted to 4.4 minutes.

In rolling country, where the average distance between instrument stations was only 80 meters or about 262 feet, an average of 16 stations was encompassed per hour, including every detail of the work.

In view of the fact that fully one-half of the time of the field party was, of necessity, devoted to walking between stations, the rate of progress above noted as the day by day average, could not have been exceeded without a lowering of the standard of accuracy required. This was particularly true when a considerable portion of the distance leveled fell within built-up sections of the city and numerous bench marks were tied into.

An almost indispensable part of the equipment of the precise level party is the Locke hand-level. When leveling up or down steep grades, much time is sometimes lost by reason of the fact that the precise level is set either too high or too low, the line of sight passing over the top of one of the leveling rods or striking the ground below the foot of the other. By sighting through the hand-level the proper height of the instrument is readily found without loss of time.

In order to insure good progress in running precise levels an active and careful recorder is requisite. The man filling this position should be capable of recording the rod readings and obtaining their mean value with speed and accuracy, and should be able to add columns of figures rapidly and with dependable results.

## AMOUNT OF FIELD WORK.

On the Board of Estimate leveling about 340 miles of double or completed level lines have been run, each line being leveled over at least twice in opposite directions. In view of the fact that practically no precise leveling can be done during the winter months and that a study of the record shows that the actual number of working days in the field was less than three per week, the showing appears to be a very creditable one when it is remembered that it represents the work of a single field party. The level lines run more than twice amount to about 3 per cent. of the total distance leveled. It is believed that the stable supports for both rods and instruments on the Board of Estimate leveling account very largely for the small divergence (B–F) between the two runnings of each line.

There were 1,186 bench marks touched on, or about four to each mile of completed level line. The elevations and descriptions of these bench marks are given on pages 93 to 263. The number of bench marks includes all with which the leveling was directly connected, regardless of whether they are new bench marks or bench marks previously established by some other organization.

## Conclusions.

In concluding this portion of the report, the writer wishes to express entire satisfaction with the present Coast Survey type of precise level and also with the method of observation, as set forth in the "General Instructions for Precise Leveling."

The accuracy attained in precise leveling is due quite as much to the methods used and precautions taken in making the observations as to the instrumental means employed. The use of first-class instruments does not, in itself, insure good work in precise leveling. While such instruments are conducive to accurate results, fully as much depends upon the employment of a systematic method and upon painstaking care and unremitting vigilance on the part of the observer.

The precise levels of the Coast Survey type were designed with the object of obtaining the highest possible degree of precision in combination with facility and convenience of manipulation. The methods used with this instrument are as simple as those employed in ordinary wye leveling, the duties of the observer at each instrument station consisting solely in obtaining the three wire readings on each rod and keeping the bubble exactly centered while reading each wire. The time spent at each instrument station is much less than in the case of work done with the European type of precise level or with the wye level and target rods.

The "General Instructions for Precise Leveling" as set forth on pages 26, 27 and 28 were intended to secure a standard degree of accuracy combined with the greatest practicable economy of operation. The method of observation corresponds closely with that used by the Missouri and Mississippi River Commissions.

Precise leveling carried on by the national surveys and by the European governments has conclusively demonstrated the superiority of the direct-reading rod over the target rod. With a good instrument the thread readings can be accurately taken in much less time than is required to set a target and the mean of the three readings is much more accurate than any target reading.

Adherence to the method of observation hereinbefore described and the use of direct-reading rods, have, in the opinion of the writer, contributed more than any other factors to rapidity of work and accuracy of results. Direct-reading rods are, for many reasons, superior to target rods, and might to advantage be more generally used than is at present the case.

During the progress of the work the writer was frequently called upon to furnish elevations and differences of elevation to many of the city departments and to other organizations, and more than one-half of the final results of the leveling were in use throughout the city before the preparation of this report was undertaken.

It is a great convenience to have fixed elevations for bench marks throughout the entire city—standard elevations, so to speak—all based on a universal datum, and to which the departmental elevations in the various boroughs can be referred. A large number of benches have now been established with elevations referred to mean sea level, and are available for the use of all whose work depends in any way on elevation. It is to be hoped that New York City will, in the near future, range itself on the side of progress and adopt a datum plane not merely universal as respects its own territory, but scientific in its origin and in harmony with the best practice throughout the civilized world.

Reviewing the work of the Board of Estimate and Apportionment level survey, it appears that no plan could have been devised which would have led to the securing of more satisfactory results. Had a larger appropriation been available at the outset the work could have been materially expedited by placing several parties in the field. The services of additional computers would likewise have greatly hastened the completion of the task, as the necessity is apparent, in work of this kind, for very thorough checking of the field notes and office computations.

The examination of all the circuit closures, of the divergence (B–F) between the two runnings of each section and of the corrections applied, and a consideration of other evidence bearing upon the accuracy of the work, justify the conclusion that the leveling done by the Board of Estimate and Apportionment in New York City and vicinity compares very favorably with the highest standard heretofore attained in precise leveling in the United States.

## HISTORY OF THE DATUM PLANES NOW IN USE IN NEW YORK CITY.

When New York City consisted only of Manhattan Island, the city's engineers referred all their elevations to one plane known as the City Datum or Public Works Datum. The consolidation of 1898, however, brought into the engineering work of the greater city, the datum planes of the various departments of the new boroughs, so that to-day we have ten or more distinct datum planes to which our elevations are referred.

As the work of some of the city departments overlaps, the engineer who has to use the elevations of streets and structures determined by the different departments is put to considerable trouble to reconcile these elevations and this condition very frequently results in much confusion and loss of time in making the adjustments required to secure uniformity.

On account of this multiplicity of datums, many entirely new systems of levels have been established, and a great amount of computation of use only to the particular department concerned, has been required. Because of the unsatisfactory and more or less imaginary character of many of the datum planes, there has been during recent years a growing desire on the part of the city engineers of all boroughs to secure the official adoption of one datum plane for the whole city.

Prior to 1870, the municipal topographical work of the City of New York was done by engineers and surveyors in private practice, each one of whom had his own system of bench marks, referred to some mythical high water. The history of the more important departmental datum planes now in use in New York City will be briefly outlined.

*The Public Works Datum:* The oldest datum of which we have any record is the City Datum or Public Works Datum of Manhattan. The original bench mark of this datum was the water-table at the base of a column at the southwest corner of the old almshouse at Bellevue Hospital, and its elevation was fixed at "20.558 feet above high water mark."

This elevation is said to have been established about 1790 by a surveyor named John Randel, Jr. The datum plane based upon this determination is still in use as a plane of reference.

By reason of the frequent references to the name of Randel in association with the city plan and the establishment of the first recorded datum plane, it may be of interest to briefly sketch the origin and character of his connection with this work.

On April 3, 1807, the Legislature of the State of New York passed an act entitled "An act relative to improvements touching the laying out of Streets and Roads in the City of New York and for other purposes." Upon the recommendation of the commission appointed in pursuance of this act, the services of John Randel, Jr. were retained. On December 31, 1810, Mr. Randel signed a contract with the commission "to complete the survey and measurements already begun in conformity to their plans and place the necessary monuments." Active work on this contract was commenced on September 30, 1811, and was finished in 1820. In 1817, as a part of his contract, Mr. Randel established elevations on all of the monuments on five of the avenues, and presented to the commission a bill of $2,500 for this work.

In carrying out his work for the city, Mr. Randel referenced his elevations to the bench mark established by himself in 1790, referred to in a previous paragraph. It is said that this location was determined on by Mr. Randel in the belief that this point would be at or near the center of the future city. A note on the commissioner's map submitted on March 22, 1811, states that "Elevations above high water are shown as ┼ v.5., the Roman characters standing for feet and the Arabic for inches." These elevations are shown at a few of the street intersections.

On account of the inaccessibility of the original bench mark a monument was established in the hospital grounds at 27th Street and First Avenue, between the years 1830 and 1840. This bench mark is still in existence and its elevation as then fixed and still in use is 17.797 feet. No information is at hand to show at what point of the city shore line were made the tidal observations to which this monument is referenced, but it is the accepted belief that these observations took place in the immediate vicinity of Bellevue Hospital.

From this initial bench mark at Bellevue Hospital, several level lines were run to the New York City Hall and a bench mark in the form of a large **T** was cut deep into the stone at the southwest corner of the building, just below the water-table. The middle of the top bar of the **T** is the accepted bench mark, and by an agreement entered into about the year 1840 between the city surveyors its elevation was fixed at 42.21 feet.

Remarkable as it may seem, these two points were with the exception of those hereinbefore referred to as established by Randel in 1817, the only official public works bench marks in New York City until the year 1880, when the Hon. Allen Campbell, Commissioner of Public Works, directed "that permanent bench marks be established with a view to standardizing engineering work done by the various departments in the City of New York."

This work was done in 1880 by Mr. A. G. Culver, then an assistant engineer in the Department of Public Works, who ran a line of levels from the original Bellevue Hospital bench mark to Highbridge Tower, establishing bench marks five blocks apart and at every avenue intersecting these blocks between the Hudson River and the East River. No level lines were run and no bench marks were established south of Twenty-sixth Street.

The Public Works Datum was adopted by the Sewer, Water and Highway Departments in the Borough of Manhattan and was used by the Rapid Transit Railroad Commission in its original surveys. The work done in this borough by the Pennsylvania Railroad and the New York Central Railroad has likewise been referenced to the Public Works Datum, which is now universally used by local surveyors.

*The Bronx Borough Datum:* In the Borough of The Bronx all elevations have been referred to a bench mark cut in the southeasterly face of the north abutment of the old Third Avenue Bridge over the Harlem River. This bench mark was a horizontal arrow marked "10 feet above mean high water," and was established from a bench mark of the Public Works system on the Harlem side of the same bridge, the elevation of which was 11.085 feet.

The 10-foot mark on the north side of the bridge was used by the Commissioners of the Town of Morrisania, by the Park Department and by the Department of Street Improvements of the 23d and 24th Wards. From it a bench mark in the shape of a cross was established at the entrance to the Old Town Hall of the Town of Morrisania on the site now occupied by the Police Station at One Hundred and Sixtieth Street and Third Avenue. The elevation of the latter bench mark was fixed at 25.935 feet above mean high water. The Third Avenue Bridge was removed in 1894, thus destroying the original Bronx bench mark.

In 1901 it was decided to establish a complete system of bench marks in The Bronx for use by all the bureaus and thus avoid the discrepancies which had heretofore existed in the work of the various bureaus. A series of level runs was therefore made, consisting of closed circuits all based on the Town Hall bench mark and radiating from it. This is the system now used in The Bronx, and the datum referred to should be identical with the Public Works Datum of Manhattan.

*United States Navy Yard Datum:* The datum next in importance to that of the Public Works Datum of Manhattan and to that of The Bronx is the Navy Yard Datum of Brooklyn. The original bench mark of this system is a square (□) cut in the top of the coping at the southwest corner of Dry Dock No. 1, at the Brooklyn Navy Yard.

The elevation of this bench mark above mean high water was fixed by tidal observations made about 1840, and has always been considered by Brooklyn engineers to be 5 feet above mean high water.

From an inspection of the old plans on file at the Brooklyn Navy Yard, it is seen that as a result of the tidal observations made about 1840, a bench mark was established on one of the steps at the shops, and from it the sill of the Dry Dock was fixed at 26 feet below mean high water, and the coping of the Dry Dock was fixed 31 feet above it, thus making its elevation 5 feet above mean high water. The construction of the Dry Dock began in 1841, and the laying of masonry was commenced in 1847. The last stone was laid in 1850.

On page 6 of the "Report Descriptive of the Construction of the Brooklyn Water Works," by James P. Kirkwood,* Chief Engineer, and published in 1867, it is stated:

"The heights above tide always refer to mean high water at the 'Brooklyn Navy Yard as conventionally fixed at 5 feet below the coping of the Brooklyn Dry Dock.'" However, a series of tidal observations made at the Brooklyn Navy Yard by the Bureau of Yards and Docks from June, 1902, to June, 1903, show the old bench mark on Dry Dock No. 1 to be only 4.43 feet above mean high water; this value may be in error as much as ± 0.2 foot on account of the short period of observation.

The Navy Yard or Dry Dock bench mark is the origin of the elevations established by the Division of Water Supply of the City of Brooklyn. The original elevations of this system were established between 1850 and 1860, when a line of levels was run from the Dry Dock bench mark to the Ridgewood Engine House and continued to a bench mark on the east rim of the pump well at Smith's Pond, at the end of the old conduit line. In 1895, a return line of levels was run from the Smith's Pond bench mark, the elevation of which had been fixed by the previous line at 10.992 feet, to the original bench mark at Ridgewood which indicated that the latter was 0.429 foot lower than the elevation established in 1858-60. The Smith's Pond elevation was accepted as correct and the bench marks west of Smith's Pond were adjusted to this new line of levels. The elevation of the Dry Dock bench mark as used by the Division of Water Supply of the City of Brooklyn is 4.997 feet.

This bench mark is also the origin of all elevations determined by the Brooklyn

* Formerly Chief Engineer of the Brooklyn Navy Yard.

Sewer Bureau, which has established its own system of bench marks. The elevation of the Dry Dock bench mark as used by the Brooklyn Sewer Bureau is 4.999 feet.

The Department of Bridges has used the Navy Yard Datum in the construction of the Manhattan Bridge and has accepted 5.000 feet as the elevation of the Dry Dock bench mark.

The elevation of this bench mark as established by the Brooklyn Highway Department is 4.167 feet. In addition, elevations of 6.677 feet and 6.762 feet were determined for this bench mark by the Board of Estimate and Apportionment and by the Board of Water Supply, respectively. It therefore appears that no less than six elevations have been determined by six different departments for this one bench mark.

In 1887 the United States Coast and Geodetic Survey established primary bench mark "D" on machine shop No. 28 in the Brooklyn Navy Yard. This machine shop was destroyed in 1899. Primary bench mark "D" is the only bench ever established in the Navy Yard by the Coast Survey, which, contrary to a widespread belief, had nothing to do with the establishment of the Dry Dock bench mark.

*Brooklyn Highway Datum:* By an act of the Legislature passed in 1835 a commission was appointed to lay out the streets of the City of Brooklyn. J. S. Stoddard, a surveyor who had been associated with John Randel, Jr., in the laying out and monumenting of New York City, was selected by the commission to execute the work and a system of monuments was used to mark the streets. Elevations were established for most of these monuments. Mr. Stoddard's own statement as made in a memorandum now in possession of Austin Ludlam is as follows:

"These monuments all appear on the several maps. The level has been taken from 827 of these monuments to the highest tidewater mark and marked in feet and hundredths of a foot opposite each monument upon the map. This will essentially aid in the future pitching and grading of the streets as it furnishes at one view the relative height of the ground in every part of the city."

No primary bench mark or origin is mentioned in connection with these elevations. In establishing a grade chart for the adjoining district on the south along the shore known as Bay Ridge, George Ingraham, surveyor, accepted as correct the elevation of the monument on the northwest corner of Fifty-eighth Street and Seventh Avenue. About the year 1905, the Bureau of Highways ran a new series of levels from this monument, establishing elevations on nearly all of the monuments in the borough.

These elevations are now used by the Highway and Topographical Bureaus of Brooklyn, and are based on the elevation established for the monument by Mr. Stoddard, which is "98.48 feet above high water." In this case, also, there is no record of tidal observations having been taken at any specified point on the city's shore line.

A short time subsequent to the erection of the Wallabout Bridge at Kent Avenue in 1891, the Highway Bureau of the City of Brooklyn made use of a mark established on this bridge from the bench mark on Dry Dock No. 1 at the Brooklyn Navy Yard. By reason of a discrepancy later discovered in the elevation fixed for the Wallabout Bridge bench mark, the use of this bench mark was discontinued and all work was referenced directly to the Navy Yard bench mark, up to the time of the running in 1905 of a new series of levels based on the original Stoddard determinations. As in the case of the Dry Dock bench mark, no elevation for the Wallabout Bridge bench mark was ever established by the United States Coast and Geodetic Survey.

In the old towns of Flatbush, Flatlands, Gravesend, New Utrecht, New Lots and Bath Beach, the individual datums were established by the local surveyors, who were called upon from time to time to furnish grade charts for the various towns. Each surveyor established his own datum in his own way and at different times, but the evidence at hand does not show any radical departure from the Highway Datum.

The old town Datums were no doubt derived from the elevations established for the old Brooklyn monuments by J. S. Stoddard, but on account of discrepancies having crept in, the datums failed to preserve the uniformity which, no doubt, originally characterized them.

*Queens Borough Datum:* By an act of the Legislature passed in 1869 the old Town of Newtown was divided into two parts, one of which was called Long Island City. Pursuant to the provisions of this act a commission was appointed to lay out this city, and as part of their work a bench mark was established on the old Alms House on Blackwells Island, this mark being referred to mean high water.

From this initial bench mark a series of levels was run and bench marks were established in Long Island City. One of the latter was located on Schwalenberg's Hotel, at Borden Avenue and Vernon Avenue.

In July, 1900, W. S. Dalrymple, Assistant Engineer in charge of the Queens Borough Topographical Bureau, had a new series of levels extended over the borough. The origin used was primary bench mark B. on Miller's Hotel at Borden Avenue and Front Street in Long Island City, established by the United States Coast and Geodetic Survey in 1887. The elevation of this bench mark was accepted by Mr. Dalrymple as 5.770 feet above mean high water and the elevations resulting from his new levels were based upon this figure.

In October, 1902, the datum was changed by Ernest Ankener, then Assistant Engineer in charge, to agree with the datum used in laying out the Long Island City street system, the old bench mark on Schwalenberg's Hotel at Borden Avenue and Vernon Avenue being taken as the new initial bench mark.

The elevations of all bench marks previously established were changed to conform to the Long Island City Datum by subtracting 0.578 foot from the previously determined elevations. All bench marks subsequently established were based on this datum, accepting 10.970 feet as the elevation of the bench mark on Schwalenberg's Hotel.

*Richmond High Water Datum:* The first general system of bench marks for Richmond County was begun in 1890-91 by W. S. Bacot, Engineer of County Roads. The datum was "approximate high water," and was referred to a spike driven in the outlet of the Water Street sewer at Stapleton.

This datum was generally used in the County Road system, though some elevations were referred to an assumed datum when no bench mark was convenient. The levels on the Northfield town roads were connected with the Coast Survey tidal bench mark at the foot of Morningstar Road, Elm Park.

In 1899, John T. Fetherston, of the Richmond Sewer Bureau, ran a series of levels from the Coast Survey bench mark at Elm Park to Oakwood, the elevation of this bench being taken as 13.152 feet. This value was given by the Department of Docks and Ferries, and was obtained by adding 2.186 to the Coast Survey elevation, thus referring the Elm Park bench mark to the Dock Department Datum.

In 1900, the Board of Public Improvements of New York City made a survey of the Arrietta Watershed, using the County Road bench marks at Tompkinsville. The mean difference between these high water bench marks on Arrietta Street and John T. Fetherston's bench mark No. 9 at Tompkinsville was found to be 5.309 feet.

In May, 1900, all the bench marks established by the 1899 leveling were referred to high water by subtracting the above amount, and the elevations so obtained have since been used.

The bench marks established by the Topographical Survey of the Borough of Richmond have all been referred to high water as given by the recorded elevation 9.586 feet for bench mark No. 12 at Clifton. This bench is one of those included in the above system, its elevation having been determined as before mentioned and frequently checked with others of the same run. The datum now in use is, consequently, as near as can be determined, the same as that used in 1890-91 on the County Roads, and is now known as "Richmond High Water Datum."

Investigations made in connection with the Board of Estimate leveling show "Richmond High Water Datum" to be 0.803 foot higher than mean high water at the foot of South Street, St. George, as determined by tidal observations made by the Department of Docks and Ferries from 1908 to 1913 inclusive.

*Dock Department Datum:* The datum of the Department of Docks and Ferries is mean low water at Pier A on the Hudson River, and is known as the "Battery Datum." The elevation of the initial bench mark at Pier A was at first determined from a short series of tidal observations made by an automatic tide-gauge at this point, but in 1898, after 12 years of continuous tidal observations, the datum was raised 0.24 foot and called the "New Battery Datum." This is the plane now used in all boroughs by the Dock Department, and elevations previously determined were corrected to agree with the same.

Prior to 1886 the elevations of all Dock Department bench marks in Manhattan were referred to a bench mark in the form of a large T cut in the south side of the east entrance to Castle Garden, the elevation of which was taken as 14.74 feet above mean low water at this point, and 47.09 feet below the T cut bench mark at the southwest corner of the City Hall.

*Public Service Commission Datum:* The origin of this system of levels is a bench mark on the old Produce Exchange at Bowling Green, established by the Rapid Transit Railroad Commission, which ran a line of levels from the T cut bench mark on the City Hall. Its elevation as then fixed is 19.096 feet, and is the base of the extensive level system of the Public Service Commission.

*Board of Water Supply Datum:* In 1907 the Headquarters Department of the Board of Water Supply ran a series of levels through New York City, which was part of a continuous level line from Schoharie County to Suffolk County.

Most of the resulting elevations in New York City were based upon the elevations of primary bench marks established by the Coast Survey in 1887, and referred to mean sea level at Sandy Hook. The remaining elevations in the south part of the city were based upon the elevation of Coast Survey bench mark No. 12, at Fort Hamilton, which was accepted as being 39.25 feet above mean tide level as determined by the Coast Survey.

As only a small portion of the Board of Water Supply elevations are referred to mean tide at Fort Hamilton instead of to mean sea level, the datum here assumed for that department is the same as that for the Board of Estimate. It is readily seen, however, that the elevations established for any common point by each of these departments will not be identical partly for this reason.

The differences in elevation between the systems of the Board of Estimate and Apportionment and the Board of Water Supply increase toward the northerly end of the city, attaining a maximum at Dobbs Ferry, by reason of the fact that the latter department accepted as correct the Coast Survey bench marks at Dobbs Ferry, which bench marks the work of the Board of Estimate and Apportionment has since shown to have undergone changes.

*Bridge Department Datums:* The Department of Bridges has no one general datum. In the construction of the various bridges in New York City this department has made use of bench marks supplied by the various other departments. In the construction of the Brooklyn Bridge the bench marks made use of were referred to a datum which is 1.929 feet above mean sea level. In the construction of the Williamsburg Bridge a bench mark furnished by the Coast and Geodetic Survey was used and the construction work on the Manhattan Bridge was referred to the Dry Dock bench mark in the Navy Yard. For the construction of the Queensboro Bridge the elevation of a bench mark obtained from the Department of Docks and Ferries was taken, to which was added the range of tide at the bridge site. For the construction of bridges in the Borough of The Bronx the datum made use of is The Bronx Borough Datum, known also as The Bronx Highway Datum.

*The Croton Datum:* This datum, known also as the Croton Aqueduct Datum, is not in use in New York City, but on account of the frequent references to it in connection with the construction of the Old and New Croton Aqueducts, a brief statement concerning the origin of this plane of reference is here given.

In the description of the New Croton Aqueduct contained in the Report to the Aqueduct Commissioners by the President, James C. Duane, 1887-1895, the following statement concerning the Croton Datum appears on page 103: "In this connection, and for the purpose of avoiding confusion in the reading of elevations indicated on several of the plans, it should be stated that the datum plane used for the construction of the Aqueduct was the Croton Datum, i.e., mean tide at Sing Sing, the same that was used for the old Croton Aqueduct and its appurtenances.

For the part of the work built in the city, beginning at the One Hundred and Thirty-fifth Street Gate-house, in order to conform with the benches adopted by the City Department, the city base was adopted, which is mean high water in New York. The same datum is used for the construction of the Jerome Park Reservoir.

Croton datum * being 0, the city datum, after careful comparison with a number of reliable benches, has been found to be + 3.61."

The Board of Estimate precise level line along Jerome Avenue connects with Aqueduct Commission B. M. 47, the location of which is given on page 259 in the description of B. M. 1160. At this point the Croton Datum is 0.786 foot below the U. S. Standard Datum.

*Miscellaneous:* The datum planes in use by the United States Engineers in New York City and vicinity are largely local in character, and are commonly determined from tidal observations covering only one lunation. The plane of reference is usually mean low water. These determinations are made in various localities as occasion arises, and, as the level system is not a connected one, there is no common reference plane.

The numerous bench marks found in various parts of the city and marked U. S. have been, in general, established by the United States Engineers, and not by the United States Coast and Geodetic Survey, the bench marks established by the latter being few in number and their locations being all recorded and well known.

In connection with the discussion of datum planes reference may be made to what is known as the "Willets Point Datum Plane" of the United States Geological Survey. This plane has been in general use by certain of the city departments over large areas of Long Island, including portions of Brooklyn and Queens. The datum has been supposed to be mean sea level at Willets Point, Long Island, and has been used by the Geological Survey on the Topographical Surveys of Long Island.

The origin of the "Willets Point Datum" was a cross (+) cut in the face of the retaining wall at the "long dock" in the military reservation at Willets Point, with the figures + 18.0555 cut in the stone alongside of the cross.

This bench mark was established by the United States Engineers and the figure + 18.0555 was the elevation in feet of the cross above the zero of the Engineers' tide-gauge.

In 1887, the United States Coast and Geodetic Survey fixed the elevation of this

* In the " Description of the Old Croton Aqueduct," contained in the history of The Water Supply of the City of New York, 1658-1895, by Edward Wegman, the datum plane is stated to be (page 51) "mean tide in the Hudson River at the mouth of the Croton River." On page 117, the following statement appears: " The elevations for the old and the new aqueducts refer to Croton Datum, which is mean tide at Sing Sing."

On page 69 of the Report to the Aqueduct Commissioners, by The President, John F. Cowan, 1895-1907, the Croton Datum is stated to be " Mean tide at the mouth of the Croton River." On page 73 of the same report, describing The New Croton Aqueduct, the following statement appears: " The elevations refer to Croton Datum, which is mean tide at the mouth of Croton River."

bench mark at 14.135 feet above mean sea level at Sandy Hook. This elevation has been accepted by the Geological Survey and is the figure used by them on the Topographical Surveys of Long Island. The "Willets Point Datum" and the plane of mean sea level at Sandy Hook are therefore identical.

In the above discussion of datum planes in use in New York City and vicinity no reference has been made to the United States Coast and Geodetic Survey Datum. This datum, which is referred to mean sea level at Sandy Hook, N. J., is known as the Standard Datum and its determination of mean sea level forms the basis of the precise leveling of the Board of Estimate and Apportionment.

## THE QUESTION OF COASTAL SUBSIDENCE IN NEW YORK CITY AND VICINITY.

The question of coastal subsidence is one to which considerable attention has in recent years been given in various parts of the world. It appears to be the general experience of those who have given consideration to this phase of precise leveling, that the first impressions received always tend to confirm a belief in the universality of this phenomenon. More careful and accurate researches, however, lead to a contrary conclusion.

Voluminous works have been compiled by geologists all tending to show that the Atlantic coast of the United States is subsiding at the rate of from 1 to 2 feet per century. Until within the last three or four years the accuracy of the conclusions reached in this respect was seldom challenged, but in very recent times a more scientific method of research and a different interpretation placed upon the data collected, has resulted in the conclusion that a revision of opinion in this matter is necessary.

As the subject of coastal subsidence is one in which every engineer is very much interested, and the question as to whether or not there is any actual settlement of the entire Atlantic coast now going on is a very important one, a few practical proofs of the stability of the coast in New York City will now be presented.

Very often old bench marks at or near the water front are compared with bench marks further inland, and, when the change in the original difference of elevation is such as to indicate a settlement, the conclusion is frequently reached that such settlement is due to coastal subsidence. At times the departure from the original difference of elevation agrees closely with the expectable difference on an assumption of a subsidence of one foot per century or even more. When a large number of old established bench marks are connected, however, and the resulting differences of elevation are compared with those originally determined, it is seen that the amounts of departure from the original figures are not at all uniform. Moreover, a close inspection of the long established bench marks at or near the water front will usually show that they have settled owing to a sinking of the structure on which they were placed.

Many of the buildings and other structures carrying the old bench marks have been erected on filled-in ground or on piling, and it is interesting to note that when a large number of old established bench marks have been connected, as in the Board of Estimate leveling, it will be found that the bench marks showing no signs of settlement whatever are on structures founded either on sand or gravel, or on natural rock.

As the discrepancies developed between the old bench marks are not at all constant, and, as the bench marks in the best condition show no signs of settlement whatever, the conclusion reached must be that the shore line of New York City is in a condition of stability. As an illustration of what we should be led to expect were the theory of coastal subsidence well founded, let us consider the following illustration.

It has been shown on page 50 that mean sea level at Fort Hamilton, N. Y., is at the same elevation as mean sea level at Sandy Hook, N, J. This conclusion is further substantiated by the additional figures shown in the fourth and fifth columns of the table on page 80.

If we assume, therefore, a progressive subsidence along the Atlantic coast of one foot per century, the level line as brought up from Perth Amboy should show mean sea level at Fort Hamilton to be 73 millimeters or 0.24 foot higher than it really is. The expectable difference on the assumption of a two-foot subsidence per century would

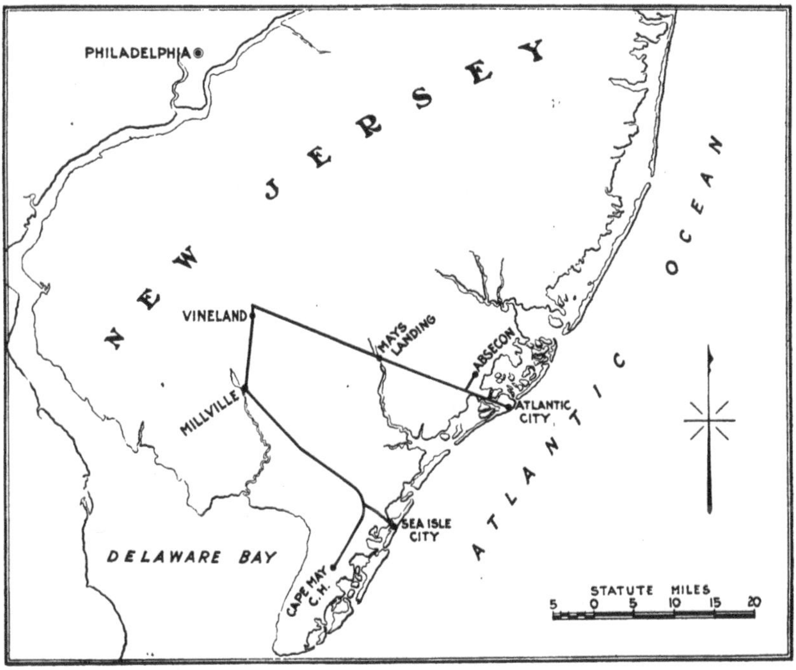


No. 19.—Precise Level Lines in New Jersey, which were Re-run in 1911.

be 146 millimeters, or 0.48 foot. The figures in the sixth and seventh columns of this table show the differences one should expect to find at each place on the assumption of a one and two-foot subsidence per century, respectively.

As these bench marks were all connected with mean sea level at Fort Hamilton, a datum based upon nineteen years of continuous tidal observations, the figures in the fourth and fifth columns not only prove conclusively that there has been no progressive subsidence of the coast in New York City and vicinity during the twenty-four years indicated, but also show that there has been no warping or tilting of the area included between the bench marks during this period.

In 1886, a large number of primary bench marks were established throughout the State of New Jersey, by the State Geological Survey. In the latter part of 1911, a number of the level lines in the southern part of the State were re-run, the work being done under the direction of C. C. Vermeule, consulting engineer. The purpose

UNITED STATES COAST SURVEY BENCH MARKS,

Established in 1887.

| Place. | Designation of Bench Mark. | Elevation Referred to Mean Sea Level at Sandy Hook in 1887. | Elevation Referred to Mean Sea Level at Fort Hamilton in 1911. | Apparent Difference of Elevation in 24 years. | Expectable Difference of Elevation, Assuming a 1-Foot Subsidence per Century. | Expectable Difference of Elevation, Assuming a 2-Foot Subsidence per Century. |
|---|---|---|---|---|---|---|
| | | meters | meters | meters | meters | meters |
| At Bay Ridge, Brooklyn.......... | J | 13.3000 | 12.2995 | —0.0005 | —0.0730 | —0.1460 |
| At Bath Beach, Brooklyn.......... | K | 8.3289 | 8.3277 | —0.0012 | —0.0730 | —0.1460 |
| At East 84th Street, Manhattan.... | No. 5 | 2.2991 | 2.2984 | —0.0007 | —0.0730 | —0.1460 |
| At Willets Point, Long Island...... | Tidal, U. S. Engineers... | 4.3083 | 4.3071 | —0.0012 | —0.0730 | —0.1460 |
| At College Point, Long Island...... | Tidal Station No. 68..... | 2.9225 | 2.9262 | +0.0037 | —0.0730 | —0.1460 |

of this work was to determine if possible whether or not any appreciable warping of the earth's crust had occurred in that region since the level lines were first run in 1886.* The 1911 leveling is shown in the accompanying sketch and the results of a comparison of the two sets of elevations are shown in the table below:

NEW JERSEY STATE GEOLOGICAL SURVEY BENCH MARKS.

| Bench Marks. | Elevations Determined in 1886. | Elevations Determined in 1911. | Difference. |
|---|---|---|---|
| | feet | feet | feet |
| Cape May Court House.................. | 19.498 | 19.498 (assumed) | 0.000 (assumed) |
| Sea Isle City............................ | 9.151 | 9.033 | —0.118 |
| Millville National Bank.................. | 33.450 | 33.453 | +0.003 |
| Vineland—West Jersey Station........... | 108.10 | 108.082 | —0.018 |
| Mays Landing—Court House............ | 19.89 | 19.790 | —0.100 |
| Absecon—M. E. Church................ | 30.660 | 30.638 | —0.022 |
| Atlantic City—Absecon Light House...... | 8.954 | 8.931 | —0.023 |

Mr. Vermeule states: "All of these bench marks are in satisfactory condition and may be considered not to have changed appreciably since the running of the original levels.

"It will be seen that the agreement between the levels of the present year and the original levels of 1886 is extremely close and entirely within the probable error of the observation.

"The agreement throughout is such that it may be assumed that there has been no important relative change of elevation between these several bench marks during the interval from 1886 to 1911."

As the New Jersey leveling of 1911 was not connected with a mean sea level determination as was the case in the Board of Estimate leveling, the conclusions of stability for the New Jersey coast are not quite so convincing. There is, however, no evidence of warping or tilting, which, it is expected, would accompany real subsidence.

From the determinations above noted, which are the result of spirit leveling of unquestioned accuracy, it is clear that from the standpoint of the geodesist or engineer there is no reliable evidence to show a general progressive subsidence of the Atlantic coast in New York City and vicinity. On the contrary, all the evidence is in favor of stability.

In 1857, Professor George H. Cook of Rutgers College, New Jersey, published in the American Journal of Science † an important paper on the subsidence of the land along the coasts of New Jersey and Long Island, in which he cited much evidence to prove that these coasts were gradually sinking at a rate of about two feet per century. Professor Cook said: "In the course of some geological examinations along the coast of Southern New Jersey, my attention was frequently called to various facts indicating a change in the relative level of the land and water, at some recent period. An attentive examination of these facts has led me to the conclusion that a gradual subsidence of the land is now in progress throughout the whole length of New Jersey

* The re-running of these lines was suggested by Professor Douglas W. Johnson, Professor of Physiography at Columbia University.

† *American Journal of Science*, s.s., vol. xxiv, pp. 341-354, 1857.

and Long Island; and from information derived from others, I am induced to think that this subsidence may extend along a considerable portion of the Atlantic coast of the United States."

Since the publication of Professor Cook's paper, a great many reports upon this interesting subject have appeared, the greater number supporting the theory of recent subsidence.

The conclusions of Professor Cook and many other authors seem to have been based mainly upon the occurrence of timber in the marshes and in the water below tide-level, upon the finding of the remains of logs, roots and stumps in such places. the rapid wearing away of the shores, the loss of land as shown by surveys, and the higher level of the water on the mill wheels along the coast. Professor Cook had also seen timber buried in the marshes of Staten Island and Long Island, and his theory of recent subsidence was upheld by many geologists of the eastern states, who had seen "swamps, now submerged at low water, containing the roots of stumps and trees:" they also found "stumps of oak trees at tide level," and many cases of "former meadow or swamp land which is now salt marsh."

Professor Cook set the average rate of subsidence "in the district where the observations were made, at two feet per century." In later years, however, particularly during the latter part of his term of office as State Geologist for New Jersey (1864-1900), Professor Cook was inclined to modify his views, and stated that the rate of subsidence might be much less.

In attempting to disprove the very generally accepted theory of coastal subsidence from a geological point of view, the writer cannot do better than present the conclusions of Douglas Wilson Johnson, Ph. D., Professor of Physiography at Columbia University, who has done more to controvert this theory than any other geologist in the United States.

Professor Johnson's conclusions are set forth in an article entitled "Fixite de la côte Atlantique de l'Amerique du Nord," published in the "Annales de Geographie," vol. XXI, pp. 193-212, 1912, at Paris. These conclusions, together with other proofs of stability as observed by Professor Johnson, will now be presented.

"It is to-day generally accepted as a well established fact that the Atlantic coast of North America is gradually subsiding, at a rate which is variously estimated from 20 to 75 centimeters per century.

"The supposed changes of level of the Atlantic coast have interested me for several years. A study of the form of Nantasket Beach, near Boston, showed that this portion of the coast could not have subsided more than a meter during the last 1,000 or 2,000 years.* An examination of certain shore line changes produced at Scituate, also near Boston, by the great storm of 1898, showed that all the appearances of a subsidence of the coast could be produced by an increased height of high tide resulting from a change in the form of the shore line.†

"I have made a study of the most important points on the Atlantic coast from the northern side of Prince Edward Island to the Florida Keys, as well as a number of places on the coasts of Sweden, England and Holland. The results of these studies seem to me to justify, for the Atlantic coast of North America, the following conclusions: (1) This coast cannot have subsided progressively 20 centimeters or more per century during the last few thousand years. It has remained relatively stable during all this period. (2) The coast cannot have subsided as much as 30 centimeters in the last century. (3) There is no satisfactory evidence of any subsidence whatever during the last few thousand years.

"Throughout this article the expression 'recent subsidence' is employed to designate

*Johnson, D. W., and Reed, W. G., The form of Nantasket Beach. Jour. Geol. 18:162-189. 1910.

†Johnson, D. W., The supposed recent subsidence of the Massachusetts and New Jersey coasts. Science N. S. 32:721-723. 1910; also The Botanical evidence of coastal subsidence. Science N. S. 33:300-302. 1911.

subsidence within the last few thousand years, and 'remote subsidence' to designate a sinking of the land which occurred more than 4,000 or 5,000 years ago. We may now consider some of the supposed proofs of recent subsidence of the coast.

"Corduroy roads discovered in the salt marshes of New Jersey and which were believed to indicate a sinking of more than 50 centimeters per century have been proven untrustworthy: even at the present time the inhabitants of the region have built such roads along the shore in the most moist portions of the marsh and find that these gradually sink into the marsh on account of its softness and the weight of the roadway itself. Even when they are not used these roadways continue to sink into the marsh due to their own weight.

### Fictitious Appearance of Changes of Level.

"At many points on the Atlantic coast one may observe large numbers of trees killed by salt water so recently that they still stand erect, and even retain their branches. These trees have often been cited as a convincing proof of the recent progressive subsidence of the land. The dead forests along the coasts of New Jersey, the Carolinas, and Georgia, many of which I have examined, are most frequently to be explained as the result of fluctuations in the plane of local high water due to a change in the outline of the coast, or to the attack of the waves undermining the roots of the trees and exposing them to salt water. I have seen no case where the killing of the trees could safely be ascribed to a sinking of the coast.

### Submerged Stumps.

"Closely allied to the foregoing evidence of subsidence is that furnished by submerged stumps. These are found along all parts of the Atlantic coast, at depths varying from a few inches below high tide to ten feet or more below low tide level, and have been repeatedly cited by both botanists and geologists as conclusive proofs of recent subsidence. These submerged stumps are in many cases the result of undermining of trees by the waves, allowing them to slip down into the salt water; in other cases they are the tap roots of certain trees which descend to a great depth. The loblolly pine has a tap root as large as its trunk which runs down 2 or 3 meters, and then sends off smaller roots. A forest of such trees growing on a low coast may be attacked by the waves, and as the earth is removed the trees die and finally break off at or below water level. In this way deeply submerged 'stumps' are produced which will seem to the ordinary observer a convincing proof of subsidence.* Illustration No. 20 represents several stages in this process.

"Submerged stumps, due to a local rise of the high tide level, to the compression of peat bogs caused by a lowering of the ground-water level as the waves cut into the shoreward side of such bogs (illustration No. 21), to the compression of peat deposits under the weight of barrier beaches (illustration No. 22), and to other causes, have been observed at many points along the coast. The more one sees of this type of evidence the more does he realize its unreliability.

### Submerged Peat.

"Deposits of submerged peat, like submerged stumps, are abundant all along the coast. These deposits are sometimes a little below high tide, often a considerable distance below low tide. It is argued that fresh water peat must have formed above sea level and that it could not reach a point below sea level without a sinking of the land. There are many objections to this line of argument. In the first place floating

*Lyell, Charles, A second visit to the U. S. of North America. 2d ed. London, 1850. 1:316-317.

peat bogs form on ponds and lakes often of considerable dimensions, and support large trees. As the trees increase in size and the bog increases in weight by the addition of new material, the lower part of the bog is forced downward, sometimes below sea level. This would give submerged stumps in fresh water peat below sea level without a change in general level. In the second place, one must remember that the lower portions of these peat bogs may be very ancient, and even if they were

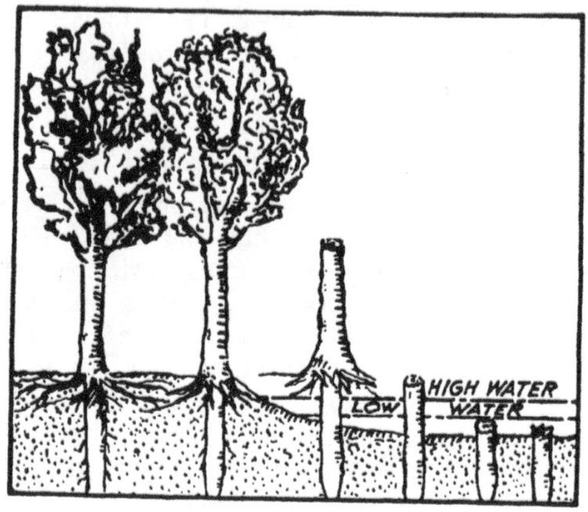


No. 20.—Submerged Stumps Resulting from Normal Retrogression of the Shore Line.

formed above sea level and later depressed, that depression may have occurred many thousands of years ago. In the third place, when such a bog is encroached upon by the sea, the level of the ground-water table in the bog, formerly at or near its surface, is rapidly lowered. Near the seaward margin of the bog the ground-water table may decline to mean sea level: and, right at the margin, to low tide level when the tide is out. As a result of this removal of water, the surface of the bog is

No. 21.—Submerged Peat and Stumps Produced by an Invasion of Peat Bogs by the Sea; H. W., High Water; L. W., Low Water.

rapidly lowered, carrying down with it trees which are killed by exposure to high tide. This is shown in illustration No. 21. How extensive such a settling of the surface may be is suggested by illustration No. 23, which represents the result of an artificial lowering of the ground-water level.* Furthermore, the alternate submerg-

* Fens of Eastern England, near Peterborough. When it was proposed to drain this land in 1848 the cast iron post shown in the illustration was driven through the peat into the stiff clay beneath until the cap on top of the post became flush with the ground. An inscription on the cap states that it marked the surface of the ground in 1848. The cap is now high above the ground and the amount can be estimated by the size of the men shown alongside. Photo by D. W. Johnson.

ence and draining of the bog removes so much of its content that the surface may even slope down to a level considerably below that of high tide. Submerged deposits of fresh peat containing upright stumps, therefore, are not to be regarded as a conclusive proof of subsidence, either remote or ancient.

"It is well known that the attack of the waves often drives a barrier beach inward over the salt marsh. The enormous weight of the beach necessarily results in a compression of the peat deposit, so that the surface of the latter is exposed near or below

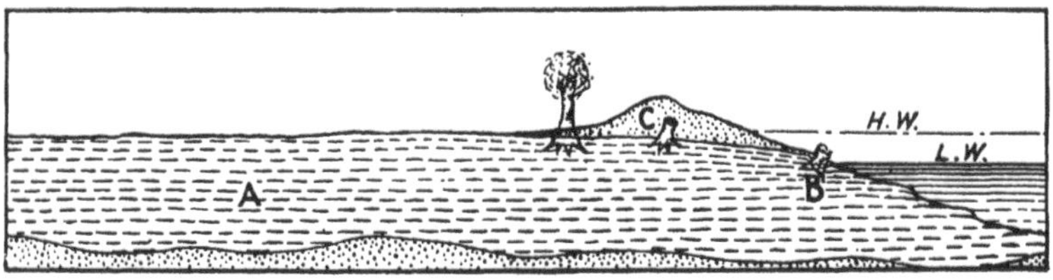


No. 22.—Submerged Peat Outcropping at Low Tide (B) Compressed by Weight of Barrier Beach, which is Encroaching on Salt Marsh Peat Deposit (A); H. W., High Water; L. W. Low Water.

low tide level on the seaward side of the beach (illustration No. 22). On the coast near Boston a barrier beach has been driven back over a salt marsh more than 70 meters in twelve years. To-day the former surface of the meadow, with the wheel tracks of an old road, impressions of horses' hoofs, and the stumps of trees which had gained a foothold on the marsh inside of the beach, are all exposed at low tide on the seaward side of the beach. Those who would interpret this as a result of coastal subsidence must admit a subsidence of perhaps two meters in twelve years, of which there is no corroborating evidence whatever. Good examples of this compression due to the enormous weight of the barrier beaches are shown all along our coast.

No. 23.—Showing Soil Shrinkage, Due to Draining of Peaty Soil; in 1848 the Surface of the Land was Even with the Top of the Post.

### Phenomena Produced by a Local Rise in the High Tide Level.

"On a tidal coast, if we have a bay like that shown in illustration No. 24, almost separated from the ocean by a barrier beach, but connected with it by a narrow tidal inlet, the waters of the rising tide in the ocean will pass through the tidal inlet with so much difficulty that the surface of the bay will rise much more slowly than the surface of the ocean. When the tide in the ocean has reached its maximum, and has begun to fall, the surface of the bay will still remain much lower. When it is low tide in the ocean, the water in the bay will remain at a higher level, because this water cannot escape fast enough to maintain equality of levels between the two water bodies. Hence, high tide level in the bay is lower than high tide level in the ocean. This is shown in illustration No. 25, which represents a cross-section of such a bay as shown in illustration No. 24, in the direction AB. It is evident that around the

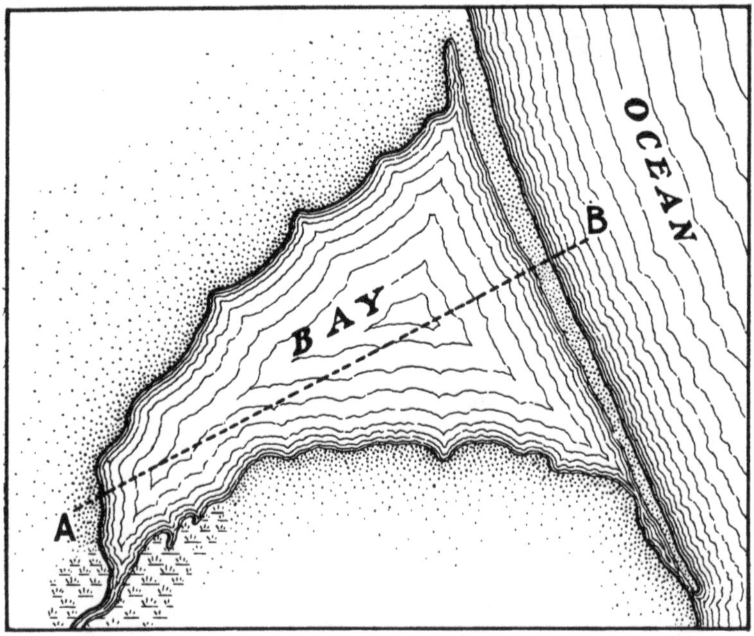


No. 24.—Bay Separated from the Open Ocean by a Barrier Beach.

shores of the bay, trees and other fresh water vegetation will grow down to the level of the high tide of the bay, and thus *below* the high tide level of the adjacent ocean. Salt marshes in the bays will likewise grow up to the high tide level of the bay, farmers will build dikes to reclaim their marshes at this same level, and in other ways the level will be so marked as to render readily perceptible any increase in the height of the tides.

"Now let us consider the consequences which must follow if storm waves make a large breach in the barrier beach. With free access to the bay through the larger opening, the tidal waters will at once rise as high in the bay as in the open ocean (CD, illustration No. 25). All trees whose bases are below the line CD will be washed by the tides and killed. The standing forests of dead trees will later be represented by submerged stumps. Dikes raised by the farmers will be overflowed by the tides. The surface of the salt marsh will build up to the new high tide level, enveloping both stumps and dikes. Fresh water peat, formerly beyond the reach of salt water, may now be buried under a layer of salt peat. In short, most of the phenomena usually cited as proofs of general coastal subsidence will be produced by a local rise of the high tide caused by a change in the form of the shore line. If the bay narrows

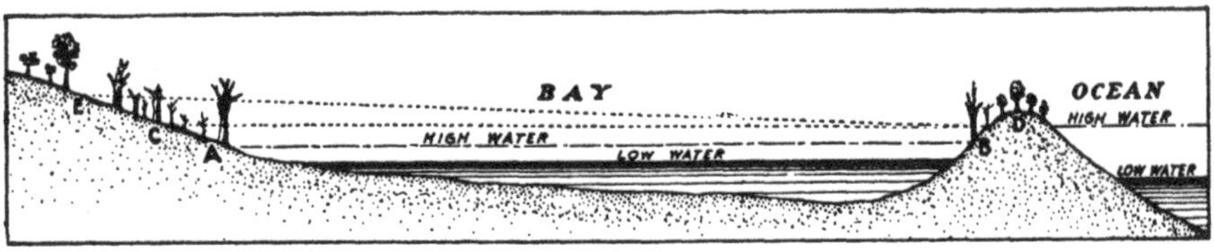


No. 25.—Diagram showing fictitious subsidence of the coast; as long as the barrier beach (D) nearly closes the mouth of the bay, high water in the bay is lower than high water in the open sea; trees grow down to this lower level (AB) along the shores of the bay; when the barrier beach is broken through or removed, high water in the bay rises as high (CD) as the open sea, and all the trees between the levels AB and CD are killed by the salt water; if the bay narrows going inland, the tide is forced to rise even above the level it attains in the open sea, or to the position ED, and at the head of the bay all the trees between A and E are killed; in addition to these submerged forests, other fictitious indications of subsidence are thus produced.

inland as shown in illustration No. 24, the tidal wave will increase in height as it advances, so that the level of high tide at the head of the bay will rise far above that of the open ocean (ED, illustration No. 25). In this case all the trees between A and E will be killed at the head of the bay, and the appearance of subsidence will be unusually pronounced.

"Illustration No. 26 represents the consequences of the opposite type of change.

No. 26.—Diagram showing fictitious elevation of the coast; before the barrier beach (D) was constructed, the tide in the bay rose as high as in the open sea, and the cliff (f) and bench (p) were carved by the waves; since the building of the barrier beach, high water in the bay is lower than in the ocean, the cliff and bench are no longer reached by the waves, and appear to represent an "elevated shore line"; the uniform altitude of the beach ridges on the barrier beach shows that the relative level of land and sea have long remained constant.

When the bay was open to the ocean, the waves cut a cliff (f) and bench (p). But the construction of a barrier beach (D) has so reduced the level of high tide in the bay that the waves no longer reach the cliff or the inner part of the bench. These become covered with trees and other fresh vegetation, and constitute what is usually called an 'elevated shore line.' Cliffs and benches of this origin have been cited as proofs of recent coastal elevation.

"When critically examined, neither the botanical nor other evidence of recent coastal subsidence seems to me conclusive. On the other hand, the physiographic evidence, so far as I have been able to analyze it, indicates a long period of coastal stability.

### Proofs of the Stability of the Coast.

"Among the proofs that the Atlantic coast has remained constant for a long period of time are the following:

"1. The level of the base of old cliffs separated from the sea thousands of years ago by the construction of barrier beaches after the manner indicated in illustration No. 26 is to-day exactly where it should be if the barrier beach were removed and the sea were to attack the coast again.

"2. The barrier beaches are in many cases constructed of successive parallel ridges thrown up by the waves and the average height of the oldest ridges is the same as the average size of the most recent ridges, showing that there has been no appreciable change of level during the construction of the beach. Inasmuch as some of these beaches are so extensive as to have required several thousand years at least for their construction, there can have been no pronounced change of level for several thousand years. (See D, illustration No. 26.)

"3. If there had been a progressive continuous sinking of the coast continuing up to the present time there should be a continuous line of dead trees all along the coast, showing the continuous advance of the ocean. As a matter of fact such dead trees are absent from the greater part of the coast and are only found in exceptional localities where the conditions are favorable to local tidal changes as already explained."

The work of the writer on the Board of Estimate leveling must be construed as a striking confirmation of Professor Johnson's theory of coastal stability as set forth in the preceding paragraphs. It is of especial interest because it is a proof based on engineering methods of the absolute stability during the last quarter of a century of the very part of the coast which is generally supposed to be undergoing most rapid subsidence at the present time.

## DESCRIPTIONS OF BENCH MARKS.

### General Notes Describing Different Forms and Markings of Bench Marks Connected with the Level Net.

NOTE 1.—A brass tablet bolt two inches in diameter, leaded or cemented *horizontally* into the vertical face of stone or brick, flush with the surface. It is provided with a stem at the back 2¾ inches in length, which is turned to form grooves around which the lead or cement flows, holding the bolt firmly in place. Two lugs on the back of the tablet give a bond against turning.

The face of the tablet is lettered as shown, and is marked horizontally across its center by a V-shaped indentation, the center of which is the bench mark. The bench mark line and lettering are depressed, being cut in the pattern and thus formed in casting. For details of construction see illustration No. 15.

NOTE 2.—A brass tablet bolt inscribed as described in Note 1, leaded or cemented *vertically* in stone, brick or concrete. The center of the top of the bolt is the bench mark.

NOTE 3.—The bottom of a square cavity, usually 2 inches square and ¼-inch deep, cut in stone or brick and lettered

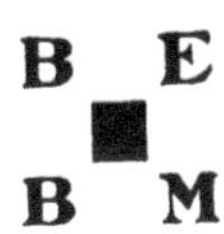


The bottom surface of the cavity is always taken as the bench mark.

NOTE 4.—The bottom of a square cavity, usually two inches square and ¼-inch deep, cut in stone and sometimes lettered.

The bottom surface of the cavity is always taken as the bench mark.

NOTE 5.—A bench mark of this type is the top of a copper bolt 3 inches long, leaded vertically in masonry.

It is usually set flush with the surface, but at times slightly above it. The letters B. M. are stamped on the top as shown:

NOTE 6.—A bench mark of this type is a copper bolt 3 inches long, leaded *horizontally* into the vertical face of masonry, with its outer end setting slightly back of the surface. It is marked horizontally across its center and has the letters B. M. stamped on the face of bolt as shown:

The center of the horizontal line is always taken as the bench mark.

NOTE 7.—The top of a ¼-inch copper bolt 2 inches long, leaded *vertically* in masonry, usually the water-table of a building.

NOTE 8.—The top of a copper bolt 3 inches long, leaded *vertically* in the horizontal surface of masonry. It is usually set flush with the surface, but at times slightly above it.

NOTE 9.—A bench mark of this type is the center of a 2-inch square cut in heavy masonry, lettered

**B E**<br>□<br>**B M**

NOTE 10.—A level surface, from 1½ inches to 3 inches square, in an inclined or irregular surface, usually a rough coping, sill, or water-table. The bottom surface of the square is always taken as the bench mark.

NOTE 11.—A bench mark of this type is one established in 1880 by Mr. A. G. Culver, then Assistant Engineer in the Department of Public Works. It is a galvanized iron bolt of ⅝-inch cross section, usually leaded horizontally into the water-table of a building and projecting about 1 inch. The top of the bolt is the bench mark.

NOTE 12.—A bench mark of this type is a ¾-inch copper bolt 3 inches long, leaded horizontally into the vertical face of masonry with its outer end setting slightly back of the surface. It is marked horizontally across its center and is lettered

**B E**<br>⊖<br>**B M**

by letters cut in the stone. The horizontal line is always taken as the bench mark.

NOTE 13.—A bench mark of this type is a galvanized iron bar of 1-inch cross-section and 3 inches long, leaded horizontally into the vertical face of masonry, with its outer end projecting 1 inch. The stem of the bolt is notched and the inner end is spread against a wedge. A solder of half lead and half tin is used in the setting, and is well tamped around the bolt. The bench mark is made of a solid piece of iron with its nose shaped to a bracket below, and is expected to be secure against accidental disturbance or violence. The iron was galvanized after shaping.

For details of construction see illustration No. 15.

NOTE 14.—A bench mark of this type is the top of a copper bolt 3 inches long, leaded vertically in masonry and lettered

**B E**

⊙

**B M**

by letters cut in the stone.

NOTE 15.—A bench mark of this type is a cross (+), usually cut with bars 2 inches long, in the vertical face of masonry. When established by the Board of Estimate and Apportionment, it has V-shaped indentations ¼-inch heavy and ¼-inch deep. The center of the horizontal bar is always the bench mark.

NOTE 16.—A bench mark of this type is one established by the United States Engineers. It is the surface within an outlined square, 1 inch or more on each side, cut in the horizontal surface of masonry and lettered "U. S."

NOTE 17.—A cross (+) cut in the horizontal surface of masonry or steel. It is usually from 1 to 2 inches long. The surface surrounding the cross is always the bench mark.

NOTE 18.—A bench mark described as a copper bolt and referred to this note is a piece of round copper 3 inches long, leaded horizontally into a masonry wall.

The horizontal line cut across its center is the bench mark.

NOTE 19.—The top of a copper bolt 3 inches long, leaded vertically in masonry that has been made level at least 1½ by 1½ inches.

NOTE 20.—A bench mark of this type is the bottom surface of a cavity 2 inches or more square, cut into an irregular block or projecting foundation stone.

NOTE 21.—This is the Board of Water Supply standard pipe casing, and is the top of an iron rod ¾-inch in diameter and about 6 feet long, which is secured to a concrete base and protected by a 3-inch wrought iron pipe casing, to the top of which is bolted a cast iron cap. The bolt and cap must be removed to use the bench mark.

NOTE 22.—The top of a brass bracket bench mark secured to the sides of public schools by means of expansion bolts and set by the Brooklyn Highway Department. The bolt projects about 5 inches, and the top of the knob at the outer end is the bench mark.

NOTE 23.—The top of a ¾-inch copper bolt 3 inches long, leaded vertically in the top of reinforced concrete city monument which is set in concrete 3'x3'x4'. Iron boxes or monument covers protect the bench mark, which is usually 6 inches below the ground surface. These bench marks were prepared by the Brooklyn Topographical Bureau. (See illustration No. 16.)

NOTE 24.—The top of a ⅝-inch copper bolt 3 inches long, leaded vertically in the top of a granite city monument. which is set in concrete 3'x3'x4'. Iron boxes or monument covers protect the bench mark, which is usually 6 inches below the ground surface. These bench marks were prepared by the Queens Topographical Bureau. (See illustration No. 16.)

NOTE 25.—The center of the top of a 3⅛-inch brass bolt marking a United States Coast and Geodetic Survey triangulation station. The bolt is set in a granite monument which is set in concrete 3'x3'x4'. On the space between the inner and outer circles of the station mark, the letters U. S. C. & G. S. are cast. (See illustration No. 17.)

NOTE 26.—A bench mark of this type is a point on a horizontal surface which was impossible for some reason to mark. This was usually the case when a bench mark was established on a polished granite surface, or in a conspicuous place.

# STANDARD ELEVATIONS OF BENCH MARKS
IN THE
# BOROUGH OF RICHMOND
AND IN
# NEW JERSEY.

| No. of Bench Mark. | Locality. | Elevation Above Mean Sea Level. | |
|---|---|---|---|
| | | Meters. | Feet. |
| 1 | Perth Amboy, New Jersey | 18.5763 | 60.946 |
| 2 | “ “ “ “ | 1.9917 | 6.534 |
| 3 | Tottenville | 2.1365 | 7.010 |
| 4 | “ | 4.1112 | 13.488 |
| 5 | “ | 12.1495 | 39.861 |
| 6 | “ | 14.8627 | 48.762 |
| 7 | “ | 23.8877 | 78.372 |
| 8 | “ | 24.3373 | 79.847 |
| 9 | “ | 23.9820 | 78.681 |
| 10 | “ | 24.3158 | 79.776 |
| 11 | Richmond Valley | 13.0193 | 42.714 |
| 12 | “ “ | 2.5414 | 8.338 |
| 13 | “ “ | 2.0892 | 6.854 |
| 14 | Pleasant Plains | 10.0724 | 33.046 |
| 15 | “ “ | 13.7425 | 45.087 |
| 16 | Princes Bay | 14.7692 | 48.455 |
| 17 | “ “ | 10.4028 | 34.130 |
| 18 | Huguenot | 21.6453 | 71.015 |
| 19 | Kreischerville | 19.2282 | 63.085 |
| 20 | Rossville | 6.0932 | 19.991 |
| 21 | “ | 3.7296 | 12.236 |
| 22 | “ | 2.9648 | 9.927 |
| 23 | “ | 6.3034 | 20.680 |
| 24 | “ | 4.0747 | 13.368 |
| 25 | “ | 22.3648 | 73.375 |
| 26 | “ | 4.4548 | 14.616 |
| 27 | “ | 4.4215 | 14.506 |
| 28 | Greenridge | 13.6077 | 44.645 |
| 29 | “ | 13.8270 | 45.364 |
| 30 | Fresh Kills | 3.6615 | 12.013 |
| 31 | “ “ | 6.3672 | 20.890 |
| 32 | “ “ | 3.0527 | 10.015 |
| 33 | “ “ | 3.1768 | 10.423 |
| 34 | “ “ | 3.3480 | 10.984 |
| 35 | “ “ | 4.3997 | 14.435 |
| 36 | “ “ | 3.8030 | 12.477 |
| 37 | “ “ | 4.2181 | 13.839 |
| 38 | Richmond | 14.8738 | 48.798 |
| 39 | “ | 12.9439 | 42.467 |
| 40 | “ | 14.9104 | 48.919 |
| 41 | “ | 23.7298 | 77.854 |
| 42 | “ | 24.2660 | 79.613 |
| 43 | New Dorp | 18.4272 | 60.457 |
| 44 | “ “ | 14.6694 | 48.128 |
| 45 | “ “ | 6.1209 | 20.082 |
| 46 | “ “ | 4.6093 | 15.122 |

| No. of Bench Mark. | Locality. | Elevation Above Mean Sea Level. | |
|---|---|---|---|
| | | Meters. | Feet. |
| 47 | New Dorp | 1.7505 | 5.743 |
| 48 | " " | 1.9472 | 6.388 |
| 49 | " " | 2.7259 | 8.943 |
| 50 | " " | 27.9867 | 91.820 |
| 51 | Oakwood | 17.4096 | 57.118 |
| 52 | " | 17.7622 | 58.275 |
| 53 | " | 11.8789 | 38.973 |
| 54 | New Dorp | 14.8404 | 48.689 |
| 55 | " " | 36.8989 | 121.059 |
| 56 | Manor Road | 57.7881 | 189.593 |
| 57 | New Springville | 6.8375 | 22.433 |
| 58 | Union Avenue | 3.1141 | 10.217 |
| 59 | Linoleumville | 11.9856 | 39.323 |
| 60 | " | 14.0195 | 45.996 |
| 61 | " | 11.9646 | 39.254 |
| 62 | " | 8.5524 | 28.059 |
| 63 | Travisville | 7.4936 | 24.585 |
| 64 | Bloomfield | 4.8996 | 16.075 |
| 65 | Watchogue | 2.8178 | 9.245 |
| 66 | " | 2.7589 | 9.052 |
| 67 | Summerville | 3.1354 | 10.287 |
| 68 | Arlington | 16.4668 | 54.025 |
| 69 | " | 14.0082 | 45.959 |
| 70 | " | 14.2148 | 46.636 |
| 71 | Summerville | 6.4443 | 21.143 |
| 72 | Old Place | 3.6535 | 11.987 |
| 73 | New York City Farm Colony | 70.2939 | 230.623 |
| 74 | " " " " " | 71.1514 | 233.436 |
| 75 | Manor Road | 79.7884 | 261.773 |
| 76 | " " | 92.6109 | 303.841 |
| 77 | " " | 99.1763 | 325.381 |
| 78 | " " | 89.8942 | 294.928 |
| 79 | Todt Hill | 118.9792 | 390.351 |
| 80 | " " | 118.9808 | 390.356 |
| 81 | " " | 118.6407 | 389.240 |
| 82 | " " | 117.3436 | 384.985 |
| 83 | " " | 124.7363 | 409.239 |
| 84 | Castleton Corners | 74.4794 | 244.354 |
| 85 | " " | 56.2464 | 184.535 |
| 86 | " " | 55.8315 | 183.174 |
| 87 | " " | 55.9593 | 183.593 |
| 88 | Westerleigh | 37.0667 | 121.610 |
| 89 | " | 30.8529 | 101.223 |
| 90 | " | 31.3147 | 102.738 |
| 91 | " | 30.6005 | 100.395 |
| 92 | " | 12.1348 | 39.812 |
| 93 | Port Richmond | 6.2611 | 20.542 |
| 94 | Grant City | 10.9086 | 35.789 |
| 95 | " " | 11.8618 | 38.917 |
| 96 | " " | 11.8855 | 38.995 |
| 97 | Garretsons | 19.3774 | 63.574 |
| 98 | " | 18.7179 | 61.410 |
| 99 | " | 13.4481 | 44.121 |
| 100 | Grasmere | 14.0751 | 46.178 |
| 101 | " | 16.2887 | 53.441 |
| 102 | " | 27.9899 | 91.830 |
| 103 | " | 29.2171 | 95.857 |
| 104 | Arrochar | 20.7814 | 68.180 |
| 105 | " | 20.1159 | 65.997 |

| No. of Bench Mark. | LOCALITY. | Elevation Above Mean Sea Level. | |
|---|---|---|---|
| | | Meters. | Feet. |
| 106 | Arrochar | 18.9737 | 62.250 |
| 107 | Fort Wadsworth | 29.5896 | 97.079 |
| 108 | " " | 30.5477 | 100.222 |
| 109 | " " | 32.6589 | 107.148 |
| 110 | " " | 31.0960 | 102.021 |
| 111 | " " | 22.2512 | 73.002 |
| 112 | Elizabethport, N. J. | 3.7787 | 12.397 |
| 113 | " " | 3.3134 | 10.873 |
| 114 | " " | 2.1586 | 7.082 |
| 115 | Howland Hook | 2.4128 | 7.916 |
| 116 | " " | 6.1894 | 20.306 |
| 117 | " " | 7.2459 | 23.773 |
| 118 | " " | 8.2440 | 27.047 |
| 119 | North Shore | 5.9105 | 19.391 |
| 120 | Arlington | 3.2997 | 10.826 |
| 121 | Mariners Harbor | 3.0728 | 10.081 |
| 122 | " " | 3.2284 | 10.592 |
| 123 | " " | 2.9664 | 9.732 |
| 124 | " " | 3.0699 | 10.072 |
| 125 | Elm Park | 4.3776 | 14.362 |
| 126 | " " | 3.7476 | 12.295 |
| 127 | Port Richmond | 3.2830 | 10.771 |
| 128 | " " | 3.5613 | 11.684 |
| 129 | " " | 4.8404 | 15.881 |
| 130 | " " | 2.9983 | 9.837 |
| 131 | " " | 3.1988 | 10.495 |
| 132 | " " | 7.7530 | 25.436 |
| 133 | " " | 7.2691 | 23.849 |
| 134 | Livingston | 6.0889 | 19.977 |
| 135 | Sailors Snug Harbor | 11.3807 | 37.338 |
| 136 | " " " | 7.4507 | 24.445 |
| 137 | " " " | 3.8294 | 12.564 |
| 138 | " " " | 2.0661 | 6.779 |
| 139 | New Brighton | 8.8838 | 29.146 |
| 140 | " " | 10.6882 | 35.066 |
| 141 | St. George | 17.4296 | 57.184 |
| 142 | " | 17.6614 | 57.944 |
| 143 | " | 13.2037 | 43.319 |
| 144 | " | 2.8597 | 9.382 |
| 145 | " | 22.7114 | 74.512 |
| 146 | Tompkinsville | 10.5764 | 34.699 |
| 147 | " | 3.0196 | 9.907 |
| 148 | " | 3.6283 | 11.904 |
| 149 | " | 12.4980 | 41.004 |
| 150 | " | 10.8574 | 35.621 |
| 151 | Stapleton | 2.9596 | 9.710 |
| 152 | " | 10.6794 | 35.037 |
| 153 | " | 10.5068 | 34.471 |
| 154 | Clifton | 3.8297 | 12.565 |
| 155 | " | 3.3737 | 11.069 |
| 156 | Rosebank | 12.8584 | 42.186 |
| 157 | " | 12.4459 | 40.833 |
| 158 | " | 13.5654 | 44.506 |
| 159 | Richmond Valley | 6.4375 | 21.120 |
| 160 | " " | 6.4234 | 21.074 |
| 161 | " " | 6.4393 | 21.126 |
| 162 | Kreischerville | 14.6171 | 47.956 |
| 163 | " | 18.3763 | 60.290 |
| 164 | " | 7.3490 | 24.111 |

| No. of Bench Mark. | Locality. | Elevation Above Mean Sea Level. | |
|---|---|---|---|
| | | Meters. | Feet. |
| 165 | Richmond | 4.2525 | 13.952 |
| 166 | Annadale | 24.4510 | 80.220 |
| 167 | " | 24.0360 | 78.858 |
| 168 | " | 28.9866 | 95.100 |
| 169 | Huguenot | 21.9483 | 72.009 |
| 170 | Eltingville | 18.8112 | 61.716 |
| 171 | Annadale | 26.0140 | 85.348 |
| 172 | Giffords | 21.6607 | 71.065 |
| 173 | " | 22.1176 | 72.564 |
| 174 | Fort Wadsworth | 7.4137 | 24.323 |
| 175 | " " | 2.2968 | 7.535 |
| 176 | " " | 4.1734 | 13.692 |
| 177 | " " | 2.2909 | 7.516 |

# DESCRIPTIONS OF BENCH MARKS

IN THE

# BOROUGH OF RICHMOND

AND IN

# NEW JERSEY.

B. M. 1.—*State Geological Survey Bench Mark, Perth Amboy, N. J.* The center of a granite monument with a rounded top 8 inches in diameter, located in a triangular grass plot in the public park at the intersection of High Street and Market Street. The monument, the elevation of which was established by the United States Coast and Geodetic Survey, is on the center line of High Street, 97.75 feet southwesterly from its intersection with the center line of Market Street, and the top is 3 inches above the ground.

Elevation, 60.946 feet.

B. M. 2.—*Perth Amboy, N. J.* The intersection of a cross (+) cut in the top of a spike with a head 1¼ inches square, in the top of the backing log on the north side of the dock at the foot of Fayette Street. The spike is near the northeast corner of the dock and 19 feet north of the south line of Fayette Street.

Elevation, 6.534 feet.

B. M. 3.—*Tottenville.* The intersection of a cross (+) cut in the top of a spike with a head ¾-inch square in the top of the fourth pile from the northwest corner, on the north side of the dock at the foot of Amboy Road. The pile containing the bench mark is on line with the north line of Amboy Road.

Elevation, 7.010 feet.

B. M. 4.—*Tottenville.* The top of a railroad spike driven vertically in the blazed root of a tree stump on the south side of Amboy Road about 95 feet east of the Arthur Kill. The bench mark is on the north side of the tree, which is opposite a sewer manhole at the foot of Amboy Road.

Elevation, 13.488 feet.

B. M. 5.—*Tottenville.* The bottom of a square cut in the top of the east curb on the northeast corner of Hopping Avenue and Amboy Road. The square is at the northwest corner of a curved section of the curb and 3 inches south of the north line of the concrete sidewalk on Amboy Road. (Note 10, p. 90.)

Elevation, 39.861 feet.

B. M. 6.—*Tottenville.* The bottom of a square cut in the top of the northwest corner of the bluestone sill of bay window at the west side of the two-story and attic frame house of Mr. Barnes, on the north side of Amboy Road, about 150 feet east of Hopping Avenue. The bench mark is 1 foot above the ground, and an arrow head on the west side of the sill points to the spot. (Note 10, p. 90.)

Elevation, 48.762 feet.

B. M. 7.—*Tottenville.* The center of a 1½-inch square (⊓) cut in the top of the west end of the first or lowest step at the Summit Street entrance to Public School No. 1, between Garretson Avenue and Prospect Avenue.

Elevation, 78.372 feet.

B. M. 8.—*Tottenville.* A horizontal line on the face of a 2-inch brass tablet bolt leaded horizontally into the north side of Public School No. 1, on the south side of Summit Street between Garretson Avenue and Prospect Avenue. The bolt is in the east side of the west buttress at the northwest corner of the building and 1.75 feet above the ground. (Note 1, p. 89.)

Elevation, 79.847 feet.

B. M. 9.—*Tottenville.* A cross (+) cut in the bluestone near the northwest corner of Public School No. 1, on the south side of Summit Street between Garretson Avenue and Prospect Avenue. The cross is in the east side of the west buttress on the north side of the building, below B. M. No. 8. (Note 15, p. 91.)

Elevation, 78.681 feet.

B. M. 10.—*Tottenville.* The center of a 1½-inch square (⊓) cut in the top of the west end of the sixth or top bluestone step at the north or main entrance to the Church of Our Lady, on the southeast corner of Prospect Avenue and Amboy Road.

Elevation, 79.776 feet.

B. M. 11.—*Richmond Valley.* A square (⊓) cut in the top of the east end of the third step of the east entrance to the Bethel Methodist Episcopal Church, on the northeast corner of Church Street and Amboy Road.

Elevation, 42.714 feet.

B. M. 12.—*Richmond Valley.* A cross (+) cut in the foundation stone at the northwest corner of the one-story and attic frame house on the northeast corner of Richmond Valley Road and Amboy Road. The cross is on the west side of the building and 9 inches above the ground. The letter B is cut in the stone above the cross. (Note 15, p. 91.)

Elevation, 8.338 feet.

B. M. 13.—*Richmond Valley.* The center of a 2-inch square (⊔) cut in the top of the east end of the south wall of the stone culvert crossing Richmond Valley Road about 25 feet south of the center line of the Staten Island Rapid Transit Railway. The bench mark is about 3 feet east of the west side of the frame house on the northeast corner of Amboy Road and Richmond Valley Road. An arrowhead cut in the face of the wall points to the bench mark.

Elevation, 6.854 feet.

B. M. 14.—*Pleasant Plains.* A 2-inch square (⊓) cut in the top of the south end of the bluestone threshold step at the house entrance to three-story brick tenement at No. 6343 Amboy Road, about 30 feet south of Station Avenue.

Elevation, 33.046 feet.

B. M. 15.—*Pleasant Plains.* The bottom of a square cut in the top of the south end of the most northerly bluestone window-sill on the east side of Public School No. 3. The bench mark is about 12 feet from the north side of the building and 1 foot above the ground. (Note 10, p. 90.)

Elevation, 45.087 feet.

B. M. 16.—*Princes Bay.* The center of the top of a small topographical monument on the southwest corner of Princes Bay Avenue (Seguine Avenue) and the Staten Island Rapid Transit Railway. The monument is on the prolongation of the center line of the sidewalk on the west side of Princes Bay Avenue, is about 10 feet north of the post and wire fence at the corner, and 16 inches below the ground. It is protected by a cylindrical iron casing with cover.

Elevation, 48.455 feet.

B. M. 17.—*Princes Bay.* A cross (+) cut in the top of the south end of the fourth or top granite step at the main entrance to the two-story and attic red brick house of Silas Miller, on the east side of Princes Bay Avenue (Seguine Avenue) about 1,500 feet south of the Staten Island Rapid Transit Railway. (Note 17, p. 91.)

Elevation, 34.130 feet.

B. M. 18.—*Huguenot.* The bottom of a square cut in the top of the center of the bluestone window-sill at the southwest corner of Public School No. 5, on the north side of Amboy Road about 510 feet east of Huguenot Avenue. The sill containing the bench mark faces Amboy Road, is between the two cellar windows and 8 inches above the ground. (Note 10, p. 90.)

Elevation, 71.015 feet.

B. M. 19.—*Kreischerville.* The center of a 1-inch square (⊔) cut in the top of the brick water-table at the southwest corner of Public School No. 4, on the east side of Fresh Kills Road, about 200 feet north of Sharrots Lane. The square is 4 feet above the ground and a cross is cut in the brick below the bench mark.

Elevation, 63.085 feet.

B. M. 20.—*Rossville.* The highest point within a 1-inch square (⊔) cut in the top of the south corner of the stone abutment at the northwest corner of St. Luke's Church, on the southeast corner of St. Luke's Avenue and Fresh Kills Road.

Elevation, 19.991 feet.

B. M. 21.—*Rossville.* The center of the horizontal bar of a **T** cut in the brick at the northeast corner of Frank Engert's three-story brick hotel on the south side of Fresh Kills Road, about 86 feet west of the center line of Prospect Avenue. The **T** is 2.1 feet above the sidewalk and 2 inches west of the northeast corner of the building.

Elevation, 12.236 feet.

B. M. 22.—*Rossville.* A 1-inch square (⊔) cut in the top of the west end of the bluestone curb on the north side of Fresh Kills Road, between Rossville Avenue and Prospect Avenue.

Elevation, 9.927 feet.

B. M. 23.—*Rossville.* A square (⊓) cut in the top of the bluestone coping of retaining wall on the south side of Fresh Kills Road, just east of St. Luke's Church and west of Prospect Avenue. The bench mark is at the point of curve on the north edge of the coping, is 2.5 feet above the sidewalk, and about 6 feet east of the west end of the wall.

Elevation, 20.680 feet.

B. M. 24.—*Rossville.* A square (⊔) cut in the top of the southeast corner of the New York and New Jersey boundary monument near the southwest corner of Fresh Kills Road and St. Luke's Avenue. The monument is granite with a square top which is 12 inches above the ground. It is on the south side of Fresh Kills Road 3 feet south of the fence line and about 80 feet west of the center line of St. Luke's

Avenue, has the inscription N. Y. & N. J. on the west side, and B. M. 1889 on the east side.

Elevation, 13.368 feet.

B. M. 25.—*Rossville.* A horizontal line on the face of a 2-inch brass bolt leaded horizontally into the limestone front of Public School No. 6, on the east side of Rossville Avenue, about midway between Fresh Kills Road and Woodrow Road. The bolt faces Rossville Avenue, is in the center of the building and 6 inches above the ground. (Note 1, p. 89.)

Elevation, 73.375 feet.

B. M. 26.—*Rossville.* The center of a square (⊓) cut in the top of the southwest corner of the north parapet wall of culvert crossing Fresh Kills Road at Huguenot Avenue (Swains Lane). Bench mark is 3.5 feet above the ground.

Elevation, 14.616 feet.

B. M. 27.—*Rossville.* The center of a square (⊓) cut in the top of the northwest corner of the south parapet wall of culvert crossing Fresh Kills Road at Huguenot Avenue (Swains Lane). The bench mark is 3.5 feet above the ground.

Elevation, 14.506 feet.

B. M. 28.—*Greenridge.* A cross (+) between the letters B. M., cut in the brick at the northwest corner of Public School No. 7, on the south side of Fresh Kills Road between Washington Avenue and Greenridge Avenue. The cross is on the third course of brick above the ground on the north side of the building and 10 inches from the west side. (Note 15, p. 91.)

Elevation, 44.645 feet.

B. M. 29.—*Greenridge.* A 1-inch square (□) cut in the top of the west end of the third bluestone step at the north or main entrance to Public School No. 7, on the south side of Fresh Kills Road between Washington Avenue and Greenridge Avenue.

Elevation, 45.364 feet.

B. M. 30.—*Fresh Kills.* A cross (+) cut in the north face of the retaining wall on the southwest corner of Fresh Kills Road and Bridge Avenue. The cross is 7.3 feet west of Bridge Avenue and 1.4 feet above the ground. (Note 15, p. 91.)

Elevation, 12.013 feet.

B. M. 31.—*Fresh Kills.* A square (⊓) cut in the top of the east end of the bluestone sill of the most northerly cellar window at the northeast corner of Emma Banker's two-story stone and frame dwelling, near the southwest corner of Fresh Kills Road and Bridge Avenue. The point is on the north side of the house and about 13 feet from the east side. It is 170 feet west of Bridge Avenue and about 490 feet south of Fresh Kills Road.

Elevation, 20.890 feet.

B. M. 32.—*Fresh Kills.* A ¾-inch copper bolt in a small topographical monument near the southwest corner of Fresh Kills Road and Bridge Avenue (monument station No. 8). The monument is 11.5 feet northeast of the retaining wall at the corner and 6.3 feet north of the telegraph pole nearest the corner. The top of the monument is 1.3 feet below the ground and is protected by a cylindrical iron casing with cover.

Elevation, 10.015 feet.

B. M. 33.—*Fresh Kills.* A square (□) cut in the west end of the old stone wall on the northeast corner of Fresh Kills Road and Bridge Avenue. The bench mark is 4 inches above the ground. Established by the Department of Docks and Ferries.

Elevation, 10.423 feet.

B. M. 34.—*Fresh Kills.* A square (□) cut in the top of the east end of the second bluestone step leading to the two-story frame house on the north side of Fresh Kills Road about 1,800 feet east of Bridge Avenue.

Elevation, 10.984 feet.

B. M. 35.—*Fresh Kills.* A cross (+) between the letters B. M., cut in the face of retaining wall on the north side of Fresh Kills Road, in front of the Breen estate. The cross is 3.5 feet west of the east end of the wall, is 8 inches above the ground and about 600 feet west of Gifford's Lane. (Note 17, p. 91.)

Elevation, 14.435 feet.

B. M. 36.—*Fresh Kills.* A ¾-inch copper bolt in the small topographical monument on the southeast corner of Fresh Kills Road and Gifford's Lane. The top of the monument is 6 inches below the surface and is protected by a square iron casing with cover.

Elevation, 12.477 feet.

B. M. 37.—*Fresh Kills.* The highest point within a 2-inch square (□) cut in the top of the large projecting stone in the base of retaining wall on the east side of Gifford's Lane, 9 feet south of its intersection with the southerly line of Fresh Kills Road.

Elevation, 13.839 feet.

B. M. 38.—*Richmond.* The highest point within a 1-inch square (□) cut in the top of the northeast corner of the water-table near the northwest corner of the Richmond County Court House, on the southerly side of Centre Street, opposite Court Street. The bench mark is 26.5 feet east of the west side of the building and 7.5 feet south of the most westerly of four massive columns in front of the Court House. It is about 6.5 feet above the sidewalk.

Elevation, 48.798 feet.

B. M. 39.—*Richmond.* A cross (+) between the letters B. M., cut in the third course of brick above the walk on the north side of Public School No. 28, on the southwest corner of Garretson Street and Centre Street. The cross is in the center of the front of the building between the two basement windows. (Note 15, p. 91.)

Elevation, 42.467 feet.

B. M. 40.—*Richmond.* A ¾-inch copper bolt in the top of the east end of the large granite step or door-sill at the entrance to the Richmond County Court House, on the south side of Centre Street opposite Court Street. The step containing the bolt is part of the main building and about 6.5 feet above the sidewalk. (Note 5, p. 89.)

Elevation, 48.919 feet.

B. M. 41.—*Richmond.* A ¾-inch copper bolt in the top of the small topographical monument (traverse station No. 15) in the school grounds on the southeast corner of Richmond Road and Egbert Avenue (Saw Mill Road). The monument is west of the northwest corner of Public School No. 10 and is in line with the prolongation of the west curb of Egbert Avenue. The top of the monument is 8 inches below the ground and is protected by a cylindrical iron casing with cement block cover.

Elevation, 77.854 feet.

B. M. 42.—*Richmond.* A cross (+) between the letters B. M., cut in the fourth course of brick above the ground at the northwest corner of Public School No. 10, on the southeast corner of Richmond Road and Egbert Avenue (Saw Mill Road). The cross is on the westerly side of the building and 8 inches above the ground. (Note 15, p. 91.)

Elevation, 79.613 feet.

B. M. 43.—*New Dorp.* The center of a brass plate in the top of the topographical monument on the southeast corner of Rose Avenue and First Street. The top of the monument is 8 inches below the sidewalk and is protected by a cylindrical iron casing with cover.

Elevation, 60.457 feet.

B. M. 44.—*New Dorp.* A ½-inch copper bolt in the top of the southwest corner of the bluestone door-sill at the house entrance to three-story yellow brick building at No. 127 Fifth Street, between Rose Avenue and New Dorp Lane. The top of the bolt is flush with the step and 6 inches above the sidewalk. (Note 8, p. 90.)

Elevation, 48.128 feet.

B. M. 45.—*New Dorp.* A ¾-inch copper bolt in the top of the topographical monument at New Dorp Lane and Southside Boulevard. The monument is just northeast of the center line of New Dorp Lane and on the northwesterly curb line of the Boulevard. It is 6 inches below the surface and protected by a square iron casing with cover.

Elevation, 20.082 feet.

B. M. 46.—*New Dorp.* The highest point within a 1½-inch square (⊓) cut in the top of the southwesterly side of the brownstone water-table at the most westerly corner of the concrete stable on the Vanderbilt Estate, on the northerly side of New Dorp Lane opposite Vanderbilt Avenue. The bench mark is about 325 feet southeasterly from the southerly line of Mill Road and about 38 feet northeast of New Dorp Lane. It is 1.75 feet southeasterly from the corner of the stable and 1 foot above the ground.

Elevation, 15.122 feet.

B. M. 47.—*New Dorp.* The center of a 1½-inch square (⊓) cut in the top of the southerly corner of the bluestone door-sill at the southwesterly side of the one-story brick laundry building in the grounds of St. John's Guild, near the foot of New Dorp Lane. The bench mark is about 1,550 feet southwest of New Dorp Lane and about 420 feet southeast of Cedar Grove Avenue.

**Elevation, 5.743 feet.**

B. M. 48.—*New Dorp.* The center of a brass bolt marking the U. S. C. & G. S. triangulation station Elm Tree, near the foot of New Dorp Lane. The bolt is about 155 feet northeast of New Dorp Lane and about 87 feet west of Elm Tree Beacon. The bench mark is 1.5 feet below the ground surface, but exposed. (Note 25, p. 91.)

Elevation, 6.388 feet.

B. M. 49.—*New Dorp.* This bench mark is the pointed top of the most westerly of four granite monuments surrounding Elm Tree Beacon and marking the U. S. Government property. The monument is about 175 feet northeast of New Dorp Lane and about 27 feet west of the beacon. The top of the monument is 1.75 feet above the ground and a 1-inch square surrounds the bench mark. This monument is the one nearest to triangulation station Elm Tree.

Elevation, 8.943 feet.

B. M. 50.—*New Dorp.* The bottom of a square cut in the top of the southeast corner of the bluestone water-table of south buttress of the clock tower at the Amboy Road entrance to Ocean View Cemetery, at Grant Avenue. The bench mark is 6 inches above the ground and the letters B. M. are cut in the water-table below. (Note 10, p. 90.)

Elevation, 91.820 feet.

B. M. 51.—*Oakwood.* A ¾-inch copper bolt in topographical monument No. 10, on the northeasterly side of Clark Avenue and almost on the center line of Amboy Road. The top of the monument is 7 inches below the surface and is protected by a square iron casing with cover.

Elevation, 57.118 feet.

B. M. 52.—*Oakwood.* A ½-inch copper bolt in the top of the southeast corner of the bluestone basement window-sill of two-story and attic frame house on the northwest corner of Guyon Avenue and the Staten Island Rapid Transit Railway. The bolt is in the sill of the most southerly basement window, is 5.7 feet from the south corner of the building and 3 feet above the ground. (Note 8, p. 90.)

Elevation, 58.275 feet.

B. M. 53.—*Oakwood.* The highest point within a square (⊡) cut in the top of the most northerly corner of a stone culvert on the east side of Tysens Lane about 25 feet northwest of the center line of the Staten Island Rapid Transit Railway. An arrow is cut in the side of the culvert farthest from the railroad and points to the bench mark.

Elevation, 38.973 feet.

B. M. 54.—*New Dorp.* A cross (+) cut in the east side of the third brick above the bluestone door-sill at the west side of the house entrance to No. 127 Fifth Street, between Rose Avenue and New Dorp Lane. The building is three-story yellow brick. (Note 15, p. 91.)

Elevation, 48.689 feet.

B. M. 55.—*New Dorp.* The center of the top of a concrete monument on the northeast corner of Egbert Avenue and Saw Mill Road, opposite Meisner Avenue. The monument is a cylindrical tile pipe filled with concrete, and the top is 3 inches above the ground.

Elevation, 121.059 feet.

B. M. 56.—*Manor Road.* A ¾-inch copper bolt in the top of topographical monument No. 5, at Manor Road and Saw Mill Road (Rockland Avenue). The monument is on line with the southerly line of Manor Road and 15 feet from the west line of Saw Mill Road. The top of the monument is 1 foot below the ground and is protected by a cylindrical iron casing with cover.

Elevation, 189.593 feet.

B. M. 57.—*New Springville.* A ¾-inch copper bolt in the top of the topographical monument at the intersection of Old Stone Road and Union Avenue (traverse station No. 88). The top of the monument is 2 feet below the ground and is protected by a cylindrical iron casing with cover.

Elevation, 22.433 feet.

B. M. 58.—*Union Avenue.* The center of a 3-inch circle cut in the outcropping natural rock on the east side of Union Avenue about 465 feet south of Richmond Turnpike. The rock containing the bench mark is north of a small bridge over the creek crossing Union Avenue and about 465 feet south of Richmond Turnpike.

Elevation, 10.217 feet.

B. M. 59.—*Linoleumville.* A horizontal line in the end of a ¾-inch copper bolt leaded horizontally into the brick at the northeast corner of Public School No. 26. on the south side of Richmond Turnpike, between Wild Avenue and Prospect Avenue. The bolt is on the fourth course of brick above the bluestone base and faces Richmond Turnpike. It is 7½ inches from the northeast corner of the building and 1.3 feet above the sidewalk. (Note 6, p. 90.)

Elevation, 39.323 feet.

B. M. 60.—*Linoleumville.* A cross (+) cut in the bluestone coping 7.75 feet above the sidewalk at the northeast corner of Public School No. 26, on the south side of Richmond Turnpike between Wild Avenue and Prospect Avenue. The cross is directly above B. M. 59. (Note 15, p. 91.)

Elevation, 45.996 feet.

B. M. 61.—*Linoleumville.* A square (ㅁ) cut in the east end of the second bluestone step at the north or main entrance to Public School No. 26, on the south side of Richmond Turnpike between Wild Avenue and Prospect Avenue.

Elevation, 39.254 feet.

B. M. 62.—*Linoleumville.* The top of a large brass plug in the concrete monument on the south side of Richmond Turnpike, about 184 feet east of the east line of Wild Avenue. The top of the monument is 1.5 feet below the ground and 18 feet south of the south line of Richmond Turnpike. Known as Richmond No. 779.

Elevation, 28.059 feet.

B. M. 63.—*Travisville.* A 1½-inch square (ㅁ) cut in the top of the west end of the stone door-sill at the westerly entrance to Travis Hall at No. 3921 Richmond Turnpike, about 150 feet west of its intersection with Lexington Avenue. The bench mark is 4 feet east of the west side of the building and 4 inches from the west end of the sill.

Elevation, 24.585 feet.

B. M. 64.—*Bloomfield.* The center of the horizontal bar of a T cut in the brick at the northeast corner of Public School No. 25, on the west side of Chelsea Road about 1,400 feet south of Decker Avenue. The bench mark is on the north side of the building on the eighteenth course of brick above the ground, and 1¾ inches from the corner.

Elevation, 16.075 feet.

B. M. 65.—*Watchogue.* The highest point within a square (ㅁ) cut in the top of the south end of stone culvert wall on the east side of South Avenue about 1,150 feet south of Lamberts Lane.

Elevation, 9.245 feet.

B. M. 66.—*Watchogue.* The highest point within a square (□) cut in the top of the north end of stone culvert wall on the west side of South Avenue about 1,150 feet south of Lamberts Lane.

Elevation, 9.052 feet.

B. M. 67.—*Summerville.* The center of a square (□) cut in the top of the boulder on the southeast corner of South Avenue and Washington Avenue. The bench mark is on the south curb line of Washington Avenue and on the center line of the east sidewalk of South Avenue.

Elevation, 10.287 feet.

B. M. 68.—*Arlington.* The underside of the limestone coping at the southwest corner of Public School No. 23, on the east side of Mersereau Avenue, just south of Arlington Place. The point is between two parallel lines cut in the underside of the coping, which is 8.5 feet above the ground.

Elevation, 54.025 feet.

B. M. 69.—*Arlington.* A square (□) cut in the top of the south end of the first step under the limestone balustrade at the west entrance to Public School No. 23, on the east side of Mersereau Avenue, just south of Arlington Place.

Elevation, 45.959 feet.

B. M. 70.—*Arlington.* A horizontal line on the face of a brass tablet bolt 2 inches in diameter, leaded horizontally into the southwest corner of Public School No. 23, on the east side of Mersereau Avenue just south of Arlington Place. The bench mark faces Mersereau Avenue, is 6 inches from the corner of the building and 1.2 feet above the ground. (Note 1, p. 89.)

Elevation, 46.636 feet.

B. M. 71.—*Summerville.* A square (□) cut in the top of the northwest edge of the west window-sill on the north side of Public School No. 24, on the south side of Washington Avenue between Franklin Avenue and Lincoln Avenue. The bench mark is 5.4 feet from the west side of the building.

Elevation, 21.143 feet.

B. M. 72.—*Old Place.* The center of a brass bolt marking the U. S. C. & G. S. triangulation station Old Place, about 785 feet west of Western Avenue and about 175 feet northeast of Washington Avenue. The triangulation station is in the rear of a one-story frame house occupied by Frederick Denker and owned by the Standard Oil Company. The bench mark is 6 inches below the ground. (Note 25, p. 91.)

Elevation, 11.987 feet.

B. M. 73.—*New York City Farm Colony.* A ¾-inch copper bolt in the top of the south end of the stone door-sill at the main or west entrance to the male dormitory of the New York City Farm Colony, west of Manor Road and south of County House Road. (Note 5, p. 89.)

Elevation, 230.623 feet.

B. M. 74.—*New York City Farm Colony.* A cross (+) cut in the south end of bluestone window-sill on the south side of the main entrance to the male dormitory of the New York City Farm Colony, west of Manor Road and south of County House Road. The point is on the west side of the sill and is 3 feet above the ground. (Note 15, p. 91.)

Elevation, 233.436 feet.

B. M. 75.—*Manor Road.* A ¾-inch copper bolt in the topographical monument at Manor Road and Bradley Avenue. The top of the monument is 1 foot below the ground, is about 25 feet from the south side of Manor Road and 20 feet east of the west line of Bradley Avenue. It is protected by a cylindrical iron casing with cover.

Elevation, 261.773 feet.

B. M. 76.—*Manor Road.* A ¾-inch copper bolt in the top of the topographical monument on the south side of Manor Road and on the center line of an old road leading to the McAndrews estate, on the north side of Manor Road. The monument is about 800 feet west of Egbert Avenue, is 3 feet north of the post and wire fence and 6 inches below the ground.

Elevation, 303.841 feet.

B. M. 77.—*Manor Road.* A cross (+) cut in the west face of the most southerly of two large stone gate posts at the entrance to the large estate known as "The Gables," on the southeast corner of Manor Road and Egbert Avenue. The point is 1.3 feet above the ground and the letter B is cut in the stone above. (Note 15, p. 91.)

Elevation, 325.381 feet.

B. M. 78.—*Manor Road.* The center of the top of a concrete monument at Manor Road and Ocean Terrace. The monument is in line with the southerly line of Ocean Terrace and is near the center line of Manor Road. The point is marked by a brass tack, is 6 inches below the ground surface, and is protected by a square iron cover.

Elevation, 294.928 feet.

B. M. 79.—*Todt Hill.* A ¾-inch copper bolt cemented to the southwest corner of a marble monument at U. S. C. & G. S. triangulation station Bogart, near the northwest corner of Todt Hill Road and Ocean Terrace. The brass bolt which marked the triangulation station is now gone. The top of the monument containing the bench mark is flush with the ground, is about 225 feet west of Todt Hill Road and 25 feet north of Ocean Terrace.

Elevation, 390.351 feet.

B. M. 80.—*Todt Hill.* A square (□) cut in the top of the marble city monument at the U. S. C. & G. S. triangulation station Bogart, near the northwest corner of Todt Hill Road and Ocean Terrace. The monument is about 225 feet west of Todt Hill Road and about 25 feet north of Ocean Terrace. The square is on top of the east side of the monument and flush with the surface of the ground.

Elevation, 390.356 feet.

B. M. 81.—*Todt Hill.* A ¾-inch copper bolt in the top of a topographical monument on the center line of Todt Hill Road and the north side of Ocean Terrace. The top of the monument is 6 inches below the ground, is on line with the post and rail fence on the north side of Ocean Terrace and 20 feet from the corner of the fence. The monument is protected by a square iron casing with cover.

Elevation, 389.240 feet.

B. M. 82.—*Todt Hill.* A ¾-inch copper bolt in the top of a topographical monument about 10 feet east of the west line of Todt Hill Road and about 80 feet north of the north line of Ocean Terrace. The top of the monument is 10 inches below the ground surface and protected by a square iron casing with cover.

Elevation, 384.985 feet.

B. M. 83.—*Todt Hill.* A horizontal line in the end of a ¾-inch copper bolt leaded horizontally into the stone foundation on the north side of David J. Tysen's dwelling, near the southeast corner of Ocean Terrace and Todt Hill Road. The house is about 100 feet south of Ocean Terrace and about 200 feet east of Todt Hill Road. The bench mark is 8 feet from the east side of the house and 2 feet above the ground. (Note 18, page 91.)

Elevation, 409.239 feet.

B. M. 84.—*Castleton Corners.* A cross (+) cut in the south end of the south window-sill of the red brick restaurant building connected with the Monroe Eckstein Brewery, on the west side of Manor Road about 300 feet south of Smiths Lane. The bench mark is 5.5 feet above the sidewalk and 4.67 feet from the south end of the building. (Note 15, page 91.)

Elevation, 244.354 feet.

B. M. 85.—*Castleton Corners.* A square (⊔) cut in the top of the southeast corner of the granite base of watering trough on the southwest corner of Manor Road and Richmond Turnpike.

Elevation, 184.535 feet.

B. M. 86.—*Castleton Corners.* A ¾-inch copper bolt in the top of the west end of the lowest concrete step at the main entrance to grocery store at No. 1765 Richmond Turnpike, on the corner of Manor Road. The bench mark is 6.15 feet from the west side of the building and 6 inches above the sidewalk. (Note 5, page 89.)

Elevation, 183.174 feet.

B. M. 87.—*Castleton Corners.* A 1½-inch cross (+) cut in the brick at the southwest corner of dwelling at No. 1765 Richmond Turnpike, on the corner of Manor Road. The bench mark is on the south side of the building, is on the sixth course of brick above the sidewalk and 3 inches from the corner. (Note 15, p. 91.)

Elevation, 183.593 feet.

B. M. 88.—*Westerleigh.* A cross (+) cut in the top of a city monument on the southwest corner of Leonard Avenue and Jewett Avenue. The monument has a very rough top and is at the inner corner of the sidewalk, 2 inches below the surface. (Note 17, page 91.)

Elevation, 121.610 feet.

B. M. 89.—*Westerleigh.* A horizontal line on the face of a brass tablet bolt 2 inches in diameter, set horizontally into the limestone at the northeast corner of Public School No. 30, on the south side of the Boulevard, between Fisk Avenue and Warowell Avenue. The bench mark faces the Boulevard, is 1.5 feet from the corner of the building and 1.17 feet above the sidewalk. (Note 1, page 89.)

Elevation, 101.223 feet.

B. M. 90.—*Westerleigh.* The bottom of a square cut in the top of the corner of the second course of limestone at the extreme northeast corner of Public School No. 30, on the south side of the Boulevard between Fisk Avenue and Warowell Avenue. (Note 10, page 90.)

Elevation, 102.738 feet.

B. M. 91.—*Westerleigh.* A cross (+) cut in the bluestone base on the north side of Public School No. 30, on the south side of the Boulevard between Fisk Avenue and Warowell Avenue. The bench mark is 1.5 feet from the northeast corner of the building. (Note 17, page 91.)

Elevation, 100.395 feet.

B. M. 92.—*Westerleigh.* A ¾-inch copper bolt in the top of a 4″ x 4″ monument on the northwest corner of Jewett Avenue and Cherry Lane. The monument is in the center of the sidewalk and 11 inches below it, and is protected by a cylindrical iron casing with cover.

Elevation, 39.812 feet.

B. M. 93.—*Port Richmond.* A square (⊓) cut in the top of the west end of the granite door-sill at the Catherine Street entrance to the brick tower of the German Lutheran Church, on the southwest corner of Catherine Street and Jewett Avenue. The bench mark is 1 inch above the sidewalk.

Elevation, 20.542 feet.

B. M. 94.—*Grant City*. A 1½-inch square (⊓) cut in the top of the northwest corner of the large bluestone door-sill at the main entrance to the Midland Park Hotel, on the northeasterly corner of Lincoln Avenue and Railroad Avenue. The hotel is a two-story building, the lower portion being red brick.

Elevation, 35.789 feet.

B. M. 95.—*Grant City*. A square (□) cut in the top of the east end of the second bluestone step of stoop at the boys' entrance to Public School No. 33, on the northwesterly corner of Washington Avenue and Thompson Street. The bench mark is on the Washington Avenue side of the building.

Elevation, 38.917 feet.

B. M. 96.—*Grant City*. A horizontal line in the end of a ¾-inch copper bolt leaded horizontally into the fifth course of brick in the southerly corner of Public School No. 33, on the northwesterly corner of Washington Avenue and Thompson Street. The bolt faces Thompson Street and is 6 inches from the Washington Avenue side of the building. (Note 6, page 90.)

Elevation, 38.995 feet.

B. M. 97.—*Garretsons*. A cross (+) cut in the foundation on the northeasterly side of the Cromwell Hose Co. building, on the southeasterly corner of Richmond Road and Sea View Avenue. The bench mark is below the granite cornerstone, is 3.6 feet above the ground and 6 inches from the northerly corner of the building. (Note 15, page 91.)

Elevation, 63.574 feet.

B. M. 98.—*Garretsons*. A ½-inch copper bolt in the top of the north end of the bluestone sill of basement window of the Cromwell Hose Co. building, on the southeasterly corner of Richmond Road and Sea View Avenue. The bench mark is on the east side of the building and 2.7 feet above the ground. (Note 8, page 90.)

Elevation, 61.410 feet.

B. M. 99.—*Garretsons*. A square (□) cut in the top of the north end of the second step at the southeasterly entrance to Public School No. 11, on the northwesterly side of Jefferson Avenue between Garretson Avenue and Cromwell Avenue.

Elevation, 44.121 feet.

B. M. 100.—*Grasmere*. A ¾-inch copper bolt in topographical monument No. 37, on the westerly side of the Southside Boulevard about 235 feet north of Old Town Road. The monument is 18 feet southeast of the low iron railing at the gate in front of the Tocci estate, and 6 inches below the ground.

Elevation, 46.178 feet.

B. M. 101.—*Grasmere*. The bottom of a 2-inch square cut in the top of the southerly corner of stone culvert wall on the northwesterly side of the Southside Boulevard, between Parkinson Avenue and Old Town Road, just east of the Tocci estate. An arrowhead points to the bench mark. (Note 10, p. 90.)

Elevation, 53.441 feet.

B. M. 102.—*Grasmere*. A cross (+) cut in the bluestone on the northwesterly side of the most westerly of two high brick gate-posts at the lodge gate on the Cameron estate, on the south side of Fingerboard Road about ½-mile east of Clove Avenue. The bench mark is 1.5 feet above the ground and directly below the iron catch for the small gate. (Note 15, page 91.)

Elevation, 91.830 feet.

B. M. 103.—*Grasmere.* A square (⊓) with the letters B. M. alongside, cut in the top of the northeasterly edge of the bluestone coping on top of the brick wall just west of the lodge gate on the Cameron estate, on the south side of Fingerboard Road about ½-mile east of Clove Avenue. The bench mark is 1.2 feet easterly from the side of the brick column at the end of the wall and 4 feet above the ground.

Elevation, 95.857 feet.

B. M. 104.—*Arrochar.* The highest point within a 1-inch square (□) cut in the top of the southwest corner of a large stone in the retaining wall on the northeast corner of Sand Lane and Richmond Avenue. The bench mark is on the west side of the wall about 8.5 feet from the Richmond Avenue curb.

Elevation, 68.180 feet.

B. M. 105.—*Arrochar.* The intersection of a cross (+) within the chiseled quadrant of a circle on top of the northwest corner of the east abutment of the highway bridge on Richmond Avenue, over the tracks of the Staten Island Rapid Transit Railway, just west of Florida Avenue. The letters B. M. are cut alongside of the bench mark, which is on the sidewalk grade.

Elevation, 65.997 feet.

B. M. 106.—*Arrochar.* The bottom of a square cut in the north end of the bridge seat of the east abutment of highway bridge on Richmond Avenue, over the tracks of the Staten Island Rapid Transit Railway, just west of Florida Avenue. The letters B. M. B. E. are cut in the stone and surround the bench mark, which is about 4 feet below the sidewalk.

Elevation, 62.250 feet.

B. M. 107.—*Fort Wadsworth.* A 2-inch square (□) cut in the top of the east side of the concrete coping of a square duct manhole on the southeast corner of Richmond Avenue and New York Avenue. The bench mark is 2.5 feet northwest of a high granite post at the corner.

Elevation, 97.079 feet.

B. M. 108.—*Fort Wadsworth.* The top of a ¾-inch iron rod inside of a wrought-iron pipe casing in the center of the west sidewalk of New York Avenue about 475 feet north of Richmond Avenue. The bench mark is about 20 feet from the curb and 44.5 feet southeasterly from the southeast corner of the Post Exchange. (Note 21, page 91.)

Elevation, 100.222 feet.

B. M. 109.—*Fort Wadsworth.* A horizontal line on the face of a brass tablet bolt 2 inches in diameter, cemented horizontally into the first course of limestone in the northeast corner of the one-story and basement red brick Post Exchange building 30 feet west of New York Avenue and about 555 feet north of Richmond Avenue. The bench mark is on the north side of the building 5.5 feet above the ground and 4 inches from the corner. (Note 1, page 89.)

Elevation, 107.148 feet.

B. M. 110.—*Fort Wadsworth.* The center of the horizontal bar of a **T** cut in the bluestone coping at the northeast corner of the one-story and basement red brick Post Exchange building 30 feet west of New York Avenue and about 555 feet north of Richmond Avenue. The bench mark is on the north side of the building and 6 inches above the ground.

Elevation, 102.021 feet.

B. M. 111.—*Fort Wadsworth.* A square (∟) cut in the top of the northeast corner of a large granite post at the northerly entrance to the military reservation. The bench mark is 50 feet east of New York Avenue and at the south side of the road.

Elevation, 73.002 feet.

B. M. 112.—*Elizabethport, N. J.* This is the Coast and Geodetic Survey tidal bench mark. It is the intersection of a cross (+) lettered U. S., cut in the brick on the easterly side of Samuel L. Moore & Sons Co. foundry (formerly Worrall & Co.), on the northerly side of Front Street, between Franklin Street and Fulton Street. The cross is in the seventh course of brick above the ground and about 80 feet from Franklin Street. Established in 1886.

Elevation, 12.397 feet.

B. M. 113.—*Elizabethport, N. J.* A square (∟) cut in the top of the northeast corner of the lowest bluestone step at the saloon door entrance to the three-story yellow brick hotel on the southwest corner of Marshall Street and Front Street.

Elevation, 10.873 feet.

B. M. 114.—*Elizabethport, N. J.* The intersection of a cross (+) cut in the top of a 1-inch spike in the top of the backing log at the easterly end of the dock at the foot of Marshall Street. The bench mark is almost on line with the southerly curb line of Marshall Street.

Elevation, 7.082 feet.

B. M. 115.—*Howland Hook.* The intersection of a cross cut in the top of a large spike with hemispherical head on the westerly side of the dock at the ferry slip. The bench mark is about 30 feet from the outshore corner of the dock.

Elevation, 7.916 feet.

B. M. 116.—*Howland Hook.* The center of a square (□) cut in the top of the west end of the basin head on the southeast corner of Western Avenue and Richmond Terrace.

Elevation, 20.306 feet.

B. M. 117.—*Howland Hook.* The intersection of a cross (+) cut in the top of the southwest nut on top of the fire hydrant on the southeast corner of Western Avenue and Richmond Terrace.

Elevation, 23.773 feet.

B. M. 118.—*Howland Hook.* The center of the horizontal bar of a **T** cut in the northeast corner of the brick foundation of the two-story frame dwelling on the southwest corner of Catherine Street and Richmond Terrace. The bench mark is 1.25 feet above the ground.

Elevation, 27.047 feet.

B. M. 119.—*North Shore.* A square (∟) cut in the top of the northeast corner of the granite property monument on the south side of Richmond Terrace about 540 feet east of Holland Avenue. The monument is at the west side of a stone retaining wall and the figure 4 is cut in the top of the monument, which is 12 inches above the ground.

Elevation, 19.391 feet.

B. M. 120.—*Arlington.* The center of a 1½-inch square (□) cut in the top of the center of the concrete basin head on the southeast corner of Mersereau Avenue and Richmond Terrace.

Elevation, 10.826 feet.

B. M. 121.—*Mariners Harbor.* The highest point within a square (⊔) cut in the top of the northeast corner of the limestone base of the east lamp post at the northeast corner of the Mariners Harbor National Bank, on the southwest corner of Union Avenue and Richmond Terrace. The bench mark is 1.8 feet above the sidewalk.

Elevation, 10.081 feet.

B. M. 122.—*Mariners Harbor.* A square (⊓) cut in the top of the east end of the granite door-sill at the entrance to the Mariners Harbor National Bank, on the southwest corner of Union Avenue and Richmond Terrace.

Elevation, 10.592 feet.

B. M. 123.—*Mariners Harbor.* A square (⊔) cut in the top of the southeast corner of the window-sill at the southeast corner of the one-story brick boiler and engine house of the Hecker, Jones, Jewell Milling Co., about 260 feet north of Richmond Terrace at Bush Avenue. The bench mark is on the east side of the engine house and 1.5 feet above the ground.

Elevation, 9.732 feet.

B. M. 124.—*Mariners Harbor.* The center of a 2-inch square (⊓) cut in the top of the west end of the brownstone door-sill at the west entrance to Franklin Hall (three-story brick), on the north side of Richmond Terrace opposite the foot of Harbor Road.

Elevation, 10.072 feet.

B. M. 125.—*Elm Park.* The intersection of a cross (+) within a square (⊓) cut in the top of the west end of the most easterly window-sill on the south side of the Standard Varnish Works, on the north side of Richmond Terrace about 36 feet west of the west line of Granite Avenue.

Elevation, 14.362 feet.

B. M. 126.—*Elm Park.* The highest point within a 2-inch square (□) cut near the west end of the bluestone door-sill at the store entrance to three-story brick building (No. 2558) on the south side of Richmond Terrace about 140 feet east of Granite Avenue.

Elevation, 12.295 feet.

B. M. 127.—*Port Richmond.* A square (⊔) cut in the top of the east end of the large bluestone sill under the most easterly of two large limestone columns at the entrance to the Port Richmond National Bank, on the south side of Richmond Terrace about 25 feet west of Richmond Avenue.

Elevation, 10.771 feet.

B. M. 128.—*Port Richmond.* A square (□) cut in the top of the northwesterly end of the top step or sill at the house entrance to the two-story red brick building at No. 2045 Richmond Terrace. The building is on the northerly side of the street and alongside of the approach to the Bergen Point Ferry.

Elevation, 11.684 feet.

B. M. 129.—*Port Richmond.* A square (□) cut in the top of the northwesterly corner of the bluestone door-sill at the southerly entrance to the office of the Staten Island Ship Building Co. at No. 1983 Richmond Terrace. The building is a two-story yellow brick and just north of the Staten Island Rapid Transit Railway. The bench mark is 13 feet from the southwest corner of the building.

Elevation, 15.881 feet.

B. M. 130.—*Port Richmond.* A 1-inch square (⊓) cut in the top of the north end of granite door-sill at the house entrance to No. 1916 Richmond Terrace, near Jewett Avenue. The building is on the south side of an alley, is three-story red and yellow brick and was formerly No. 15 and No. 17.

Elevation, 9.837 feet.

B. M. 131.—*Port Richmond.* The highest point within a square (⊔) cut in the top of the southeast corner of the projecting foundation stone at the base of the west wall of three-story red brick building at No. 1851 Richmond Terrace, opposite Columbia Street. The bench mark is 12 inches from the west side of building.

Elevation, 10.495 feet.

B. M. 132.—*Port Richmond.* A horizontal line on the face of a brass tablet bolt 2 inches in diameter, leaded horizontally into the cornerstone on the east side of the entrance to Police Station at No. 1590 Richmond Terrace, between Van Street and Tompkins Place. The bench mark is in the tower portion of the building and 3 feet above the sidewalk. (Note 1, page 89.)

Elevation, 25.436 feet.

B. M. 133.—*Port Richmond.* A square (□) cut in the top of the east end of the second step at the unused easterly entrance to No. 1590 Richmond Terrace, between Van Street and Tompkins Place. The building is a two-story yellow brick and the step containing the bench mark is in the tower portion.

Elevation, 23.849 feet.

B. M. 134.—*Livingston.* The center of a 1¼-inch circle cut in the top of the granite water-table at the southwest corner of the New York and Staten Island Electric Co.'s building on the north side of Richmond Terrace about 70 feet east of Davis Avenue. The figure 5 is chiseled alongside of the bench mark, which is 4.5 feet above the ground.

Elevation, 19.977 feet.

B. M. 135.—*Sailor's Snug Harbor.* On the administration building, known as building "C," about 500 feet west of Tysen Avenue and about 100 feet south of Richmond Terrace, opposite the main entrance to the grounds of Sailor's Snug Harbor. The bench mark is a square (⊓) cut in the top of the east edge of a massive granite block at the northeast corner of the steps and supporting the lamp-post. The point is 4.3 feet from the north end of the block and 2.9 feet above the ground.

Elevation, 37.338 feet.

B. M. 136.—*Sailor's Snug Harbor.* The highest point within a quadrant (∟) cut in the top of the northeast corner of the granite coping of the stone "quarterdeck" on the north side of Richmond Terrace opposite to and north of the main entrance to Sailor's Snug Harbor. The bench mark is about level with the sidewalk, is outside of the iron railing and about 16 feet northeast of the flagpole.

Elevation. 24.445 feet.

B. M. 137.—*Sailor's Snug Harbor.* The center of a square (⊔) cut in the north-east corner of the long granite platform at the foot of the stone steps leading from the north side of Richmond Terrace to the tracks of the Staten Island Rapid Transit Railway, opposite the main entrance to Sailor's Snug Harbor. The bench mark is about 12 feet below the sidewalk and 2.5 feet above the railroad tracks.

Elevation, 12.564 feet.

B. M. 138.—*Sailor's Snug Harbor.* The bottom of a square cut in the southwest corner of the copingstone of the sea wall north of Sailor's Snug Harbor. The bench mark is about 260 feet west of the flagpole opposite the main entrance and about 15 feet east of the one-story boat house at the sea wall. (Note 10, page 90.)
Elevation, 6.779 feet.

B. M. 139.—*New Brighton.* The highest point within a quadrant (ᒷ) cut in the northeast corner of the bluestone door-sill at the store entrance to the Richmond Building, on the southwest corner of York Avenue and Richmond Terrace. The bench mark is about 22 feet west of the northeast corner of the building.
Elevation, 29.146 feet.

B. M. 140.—*New Brighton.* The center of a square (⊡) cut in the top of the west end of the granite door-sill at the entrance to the two-story brick building at No. 441 Richmond Terrace, at Westervelt Avenue. The building containing the bench mark is the New Brighton Post Office.
Elevation, 35.066 feet.

B. M. 141.—*St. George.* A 1½-inch square (□) cut in the top of the north end of the lowest granite step at the southeast entrance to the Richmond Borough Hall, on the northwest corner of Jay Street and South Street. The bench mark is on the Jay Street side of the building and 6 inches above the walk.
Elevation, 57.184 feet.

B. M. 142.—*St. George.* A horizontal line on the face of a brass tablet bolt 2 inches in diameter, set horizontally into the southeast corner of the Richmond Borough Hall, on the northwest corner of Jay Street and South Street. The bench mark is in the limestone block under the inscribed cornerstone, faces Jay Street and is 1.75 feet above the walk. (Note 1, page 89.)
Elevation, 57.944 feet.

B. M. 143.—*St. George.* A 2-inch square (□) cut in the top of the west end of the granite base block of electric lamp at the west end of the concrete wall between the two wagon approaches to the New York Ferry. The bench mark is about 35 feet north of the grounds of the U. S. Lighthouse Department and about 5 feet east of the center line of Jay Street.
Elevation, 43.319 feet.

B. M. 144.—*St. George.* A 2-inch square (⊡) cut in the top of the sea wall at the foot of South Street, just east of the grounds of the U. S. Lighthouse Department. The bench mark is on the outshore edge of the coping, is 25.35 feet east of the crib bulkhead at the foot of South Street and 7 feet east of the east end of the brick wall on the south side of the street.
Elevation, 9.382 feet.

B. M. 145.—*St. George.* A square (□) cut in the top of the southeast corner of the lowest granite step just north of the south lamp at the west entrance to the Richmond Borough Hall. This is Richmond topographical bench mark No. 456.
Elevation, 74.512 feet.

B. M. 146.—*Tompkinsville.* A 1-inch square (□) cut in the top of the south end of the granite curb at the corner of the retaining wall on the east side of Central Avenue, about 43 feet south of the north line of Arrietta Street.
Elevation, 34.699 feet.

B. M. 147.—*Tompkinsville.* A 1-inch square (⊾) cut in the top of the southeast corner of the first step at the store entrance to the four-story brick house at No. 65 Arrietta Street, formerly No. 1. The corner of the step is lettered B. W. S. and is about 40 feet west of the inshore end of the pier.

Elevation, 9.907 feet.

B. M. 148.—*Tompkinsville.* A square (□) cut in the top of the west end of the top step at the house entrance to No. 65 Arrietta Street, formerly No. 1. The building is four-story brick, is on the north side of the street and near the pier.

Elevation, 11.904 feet.

B. M. 149.—*Tompkinsville.* The center of the surface within a quadrant (⊾) cut in the top of the southeast corner of the lowest granite step at the store entrance to the three-story brick house at No. 12 Griffin Street. This point is Richmond topographical bench mark No. 9, and the building was formerly the Post Office.

Elevation, 41.004 feet.

B. M. 150.—*Tompkinsville.* A square (□) lettered B. W. S., cut in the top of the southwest corner of the concrete retaining wall on the southeast corner of Arrietta Street and the Boulevard, about 50 feet west of the center line of Minthorne Street. The bench mark is 9½ inches above the sidewalk and just west of the southwest corner of Arrietta Street and Minthorne Street.

Elevation, 35.621 feet.

B. M. 151.—*Stapleton.* The top of a ¾-inch iron rod inside of a wrought-iron pipe casing in Stapleton Park, near the northeast corner. The bench mark is on line with the easterly curb line of Beach Street, is about 43 feet west of the west curb line of Bay Street, and about 20 feet south of the south curb line of Water Street. The top of the pipe casing is 10 inches above the ground. (Note 21, page 91.)

Elevation, 9.710 feet.

B. M. 152.—*Stapleton.* The bottom of a square cut in the top of the granite water-table at the extreme southeast corner of Public School No. 14, on the northwest corner of Broad Street and Brook Street. The bench mark is 1.25 feet above the sidewalk. (Note 10, page 90.)

Elevation, 35.037 feet.

B. M. 153.—*Stapleton.* A cross (+) cut in the granite foundation at the southeast corner of Public School No. 14, on the northwest corner of Broad Street and Brook Street. The bench mark faces Brook Street, is 9½ inches above the sidewalk and 7 inches from the corner of the building. (Note 15, page 91.)

Elevation, 34.471 feet.

B. M. 154.—*Clifton.* The center of the surface within a quadrant (⊾) cut in the top of the northeast corner of the bluestone water-table of the two-story and attic brick and frame house on the southwest corner of Norwood Avenue and Bay Street. The point is 6 inches above the sidewalk. Richmond topographical bench mark No. 12.

Elevation, 12.565 feet.

B. M. 155.—*Clifton.* A cross (+) cut in the top of the southwest corner of the water-table of Scott's stables, on the northeast corner of Bay Street and Norwood Avenue (Amos Street). The water-table is flush with the sidewalk. (Note 17, page 91.)

Elevation, 11.069 feet.

B. M. 156.—*Rosebank.* A horizontal line on the face of a brass table bolt 2 inches in diameter set horizontally into the limestone at the south side of the northeast buttress of Public School No. 13, on the southwest corner of Anderson Avenue and Clifton Avenue. The bench mark is about 40 feet south of the north end of the building, is 1.5 feet from the east side of the buttress and 2 feet above the ground. (Note 1, page 89.)

Elevation, 42.186 feet.

B. M. 157.—*Rosebank.* A cross (+) cut in the granite foundation on the east side of Public School No. 13, on the southwest corner of Anderson Avenue and Clifton Avenue. The bench mark is on the south side of the northeast buttress, is 8 inches above the ground and about 40 feet from the north end of the building. (Note 15, page 91.)

Elevation, 40.833 feet.

B. M. 158.—*Rosebank.* A 2-inch square (⊔) cut in the top of the northeast corner of the bluestone coping of retaining wall on the west side of Anderson Avenue, in front of Public School No. 13. The bench mark is about 65 feet south of the Clifton Avenue corner of the wall and 5 feet above the sidewalk.

Elevation, 44.506 feet.

B. M. 159.—*Richmond Valley.* The bottom of a square cut in the top of the east end of the first bluestone window-sill west of the north or main entrance to O. H. Barnard's silk mill, on the south side of Richmond Valley Road about 250 feet east of Fresh Kills Road. An arrowhead cut in the face of the sill points to the bench mark. (Note 10, p. 90.)

Elevation, 21.120 feet.

B. M. 160.—*Richmond Valley.* The bottom of a square cut in the top of the east end of the west window-sill on the north side of O. H. Barnard's silk mill on the south side of Richmond Valley Road about 250 feet east of Fresh Kills Road. The bench mark is 7.5 feet from the west side of the building. (Note 10, page 90.)

Elevation, 21.074 feet.

B. M. 161.—*Richmond Valley.* A square (⊓) cut in the top of the west end of the east window-sill on the north side of O. H. Barnard's silk mill on the south side of Richmond Valley Road about 250 feet east of Fresh Kills Road. The bench mark is 7.75 feet from the east end of the building.

Elevation, 21.126 feet.

B. M. 162.—*Kreischerville.* A square (□) cut in the top of the north end of the second bluestone step at the entrance to Killmeyer's estate, on the west side of Fresh Kills Road about 100 feet south of Kreischer Street.

Elevation, 47.956 feet.

B. M. 163.—*Kreischerville.* A horizontal line in the end of a ¾-inch copper bolt leaded horizontally into the brick under the middle window on the west side of Public School No. 4, on the east side of Fresh Kills Road about 200 feet north of Sharrots Lane. The bolt is in the sixth course of brick above the ground. (Note 6, page 90.)

Elevation, 60.290 feet.

B. M. 164.—*Kreischerville.* The center of the horizontal bar of a **T** cut in the south face of the brick buttress at the southeast corner of the two-story brick building of the Richmond Brick & Tile Company on the west side of Kreischer

Street near Fresh Kills Road. The bench mark is about 4 feet above the ground and on the largest building of the group.

Elevation, 24.111 feet.

B. M. 165.—*Richmond.* A square (□) cut in the top of the east end of the second bluestone step at the retaining wall in front of No. 74 Fresh Kills Road. The bench mark is on the southerly side of the road and about 155 feet westerly from Centre Street. The house is occupied by W. F. Lake.

Elevation, 13.952 feet.

B. M. 166.—*Annadale.* The highest point within a 1-inch square (□) cut in the top of the south end of the brick and cement culvert wall on the west side of Annadale Road and 30 feet north of the center line of the Staten Island Rapid Transit Railway.

Elevation, 80.220 feet.

B. M. 167.—*Annadale.* A cross (+) cut in the fifth course of brick above the ground at the southwest corner of the two-story and attic frame house at No. 847 Annadale Road. The building is on the east side of the road about 160 feet north of the center line of the Staten Island Rapid Transit Railway. The bench mark is on the south side of the building and 9 inches from the corner. (Note 15, p. 91.)

Elevation, 78.858 feet.

B. M. 168.—*Annadale.* The top of a ¾-inch copper bolt in a topographical monument at the intersection of Annadale Road and Washington Avenue. The monument is near the easterly side of Annadale Road and just north of the center line of Washington Avenue. The top is 10 inches below grade and is protected by a square iron cover.

Elevation, 95.100 feet.

B. M. 169.—*Huguenot.* The intersection of a 1-inch cross (+) cut in the face of a white marble tablet set in the wall at the southeast corner of Public School No. 5, on the north side of Amboy Road about 510 feet east of Huguenot Avenue. The builder's name, David C. Keller, is cut in the face of the tablet. The bench mark is 2 feet above the ground and 3.5 feet from the corner of the building.

Elevation, 72.009 feet

B. M. 170.—*Eltingville.* A 1½-inch square (□) cut in the top of the north end of the bluestone door-sill at the saloon entrance to Daniel Geil's Hotel at No. 3651 Seaside Avenue. The building is three-story frame and brick, is on the east side of the avenue and 100 feet south of the Staten Island Rapid Transit Railway.

Elevation, 61.716 feet.

B. M. 171.—*Annadale.* A ¾-inch copper bolt in the top of a topographical monument at the intersection of Amboy Road and Annadale Road. The monument is on the center line of both roads, is 1.5 feet below grade and protected by iron casing and cover.

Elevation, 85.348 feet.

B. M. 172.—*Giffords.* This is the U. S. C. & G. S. bench mark N. It is a rude cross cut in the top of the east corner of a boulder about 6.5 feet in diameter partly imbedded in the embankment on the north side of the Staten Island Rapid Transit Railway and about 350 feet west of Giffords Lane. The bench mark is 2.75 feet above the ground. Established June 10, 1887.

Elevation, 71.065 feet.

B. M. 173.—*Giffords.* The highest point within a 1½-inch square (□) cut in the top of a boulder about 6.5 feet in diameter partly imbedded in the embankment on the north side of the Staten Island Rapid Transit Railway and about 350 feet west of Giffords Lane. The bench mark is near the east side of the boulder and the letters B. M. are chiseled alongside.

Elevation, 72.564 feet.

B. M. 174.—*Fort Wadsworth.* The highest point within a square (⊔) cut in the top of the northeast corner of the 4-inch granite water-table at the northeast corner of the red brick torpedo storehouse in the military reservation at Fort Wadsworth. The building is on the west side of the steep road leading to the dock.

Elevation, 24.323 feet.

B. M. 175.—*Fort Wadsworth.* A square (⊔) cut in the top of the outshore edge of the sea wall at Fort Wadsworth, 40.8 feet southeasterly from the old iron ring-bolt opposite the large granite buttress near the north end of the old fort.

Elevation, 7.535 feet.

B. M. 176.—*Fort Wadsworth.* The center of the horizontal bar of a T cut in the extreme northerly corner of old Fort Wadsworth. The bench mark is in the granite wall at the acute angled corner and 5 feet above the ground.

Elevation, 13.692 feet.

B. M. 177.—*Fort Wadsworth.* A square (⊓) cut in the top of the outshore edge of the sea wall at Fort Wadsworth. The bench mark is 44.5 feet southeasterly from the extreme northerly corner of the old granite fort, is 20 feet from the buttress face and in front of an old iron ring-bolt in the sea wall.

Elevation, 7.516 feet.

# STANDARD ELEVATIONS OF BENCH MARKS IN THE BOROUGH OF BROOKLYN.

| No. of Bench Mark. | Locality. | Elevation Above Mean Sea Level. | |
|---|---|---|---|
| | | Meters. | Feet. |
| 178 | Fort Lafayette | 2.4784 | 8.131 |
| 179 | " " | 2.4424 | 8.013 |
| 180 | Fort Hamilton | 1.8410 | 6.040 |
| 181 | " " | 2.6990 | 8.855 |
| 182 | " " | 3.8639 | 12.677 |
| 183 | " " | 3.8407 | 12.601 |
| 184 | " " | 3.8400 | 12.598 |
| 185 | " " | 2.8494 | 9.348 |
| 186 | " " | 8.7663 | 28.761 |
| 187 | " " | 11.9540 | 39.219 |
| 188 | " " | 12.1080 | 39.725 |
| 189 | Fort Hamilton Park | 12.9230 | 42.398 |
| 190 | " " " | 11.9340 | 39.153 |
| 191 | " " " | 12.1245 | 39.778 |
| 192 | Fifth Avenue | 25.9172 | 85.030 |
| 193 | Ninety-second Street | 24.6713 | 80.943 |
| 194 | " " | 24.3927 | 80.028 |
| 195 | " " | 16.0555 | 52.675 |
| 196 | Bath Beach | 7.2479 | 23.785 |
| 197 | " " | 8.3289 | 27.326 |
| 198 | Bensonhurst | 6.1070 | 20.036 |
| 199 | " | 6.1933 | 20.319 |
| 200 | " | 5.9640 | 19.567 |
| 201 | Harway Avenue Bridge | 2.8023 | 9.194 |
| 202 | " " " | 2.8476 | 9.343 |
| 203 | " " " | 2.6073 | 8.554 |
| 204 | " " " | 2.6053 | 8.548 |
| 205 | " " " | 3.0723 | 10.080 |
| 206 | " " " | 3.0720 | 10.079 |
| 207 | Coney Island | 3.9676 | 13.017 |
| 208 | " " | 3.4408 | 11.289 |
| 209 | " " | 2.1918 | 7.191 |
| 210 | " " | 2.2406 | 7.351 |
| 211 | " " | 2.9614 | 9.716 |
| 212 | " " | 2.3345 | 7.659 |
| 213 | " " | 3.5032 | 11.493 |
| 214 | " " | 3.2214 | 10.569 |
| 215 | " " | 2.7321 | 8.964 |
| 216 | Ocean Parkway | 3.0944 | 10.152 |
| 217 | " " | 3.5515 | 11.652 |
| 218 | " " | 2.5263 | 8.288 |
| 219 | " " | 2.1000 | 6.890 |
| 220 | " " | 3.6745 | 12.055 |
| 221 | Avenue U | 6.4603 | 21.195 |
| 222 | " | 4.6537 | 15.268 |
| 223 | " | 5.3763 | 17.639 |
| 224 | Neck Road | 5.3663 | 17.606 |
| 225 | Avenue U | 2.6392 | 8.659 |
| 226 | " | 2.7652 | 9.072 |

| No. of Bench Mark. | Locality. | Elevation Above Mean Sea Level. | |
|---|---|---|---|
| | | Meters. | Feet. |
| 227 | Flatbush Avenue | 4.6633 | 15.300 |
| 228 | " " | 4.8436 | 15.891 |
| 229 | Avenue K | 5.7626 | 18.906 |
| 230 | " | 5.7903 | 18.997 |
| 231 | Flatbush Avenue | 6.0798 | 19.947 |
| 232 | " " | 10.3048 | 33.808 |
| 233 | Ocean Parkway | 6.8771 | 22.563 |
| 234 | " " | 7.2466 | 23.775 |
| 235 | " " | 8.3635 | 27.439 |
| 236 | " " | 9.5497 | 31.331 |
| 237 | " " | 10.8832 | 35.706 |
| 238 | " " | 13.0155 | 42.702 |
| 239 | " " | 13.9221 | 45.676 |
| 240 | Bath Beach | 9.6143 | 31.543 |
| 241 | " " | 10.1269 | 33.225 |
| 242 | " " | 9.2320 | 30.289 |
| 243 | " " | 8.3169 | 27.286 |
| 244 | Fifteenth Avenue | 13.3158 | 43.687 |
| 245 | " " | 13.3112 | 43.672 |
| 246 | " " | 13.3547 | 43.814 |
| 247 | New Utrecht Avenue | 12.7279 | 41.758 |
| 248 | Eighty-sixth Street | 24.9228 | 81.768 |
| 249 | " " " | 25.7571 | 84.505 |
| 250 | " " " | 23.8335 | 78.194 |
| 251 | " " " | 23.8086 | 78.112 |
| 252 | " " " | 21.7074 | 71.218 |
| 253 | Seventy-sixth Street | 22.7467 | 74.628 |
| 254 | Sixty-eighth Street | 22.1587 | 72.699 |
| 255 | Bay Ridge Avenue | 20.0606 | 65.816 |
| 256 | Seventy-first Street | 23.0267 | 75.547 |
| 257 | " " " | 23.0682 | 75.683 |
| 258 | Second Avenue | 13.3000 | 43.635 |
| 259 | " " | 16.8812 | 55.384 |
| 260 | Seventy-ninth Street | 26.0893 | 85.595 |
| 261 | " " " | 26.1708 | 85.862 |
| 262 | " " " | 24.5708 | 80.613 |
| 263 | McKinley Park | 29.4510 | 96.624 |
| 264 | Fort Hamilton Parkway | 13.8488 | 45.436 |
| 265 | Fifteenth Avenue | 10.1597 | 33.332 |
| 266 | Fifty-second Street | 11.6129 | 38.100 |
| 267 | Ocean Avenue | 7.0850 | 23.245 |
| 268 | Fort Hamilton Parkway | 19.4175 | 63.706 |
| 269 | Sixteenth Street | 20.2450 | 66.420 |
| 270 | Fifty-ninth Street | 20.7897 | 68.208 |
| 271 | Fifty-eighth Street | 30.7930 | 101.027 |
| 272 | Fifty-ninth Street | 35.6486 | 116.957 |
| 273 | Sixtieth Street | 29.7850 | 97.720 |
| 274 | Third Avenue | 21.0033 | 68.908 |
| 275 | Sixtieth Street | 28.3223 | 92.921 |
| 276 | " " | 27.6263 | 90.637 |
| 277 | Fourth Avenue | 25.4610 | 83.533 |
| 278 | " " | 23.4491 | 76.933 |
| 279 | " " | 19.6609 | 64.504 |
| 280 | " " | 15.7223 | 51.582 |
| 281 | " " | 15.9240 | 52.244 |
| 282 | " " | 16.5127 | 54.175 |
| 283 | " " | 8.1680 | 26.798 |
| 284 | Fourteenth Avenue | 13.8043 | 45.290 |
| 285 | New Utrecht Avenue | 16.8134 | 55.162 |

| No. of Bench Mark. | Locality. | Elevation Above Mean Sea Level. | |
|---|---|---|---|
| | | Meters. | Feet. |
| 286 | Fourteenth Avenue | 18.1800 | 59.645 |
| 287 | " " | 18.1214 | 59.453 |
| 288 | " " | 17.9316 | 58.831 |
| 289 | Thirteenth Avenue | 17.1765 | 56.353 |
| 290 | Fort Hamilton Parkway | 22.7703 | 74.706 |
| 291 | " " " | 22.7347 | 74.589 |
| 292 | Fortieth Street | 26.2120 | 85.997 |
| 293 | Thirty-seventh Street | 31.3263 | 102.776 |
| 294 | Seventh Avenue | 26.9205 | 88.322 |
| 295 | Fifth Avenue | 18.8689 | 61.906 |
| 296 | " " | 19.1339 | 62.775 |
| 297 | Foster Avenue | 13.3343 | 43.748 |
| 298 | Eighteenth Avenue | 15.1947 | 49.851 |
| 299 | " " | 15.1005 | 49.542 |
| 300 | Beverly Road | 15.9279 | 52.257 |
| 301 | Ocean Parkway | 17.3070 | 56.781 |
| 302 | " " | 17.1830 | 56.375 |
| 303 | Fort Hamilton Parkway | 22.7249 | 74.557 |
| 304 | Parkside Avenue | 17.1808 | 56.367 |
| 305 | " " | 18.3256 | 60.123 |
| 306 | " " | 21.5963 | 70.854 |
| 307 | " " | 18.7316 | 61.445 |
| 308 | " " | 19.1236 | 62.741 |
| 309 | " " | 18.5594 | 60.890 |
| 310 | " " | 18.1210 | 59.452 |
| 311 | Farragut Road | 7.4332 | 24.387 |
| 312 | " " | 7.4937 | 24.586 |
| 313 | Rogers Avenue | 8.1925 | 26.878 |
| 314 | " " | 8.1955 | 26.888 |
| 315 | Avenue I | 8.8463 | 29.023 |
| 316 | Ditmas Avenue | 9.1140 | 29.902 |
| 317 | Snyder Avenue | 14.2940 | 46.896 |
| 318 | Flatbush Avenue | 13.9904 | 45.900 |
| 319 | " " | 14.8132 | 48.600 |
| 320 | " " | 14.6836 | 48.174 |
| 321 | " " | 16.1738 | 53.064 |
| 322 | " " | 17.9577 | 58.916 |
| 323 | Utica Avenue | 8.6778 | 28.470 |
| 324 | " " | 8.6708 | 28.448 |
| 325 | East Fifty-third Street | 7.6098 | 24.966 |
| 326 | Rugby | 5.8489 | 19.189 |
| 327 | Canarsie | 8.4300 | 27.657 |
| 328 | " | 5.0872 | 16.690 |
| 329 | " | 5.3312 | 17.491 |
| 330 | " | 2.6179 | 8.589 |
| 331 | " | 6.8617 | 22.512 |
| 332 | " | 4.1371 | 13.573 |
| 333 | New Lots | 11.0215 | 36.160 |
| 334 | " " | 7.8105 | 25.625 |
| 335 | " " | 11.3433 | 37.216 |
| 336 | " " | 10.5770 | 34.701 |
| 337 | " " | 12.0305 | 39.470 |
| 338 | " " | 7.8340 | 25.702 |
| 339 | " " | 7.8200 | 25.656 |
| 340 | " " | 10.8198 | 35.495 |
| 341 | East New York | 20.7000 | 67.913 |
| 342 | " " " | 21.0458 | 69.048 |
| 343 | Thirty-sixth Street | 7.4287 | 24.373 |
| 344 | Fourth Avenue | 15.7006 | 51.511 |

| No. of Bench Mark. | Locality. | Elevation Above Mean Sea Level. | |
|---|---|---|---|
| | | Meters. | Feet. |
| 345 | Fourth Avenue | 15.7124 | 51.550 |
| 346 | Fourteenth Street | 18.7616 | 61.554 |
| 347 | Prospect Park | 18.9365 | 62.128 |
| 348 | Eleventh Avenue | 39.3797 | 129.198 |
| 349 | " " | 40.7173 | 133.587 |
| 350 | Prospect Park | 47.0000 | 154.199 |
| 351 | " " | 47.3482 | 155.342 |
| 352 | Sixth Avenue | 36.1187 | 118.499 |
| 353 | Fifth Avenue | 28.3097 | 92.879 |
| 354 | Third Avenue | 9.3757 | 30.760 |
| 355 | Second Avenue | 6.2410 | 20.476 |
| 356 | Seventh Avenue | 32.2737 | 105.885 |
| 357 | " " | 32.4964 | 106.615 |
| 358 | " " | 31.4966 | 103.335 |
| 359 | Prospect Park West | 49.4251 | 162.156 |
| 360 | Fourth Avenue | 9.1202 | 29.922 |
| 361 | Butler Street | 10.3232 | 33.869 |
| 362 | " " | 11.1178 | 36.476 |
| 363 | Third Street | 2.4600 | 8.071 |
| 364 | Smith Street | 13.3783 | 43.892 |
| 365 | Henry Street | 11.9902 | 39.338 |
| 366 | " " | 11.2387 | 36.872 |
| 367 | " " | 11.5974 | 38.049 |
| 368 | Hamilton Avenue | 4.5797 | 15.025 |
| 369 | Rapelyea Street | 4.6690 | 15.318 |
| 370 | India Wharf | 2.3177 | 7.604 |
| 371 | Berriman Street | 8.1094 | 26.606 |
| 372 | Lincoln Road | 21.6564 | 71.051 |
| 373 | " " | 16.7659 | 55.006 |
| 374 | Albany Avenue | 14.9894 | 49.178 |
| 375 | " " | 14.8714 | 48.791 |
| 376 | " " | 14.8021 | 48.563 |
| 377 | Schenectady Avenue | 11.9885 | 39.332 |
| 378 | Buffalo Avenue | 28.5945 | 93.814 |
| 379 | Sutter Avenue | 11.2440 | 36.890 |
| 380 | East New York Avenue | 17.5354 | 57.531 |
| 381 | The Plaza | 44.5900 | 146.292 |
| 382 | " " | 43.6326 | 143.151 |
| 383 | Flatbush Avenue | 22.0802 | 72.442 |
| 384 | Prospect Park | 27.3379 | 89.691 |
| 385 | " " | 27.2631 | 89.446 |
| 386 | " " | 27.4332 | 90.004 |
| 387 | " " | 45.0996 | 147.964 |
| 388 | Flatbush Avenue | 43.9716 | 144.263 |
| 389 | " " | 43.9523 | 144.200 |
| 390 | Vanderbilt Avenue | 37.0390 | 121.519 |
| 391 | Eastern Parkway | 46.1371 | 151.368 |
| 392 | " " | 46.6595 | 153.082 |
| 393 | " " | 49.8162 | 163.438 |
| 394 | " " | 45.1062 | 147.986 |
| 395 | Prospect Place | 37.6221 | 123.432 |
| 396 | Grant Square | 27.2395 | 89.368 |
| 397 | Boerum Place | 17.0595 | 55.969 |
| 398 | " " | 17.0424 | 55.913 |
| 399 | " " | 17.0335 | 55.884 |
| 400 | " " | 16.4769 | 54.058 |
| 401 | Livingston Street | 15.7185 | 51.570 |
| 402 | Nevins Street | 12.4929 | 40.987 |
| 403 | Bond Street | 13.2553 | 43.488 |

| No. of Bench Mark. | Locality. | Elevation Above Mean Sea Level. | |
|---|---|---|---|
| | | Meters. | Feet. |
| 404 | Bond Street | 13.2626 | 43.512 |
| 405 | Pacific Street | 12.1224 | 39.772 |
| 406 | " " | 12.6399 | 41.469 |
| 407 | " " | 11.7556 | 38.568 |
| 408 | Dean Street | 9.8754 | 32.400 |
| 409 | Flatbush Avenue | 15.9508 | 52.332 |
| 410 | Atlantic Avenue | 20.2353 | 66.389 |
| 411 | " " | 24.4403 | 80.185 |
| 412 | " " | 25.5049 | 83.677 |
| 413 | " " | 23.7971 | 78.074 |
| 414 | " " | 19.4649 | 63.861 |
| 415 | New York Avenue | 18.8902 | 61.976 |
| 416 | " " " | 18.9303 | 62.107 |
| 417 | Atlantic Avenue | 18.3238 | 60.117 |
| 418 | " " | 20.1288 | 66.039 |
| 419 | " " | 19.1651 | 62.878 |
| 420 | " " | 23.1364 | 75.907 |
| 421 | Herkimer Street | 28.7951 | 94.472 |
| 422 | " " | 28.6401 | 93.963 |
| 423 | Vermont Street | 7.9719 | 26.154 |
| 424 | New Lots Road | 6.6140 | 21.699 |
| 425 | " " " | 7.9365 | 26.038 |
| 426 | " " " | 6.5865 | 21.609 |
| 427 | " " " | 6.5928 | 21.630 |
| 428 | " " " | 6.7215 | 22.052 |
| 429 | Ashford Street | 10.1901 | 33.432 |
| 430 | " " | 10.3501 | 33.957 |
| 431 | Atlantic Avenue | 17.9803 | 58.990 |
| 432 | Berriman Street | 8.9929 | 29.503 |
| 433 | " " | 8.6526 | 28.388 |
| 434 | Belmont Avenue | 7.4723 | 24.515 |
| 435 | Crescent Street | 8.5040 | 27.900 |
| 436 | " " | 8.9569 | 29.386 |
| 437 | " " | 8.5228 | 27.962 |
| 438 | " " | 11.5893 | 38.023 |
| 439 | Sheridan Street | 9.6295 | 31.593 |
| 440 | Grant Avenue | 10.4596 | 34.316 |
| 441 | " " | 10.3130 | 33.835 |
| 442 | Court Street | 20.8224 | 68.315 |
| 443 | " " | 20.4935 | 67.236 |
| 444 | Borough Hall | 20.3337 | 66.712 |
| 445 | Court Street | 21.8772 | 71.776 |
| 446 | Fulton Street | 22.9075 | 75.156 |
| 447 | Clinton Street | 23.8423 | 78.223 |
| 448 | Prospect Street | 14.3902 | 47.212 |
| 449 | Brooklyn Bridge | 27.6873 | 90.837 |
| 450 | " " | 20.4017 | 66.935 |
| 451 | Bridge Street | 15.3928 | 50.501 |
| 452 | Hudson Avenue | 3.1057 | 10.189 |
| 453 | " " | 2.5958 | 8.516 |
| 454 | " " | 6.5692 | 21.553 |
| 455 | Brooklyn Navy Yard | 5.1402 | 16.864 |
| 456 | " " " | 2.0351 | 6.677 |
| 457 | City Park | 4.0395 | 13.253 |
| 458 | " " | 3.8828 | 12.739 |
| 459 | Cumberland Street | 22.4389 | 73.618 |
| 460 | " " | 17.2676 | 56.652 |
| 461 | Grand Avenue | 17.0090 | 55.804 |
| 462 | Taaffe Place | 17.9340 | 58.838 |

| No. of Bench Mark. | Locality. | Elevation Above Mean Sea Level. | |
|---|---|---|---|
| | | Meters. | Feet. |
| 463 | Rodney Street | 14.7211 | 48.297 |
| 464 | Lawrence Street | 13.6185 | 44.680 |
| 465 | St. Edwards Street | 14.3177 | 46.974 |
| 466 | Flushing Avenue | 4.0507 | 13.290 |
| 467 | Taaffe Place | 7.4246 | 24.359 |
| 468 | Wythe Avenue | 9.1416 | 29.992 |
| 469 | South Sixth Street | 9.1814 | 30.123 |
| 470 | " " " | 10.9737 | 36.003 |
| 471 | " " " | 9.3944 | 30.821 |
| 472 | Fort Greene Park | 21.4525 | 70.382 |
| 473 | Eastern Parkway | 31.9026 | 104.667 |
| 474 | Bushwick Avenue | 27.5419 | 90.360 |
| 475 | " " | 16.1858 | 53.103 |
| 476 | " " | 16.8449 | 55.265 |
| 477 | " " | 21.5372 | 70.660 |
| 478 | Troutman Street | 19.1703 | 62.895 |
| 479 | Willoughby Avenue | 11.8847 | 38.992 |
| 480 | " " | 11.8892 | 39.007 |
| 481 | Bushwick Avenue | 18.0911 | 59.354 |
| 482 | Vernon Avenue | 21.3991 | 70.207 |
| 483 | " " | 20.6150 | 67.634 |
| 484 | " " | 22.0869 | 72.463 |
| 485 | " " | 22.0856 | 72.459 |
| 486 | Flushing Avenue | 4.2679 | 14.002 |
| 487 | Sumner Avenue | 19.5896 | 64.270 |
| 488 | South Ninth Street | 15.3181 | 50.256 |
| 489 | Driggs Avenue | 15.0303 | 49.312 |
| 490 | Bedford Avenue | 15.7052 | 51.526 |
| 491 | " " | 15.9473 | 52.320 |
| 492 | Wythe Avenue | 5.5721 | 18.281 |
| 493 | Kent Avenue | 3.6811 | 12.077 |
| 494 | Franklin Street | 4.3778 | 14.363 |
| 495 | " " | 4.3593 | 14.302 |
| 496 | South Sixth Street | 9.9497 | 32.643 |
| 497 | " " " | 8.2219 | 26.975 |
| 498 | " " " | 10.1435 | 33.279 |
| 499 | Greenpoint Avenue | 10.7586 | 35.297 |
| 500 | Dupont Street | 5.8130 | 19.072 |
| 501 | " " | 5.7174 | 18.758 |
| 502 | " " | 4.9959 | 16.391 |
| 503 | Manhattan Avenue | 5.5939 | 18.353 |
| 504 | " " | 2.5662 | 8.419 |

# DESCRIPTIONS OF BENCH MARKS

IN THE

## BOROUGH OF BROOKLYN.

B. M. 178.—*Fort Lafayette.* A 2-inch square (⊓) cut in the top of the outshore edge of the sea wall on the south side of Fort Lafayette. The point is opposite the boiler shop and 38 feet south of the southwest corner of the carpenter shop.

Elevation, 8.131 feet.

B. M. 179.—*Fort Lafayette.* A cross (+) cut in the top of the sea wall on the east side of Fort Lafayette. The point is just east of the carpenter shop, and 10 inches from the east edge of the wall. It is 5.4 feet east of the high iron mooring post and 18 inches south of a small ring bolt in the wall. The figure 164 is cut alongside of the point. (Note 17, p. 91.)

Elevation, 8.013 feet.

B. M. 180.—*Fort Hamilton.* Coast and Geodetic Survey B. M. 1, formerly known as B. M. S. It is the top of the tide staff established by the Coast and Geodetic Survey on the north side of the tide house on the south side of the government wharf near its southwest corner. The bench mark corresponds to a reading of 12 feet on the staff. It should be noted that the tide staff is movable vertically and that the 12-foot mark is the bench only when the staff is in its lowest position. Established in May, 1905.

Elevation, 6.040 feet.

B. M. 181.—*Fort Hamilton.* Coast and Geodetic Survey B. M. 2, formerly known as B. M. A. It is a 6-inch cross (+) cut in the top of the granite capstone on the west edge of the old stone pier, and 6.85 feet from its southwest corner. It is 1.5 feet from the west edge of the pier. Established in October, 1892.

Elevation, 8.855 feet.

B. M. 182.—*Fort Hamilton.* Coast and Geodetic Survey B. M. 3, formerly known as B. M. B. It is the top of an iron mooring post at the southwest corner of the old stone pier, 3.75 feet above the wharf. Established in October, 1892.

Elevation, 12.677 feet.

B. M. 183.—*Fort Hamilton.* The top of an iron mooring post near the north edge of the old stone pier and just south of the one-story frame boathouse. The post is 3 feet from the north edge of the pier and 43 feet from the west edge. It is 4 feet from the southwest corner of the boat house and 3.75 feet above the wharf.

Elevation, 12.601 feet.

B. M. 184.—*Fort Hamilton.* Coast and Geodetic Survey B. M. 4. formerly known as B. M. C. It is the top of an iron mooring post near the southeast corner of the widened end of the old stone pier. It is 3.75 feet above the surface of the wharf. Established in October, 1892.

Elevation, 12.598 feet.

B. M. 185.—*Fort Hamilton.* Coast and Geodetic Survey B. M. 6, formerly known as B. M. I of 1904. It is the top of an iron bolt in the west end of the fourth timber from the shore on the south side of the old stone pier. Established in April, 1904.

Elevation, 9.348 feet.

B. M. 186.—*Fort Hamilton.* Coast and Geodetic Survey B. M. 9, formerly known as B. M. H. It is the highest point of a notch 4 inches long and 1.5 inches deep, cut in the vertical face of a stone in the retaining wall on the west side of Fort Hamilton, nearly in line with the axis of the old stone pier, and 6 feet south of a drain, with which it is nearly on the same level. Established in October, 1892.

Elevation, 28.761 feet.

B. M. 187.—*Fort Hamilton.* Coast and Geodetic Survey B. M. 12, formerly known as B. M. M. It is a horizontal line cut in the end of a ¾-inch brass bolt cemented horizontally into the concrete wall at the westerly corner of the fortification. The bench mark is at the intersection of The Shore Road and Fort Hamilton Avenue, and 3.1 feet above the roadway. Established in April, 1904.

Elevation, 39.219 feet.

B. M. 188.—*Fort Hamilton.* Coast and Geodetic Survey B. M. 13. A horizontal line cut in the corner of the concrete wall at the westerly corner of the fortification, at the intersection of The Shore Road and Fort Hamilton Avenue. The point is 3.6 feet above the roadway. Established in April, 1904.

Elevation, 39.725 feet.

B. M. 189.—*Fort Hamilton Park.* Coast and Geodetic Survey B. M. 14, formerly known as B. M. X. It is a square (□) cut in the surface of the chassis of the gun carriage of a 20" Parrott gun in the southerly corner of Fort Hamilton Park. The bench mark is about 60 feet northwest of the northwest curb of Fort Hamilton Parkway, and about 60 feet east of the east curb of The Shore Road. It is 1.2 feet above the ground. Established in June, 1908.

Elevation, 42.398 feet.

B. M. 190.—*Fort Hamilton Park.* A ¾-inch copper bolt in a city monument set in the lawn on the westerly side of Fort Hamilton Park. The bench mark is 26 feet east of the east curb of The Shore Road and 60 feet southeasterly from the southeasterly curb of Fourth Avenue. The top of the monument is 4 inches below the ground and protected by a square iron cover. (Note 23, p. 91.)

Elevation, 39.153 feet.

B. M. 191.—*Fort Hamilton Park.* The top of a ¾-inch iron rod inside of a wrought-iron pipe casing in the grass plot at the northwest corner of Fort Hamilton Park. The bench mark is 21 feet from the east curb of The Shore Road and 33 feet southeast of the southeast curb of Fourth Avenue. (Note 21, p. 91.)

Elevation, 39.778 feet.

B. M. 192.—*Fifth Avenue.* The bottom of a square cut in the top of the northwest corner of the granite water-table in front of Engine House No. 142, on the southeasterly side of Fifth Avenue, about 170 feet southwest of the center line of Ninety-second Street. The bench mark is 2.8 feet from the northeasterly side of the building and 1.8 feet above the sidewalk. (Note 10, p. 90.)

Elevation, 85.030 feet.

B. M. 193.—*Ninety-second Street.* A horizontal line in the end of a ¾-inch copper bolt leaded horizontally into the granite base course at the southerly corner of Public School No. 104, on the northerly corner of Ninety-second Street and Gelston

Avenue. The bench mark faces Ninety-second Street, is 1.5 feet above the walk and 6 inches from the corner. (Note 6, p. 90.)

Elevation, 80.943 feet.

B. M. 194.—*Ninety-second Street.* A cross (+) cut in the west side of the granite base course at the southwest corner of Public School No. 104, on the northeast corner of Ninety-second Street and Gelston Avenue. The cross is 6 inches above the area. (Note 15, p. 91.)

Elevation, 80.028 feet.

B. M. 195.—*Ninety-second Street.* The center of the top of a granite city monument on the southeast corner of Ninety-second Street and Seventh Avenue. The top of the monument is 6 inches above the ground.

Elevation, 52.675 feet.

B. M. 196.—*Bath Beach.* The highest point within a square (□) cut in the top of the northerly curb of Cropsey Avenue, about 52 feet northwesterly from the northwesterly line of Bay Thirteenth Street.

Elevation, 23.785 feet.

B. M. 197.—*Bath Beach.* Coast and Geodetic Survey B. M. K. The intersection of a cross cut on the end of a ½-inch copper bolt leaded horizontally into the fifth brick above the bluestone sill of the cellar window on the southeasterly side of the two-story and attic frame house at No. 1615 Cropsey Avenue, about 75 feet northwesterly from Bay Thirteenth Street. The bench mark is on the northeasterly side of the window, about 8 feet from the front of the house. Established in May, 1887.

Elevation, 27.326 feet.

B. M. 198.—*Bensonhurst.* A 2-inch cross (+) cut in the white foundation stone at the southerly corner of the Sunshine Sanatorium Home, on the northeasterly side of Cropsey Avenue between Bay Thirty-second Street and Bay Thirty-third Street. The cross faces Bay Thirty-third Street, is 4½ inches above the sidewalk and 7 inches from the corner of the building. It is lettered B. E. B. M.

Elevation, 20.036 feet.

B. M. 199.—*Bensonhurst.* The center of the top of a granite city monument on the westerly corner of Cropsey Avenue and Twenty-second Avenue. The monument is in the corner of Bensonhurst Park and the top is 6 inches below the ground.

Elevation, 20.319 feet.

B. M. 200.—*Bensonhurst.* The center of the top of a granite city monument on the northerly corner of Cropsey Avenue and Twenty-second Avenue. The top of the monument is 1.6 feet below the ground.

Elevation, 19.567 feet.

B. M. 201.—*Harway Avenue Bridge.* The center of a square (□) cut in the top of the northwest corner of the center pier on the west side of the highway bridge over the Gravesend Drainage Canal (Coney Island Creek) at West Seventeenth Street.

Elevation, 9.194 feet.

B. M. 202.—*Harway Avenue Bridge.* A square (□) cut in the top of the northwest corner of the north pier on the west side of the highway bridge over the Gravesend Drainage Canal (Coney Island Creek) at West Seventeenth Street.

Elevation, 9.343 feet.

B. M. 203.—*Harway Avenue Bridge.* A square (□) cut in the top of the northwest corner of the south pier on the west side of the highway bridge over the Gravesend Drainage Canal (Coney Island Creek) at West Seventeenth Street.

Elevation, 8.554 feet.

B. M. 204.—*Harway Avenue Bridge.* A square (□) cut in the top of the southeast corner of the south pier on the east side of the highway bridge over the Gravesend Drainage Canal (Coney Island Creek) at West Seventeenth Street.

Elevation, 8.548 feet.

B. M. 205.—*Harway Avenue Bridge.* A 1-inch square (□) cut in the top of the southeast corner of the 3-foot square capstone supporting the steel incline on top of the south pier on the east side of the highway bridge over the Gravesend Drainage Canal (Coney Island Creek) at West Seventeenth Street.

Elevation, 10.080 feet.

B. M. 206.—*Harway Avenue Bridge.* A 1-inch square (□) cut in the top of the southwest corner of the 3-foot square capstone supporting the steel incline on top of the south pier on the east side of the highway bridge over the Gravesend Drainage Canal (Coney Island Creek) at West Seventeenth Street.

Elevation, 10.079 feet.

B. M. 207.—*Coney Island.* The center of the top of a brass tablet bolt 2 inches in diameter, in the northwest corner of the top step at the main entrance to Public School No. 80, on the west side of West Seventeenth Street, between Neptune Avenue and Mermaid Avenue. (Note 2, p. 89.)

Elevation, 13.017 feet.

B. M. 208.—*Coney Island.* A brass bracket in the bluestone base course on the east side of Public School No. 80, on the west side of West Seventeenth Street, between Neptune Avenue and Mermaid Avenue. The bracket is 1 foot south of the main entrance. (Note 22, p. 91.)

Elevation, 11.289 feet.

B. M. 209.—*Coney Island.* A square (□) cut in the top of the southeast corner of a granite city monument on the northwest corner of West Seventeenth Street and Mermaid Avenue. The top of the monument is flush with the ground.

Elevation, 7.191 feet.

B. M. 210.—*Coney Island.* A square (□) cut in the top of the large bluestone basin head on the south side of Surf Avenue about 45 feet northeasterly from the corner of Henderson's restaurant on Stratton-Henderson Walk. The bench mark is opposite the Mardi Gras Hotel.

Elevation, 7.351 feet.

B. M. 211.—*Coney Island.* The top of the small center nut on top of the high pressure fire hydrant on the southwest corner of West Fifth Street and Surf Avenue.

Elevation, 9.716 feet.

B. M. 212.—*Coney Island.* A ¾-inch copper bolt in a city monument in the southwest corner of Seaside Park, on the northeast corner of Surf Avenue and West Fifth Street. The bench mark is 4 feet east of the iron railing on West Fifth Street and 75 feet north of the north curb line of Surf Avenue. (Note 23, p. 91.)

Elevation, 7.659 feet.

B. M. 213.—*Coney Island.* The bottom of a 2-inch square cut in the center of a square capstone on top of the granite wall on the south side of Surf Avenue, in front of the Public Bath House. The bench mark is 25 feet east of the West Fifth Street end of the wall and 3.7 feet above the sidewalk. (Note 10, p. 90.)

Elevation, 11.493 feet.

B. M. 214.—*Coney Island.* A square (□) cut in the top of the northeast corner of the 5½-milestone post at the south edge of the sidewalk on the north side of Surf Avenue, about 200 feet west of Ocean Parkway.

Elevation, 10.569 feet.

B. M. 215.—*Coney Island.* The top of the north nut on the top flange of the fire hydrant on the northwest corner of Neptune Avenue and Ocean Parkway. Established by the Brooklyn Highway Bureau.

Elevation, 8.964 feet.

B. M. 216.—*Ocean Parkway.* A horizontal line on the face of a brass tablet bolt 2 inches in diameter leaded horizontally into the yellow brick at the northwest corner of the Coney Island Hospital, on the southeast corner of Avenue Z and Ocean Parkway. The bench mark is 12 inches above the ground. (Note 1, p. 89.)

Elevation, 10.152 feet.

B. M. 217.—*Ocean Parkway.* A square (□) cut in the top of the northwest corner of the square limestone block at the northwest corner of the Coney Island Hospital, on the southeast corner of Avenue Z and Ocean Parkway.

Elevation, 11.652 feet.

B. M. 218.—*Ocean Parkway.* A square (□) cut in the top of the northwest corner of the 4½ milestone post on the west side of the center driveway of Ocean Parkway, between Avenue Y and Avenue Z.

Elevation, 8.288 feet.

B. M. 219.—*Ocean Parkway.* A square (□) cut in the north end of the bluestone coping of culvert wall on the east side of Ocean Parkway about 60 feet south of Avenue Y.

Elevation, 6.890 feet.

B. M. 220.—*Ocean Parkway.* A ½-inch copper bolt in section stone III of the Ocean Parkway Base Line. The bolt is in a granite block 6″ x 6″ x 3′ set in concrete 3′ x 3′ x 4′, is on the monument line on the east side of Ocean Parkway and 257 feet south of the north line of Avenue U.

Elevation, 12.055 feet.

B. M. 221.—*Avenue U.* The top of a ⅝-inch copper bolt in the large limestone block supporting the lamp at the east side of the main entrance to the 68th Precinct Police Station on the northwest corner of Avenue U and East Fifteenth Street. The bolt is 12 inches from the south face of the building. (Note 8, p. 90.)

Elevation, 21.195 feet.

B. M. 222.—*Avenue U.* The center of the top of the concrete city monument on the northeast corner of Avenue U and East Fifteenth Street. The monument is flush with the sidewalk.

Elevation, 15.268 feet.

B. M. 223.—*Avenue U.* A square (□) lettered U. S., cut in the east end of the sidewalk on the southwest corner of Avenue U and Ocean Avenue. (Note 16, p. 91.)

Elevation, 17.639 feet.

B. M. 224.—*Neck Road.* The intersection of a cross (+) cut in the end of a ⅝-inch copper bolt leaded horizontally into the brick foundation at the southwest corner of Hector Frisbie's frame house (Gerritson's Homestead), on the northwesterly side of Neck Road, about 1,200 feet east of its intersection with Avenue U. The bench mark is on the tenth course of brick above the ground. (Note 17, p. 91.)

Elevation, 17.606 feet.

B. M. 225.—*Avenue U.* The center of a brass bolt marking the U. S. C. & G. S. triangulation station Avenue U, on the north side of Avenue U, between East Thirty-fifth Street and East Thirty-sixth Street. The bench mark is on the monument line on the north side of Avenue U and 86.5 feet west of East Thirty-sixth Street. The bolt is flush with the sidewalk. (Note 25, p. 91.)

Elevation, 8.659 feet.

B. M. 226.—*Avenue U.* The center of a granite city monument on the northeasterly corner of Kimball Road and Avenue U. The top of the monument is 3 inches below the ground.

Elevation, 9.072 feet.

B. M. 227.—*Flatbush Avenue.* A square (⊡) lettered U. S., cut in the north end of the basin head on the northwest corner of Flatbush Avenue and Avenue Q. (Note 16, p. 91.)

Elevation, 15.300 feet.

B. M. 228.—*Flatbush Avenue.* A square (⊡) cut in the top of the northeast corner of a granite city monument in the center of the cement sidewalk on the northeast corner of Flatbush Avenue and Avenue N. The top of the monument is 1 inch below the sidewalk.

Elevation, 15.891 feet.

B. M. 229.—*Avenue K.* The intersection of a cross (+) cut in the end of a ⅝-inch copper bolt leaded horizontally into the granite base course at the southwest corner of Public School No. 119, on the north side of Avenue K, between East Thirty-eighth Street and East Thirty-ninth Street. The bench mark is on the south side of the building 1 foot from the southwest corner and 6 inches above the ground. (Note 18, p. 91.)

Elevation, 18.906 feet.

B. M. 230.—*Avenue K.* A brass bracket in the granite base course on the west side of Public School No. 119, on the north side of Avenue K, between East Thirty-eighth Street and East Thirty-ninth Street. The bench mark is on the west side of the building 19.8 feet from the southwest corner and about 200 feet north of Avenue K. (Note 22, p. 91.)

Elevation, 18.997 feet.

B. M. 231.—*Flatbush Avenue.* A ½-inch copper bolt in the top of the east end of the concrete pier wall along the center line of the Manhattan Beach Railroad at Flatbush Avenue. The bolt is in the center of the wall, is 4½ inches from the east end and 5 feet above the bottom of the railroad cut. (Note 8, p. 90.)

Elevation, 19.947 feet.

B. M. 232.—*Flatbush Avenue.* The center of a square (⊡), lettered B. M., cut in the west edge of the cement sidewalk on the west side of the highway bridge over the Manhattan Beach Railroad on Flatbush Avenue, just south of Avenue H. The bench mark is 32.5 feet south of the extreme northwest corner of the bridge.

Elevation, 33.808 feet.

B. M. 233.—*Ocean Parkway.* A ½-inch copper bolt in Section Stone II of the Ocean Parkway Base Line. The bolt is set in a granite block 6″ x 6″ x 3′ set in concrete 3′ x 3′ x 4′. It is on the monument line on the east side of Ocean Parkway and 126.5 feet north of Kings Highway. The top of the monument is 9 inches below the ground.

Elevation, 22.563 feet.

B. M. 234.—*Ocean Parkway.* The center of the top of a granite city monument on the northeast corner of Avenue Q and Ocean Parkway. The top of the monument is flush with the ground.

Elevation, 23.775 feet.

B. M. 235.—*Ocean Parkway.* A square (□) cut in the top of the granite 3-mile stone post on the west side of the center driveway of Ocean Parkway, about 160 feet south of Avenue P.

Elevation, 27.439 feet.

B. M. 236.—*Ocean Parkway.* A ½-inch copper bolt in Section Stone I of the Ocean Parkway Base Line. The bolt is set in a granite block 6″ x 6″ x 3′ set in concrete 3′ x 3′ x 4′. It is on the monument line on the east of Ocean Parkway and about 168 feet north of Avenue N. The top of the monument is 6 inches below the ground.

Elevation, 31.331 feet.

B. M. 237.—*Ocean Parkway.* The center of the top of a granite city monument on the northwest corner of Ocean Parkway and Avenue L. The top of the monument is flush with the ground.

Elevation, 35.706 feet.

B. M. 238.—*Ocean Parkway.* The center of the top of a granite city monument on the northwest corner of Ocean Parkway and Avenue I. The top of the monument is 2 inches below the ground.

Elevation, 42.702 feet.

B. M. 239.—*Ocean Parkway.* A ⅝-inch copper bolt lettered B. E. B. M., leaded vertically in the east end of the north abutment of highway bridge over the Manhattan Beach Division of the Long Island Railroad at Ocean Parkway, between Avenue H and Avenue I. The bench mark is 6 inches east of the east girder, and 8 inches north of the face of the abutment. (Note 8, p. 90.)

Elevation, 45.676 feet.

B. M. 240.—*Bath Beach.* A horizontal line in the end of a ¾-inch copper bolt in the granite base course at the easterly corner of Public School No. 163, on the westerly corner of Benson Avenue and Seventeenth Avenue. The bench mark faces Benson Avenue, is 8 inches from the corner of the building and is 5½ inches above the ground. (Note 6, p. 90.)

Elevation, 31.543 feet.

B. M. 241.—*Bath Beach.* A cross (+) cut in the top of the granite window-sill near the easterly corner of Public School No. 163, on the westerly corner of Benson Avenue and Seventeenth Avenue. The bench mark is on the Benson Avenue side of the building and 9.5 feet from the Seventeenth Avenue side. (Note 17, p. 91.)

Elevation, 33.225 feet.

B. M. 242.—*Bath Beach.* The center of the top of the granite city monument on the northerly corner of Benson Avenue and Seventeenth Avenue. The top of the monument is 1 inch above the ground.

Elevation, 30.289 feet.

B. M. 243.—*Bath Beach.* The center of a square (⊓) cut in the top of the brownstone water-table on the south side of the house entrance to the three-story brick tenement at No. 1721 Seventeenth Avenue, on the southerly corner of Seventy-eighth Street. The bench mark is about 45 feet southeast of Seventy-eighth Street.

Elevation, 27.286 feet.

B. M. 244.—*Fifteenth Avenue.* A brass bracket in the granite base course on the northerly corner of Public School No. 112, on the southerly side of Fifteenth Avenue, from Seventy-first Street to Seventy-second Street. The bench mark is on the Fifteenth Avenue side of the building and 1.7 feet from the corner. (Note 22, p. 91.)

Elevation, 43.687 feet.

B. M. 245.—*Fifteenth Avenue.* The intersection of a cross (+) cut in the granite base course on the northerly corner of Public School No. 112, on the southerly side of Fifteenth Avenue, from Seventy-first Street to Seventy-second Street. The cross is on the Fifteenth Avenue side of the building and 3 inches from the corner. (Note 17, p. 91.)

Elevation, 43.672 feet.

B. M. 246.—*Fifteenth Avenue.* A square (⊓) cut in the top of the granite sill of the basement window on the westerly side of Public School No. 112, on the southerly side of Fifteenth Avenue, from Seventy-first Street to Seventy-second Street. The bench mark is on the center of the sill and 9.0 feet from the northerly corner of the building.

Elevation, 43.814 feet.

B. M. 247.—*New Utrecht Avenue.* A 9-16-inch copper bolt in the concrete foundation at the southwest corner of Sub-station No. IX of the Brooklyn Rapid Transit, on the northeast corner of New Utrecht Avenue and Sixty-second Street. The bench mark is 7 inches from the south side of the building and 3 inches from the west side. (Note 8, p. 90.)

Elevation, 41.758 feet.

B. M. 248.—*Eighty-sixth Street.* A horizontal line on the face of a brass tablet bolt 2 inches in diameter, cemented horizontally into the 12-inch brownstone coping of the Seventy-first Precinct Police Station, on the northeast corner of Eighty-sixth Street and Fifth Avenue. The bench mark is on the Fifth Avenue side of the building about 70 feet northeasterly from Eighty-sixth Street, and 4.5 feet above the sidewalk. (Note 1, p. 89.)

Elevation, 81.768 feet.

B. M. 249.—*Eighty-sixth Street.* A ¼-inch copper bolt leaded vertically in the top of the 6-inch brownstone water-table at the southeast corner of the Seventy-first Precinct Police Station on the northeast corner of Eighty-sixth Street and Fifth Avenue. The bench mark is 4.5 feet above the sidewalk. (Note 7, p. 90.)

Elevation, 84.505 feet.

B. M. 250.—*Eighty-sixth Street.* The center of the top of a granite city monument on the southwest corner of Eighty-sixth Street and Fifth Avenue. The top of the monument is 2 inches above the ground.

Elevation, 78.194 feet.

B. M. 251.—*Eighty-sixth Street.* A square (□) cut in the top of the east corner of the basin head on the northeast corner of Eighty-sixth Street and Fifth Avenue.
Elevation, 78.112 feet.

B. M. 252.—*Eighty-sixth Street.* A square (□) cut in the top of the east corner of the basin head on the northeast corner of Eighty-sixth Street and Fourth Avenue.
Elevation, 71.218 feet.

B. M. 253.—*Seventy-sixth Street.* A square (□) in the top of the easterly corner of basin head on the southwest corner of Seventy-sixth Street and Fourth Avenue.
Elevation, 74.628 feet.

B. M. 254.—*Sixty-eighth Street.* The center of the top of a granite city monument in the center of the sidewalk on the northeast corner of Sixty-eighth Street and Second Avenue. The top of the monument is 4 inches below the sidewalk.
Elevation, 72.699 feet.

B. M. 255.—*Bay Ridge Avenue.* The center of the top of a granite city monument in the center of the sidewalk on the northeast corner of Bay Ridge Avenue and Second Avenue. The top of the monument is 6 inches below the sidewalk.
Elevation, 65.816 feet.

B. M. 256.—*Seventy-first Street.* A brass bracket in the granite base course on the north side of Public School No. 102, on the southeast corner of Seventy-first Street and Ridge Boulevard (Second Avenue). The bench mark is 1.6 feet from the northwest corner of the building. (Note 22, p. 91.)
Elevation, 75.547 feet.

B. M. 257.—*Seventy-first Street.* A horizontal line in the end of a ⅝-inch copper bolt leaded horizontally into the granite base course on the north side of Public School No. 102, on the southeast corner of Seventy-first Street and Ridge Boulevard (Second Avenue). The bench mark is about 128 feet east of Ridge Boulevard and 2 feet from the east corner of the buttress. (Note 18, p. 91.)
Elevation, 75.683 feet.

B. M. 258.—*Second Avenue.* Coast and Geodetic Survey B. M. J. A cross (+) cut in the top of the northwest corner of the bluestone coping on the west side of the old highway bridge at Second Avenue and Sixty-fifth Street, over the tracks of the Manhattan Beach Division of the Long Island Railroad.
Elevation, 43.635 feet.

B. M. 259.—*Second Avenue.* A 2-inch square (□) cut in the top of the northwest corner of the easterly granite curb of the highway bridge on Second Avenue, over The Shore Road.
Elevation, 55.384 feet.

B. M. 260.—*Seventy-ninth Street.* A cross, lettered B + M, cut in the granite base course on the west side of Public School No. 127, on the northeast corner of Seventy-ninth Street and Seventh Avenue. The cross is on the south side of the buttress, is 13 feet from Seventy-ninth Street and 6 inches above the ground. (Note 17, p. 91.)
Elevation, 85.595 feet.

B. M. 261.—*Seventy-ninth Street.* A brass bracket in the granite base course on the west side of Public School No. 127, on the northeast corner of Seventy-ninth Street and Seventh Avenue. The bench mark is about 38 feet from Seventy-ninth Street. (Note 22, p. 91.)
Elevation, 85.862 feet.

B. M. 262.—*Seventy-ninth Street.* A square (ㅁ) cut in the top of the northeast corner of the granite city monument on the southeast corner of Seventy-ninth Street and Seventh Avenue. The top of the monument is flush with the sidewalk.

Elevation, 80.613 feet.

B. M. 263.—*McKinley Park.* The top of a ¾-inch iron rod inside of a wrought-iron pipe casing, in the southeasterly side of McKinley Park, on the northwesterly side of Fort Hamilton Parkway at Seventy-fifth Street. The bench mark is 5 feet northwest of the park wall, and about 46 feet northeast of Seventy-fifth Street. (Note 21, p. 91.)

Elevation, 96.624 feet.

B. M. 264.—*Fort Hamilton Parkway.* The center of the horizontal bar of a **T** cut in the northwesterly face of the concrete pier wall on the center line of the Manhattan Beach Division of the Long Island Railroad at Fort Hamilton Parkway. The bench mark is 2.8 feet above the bottom of the railroad cut.

Elevation, 45.436 feet.

B. M. 265.—*Fifteenth Avenue.* A 2-inch cross (+) cut in the top of the northeast corner of the concrete pier wall on the center line of the Manhattan Beach Railroad at Fifteenth Avenue, about 225 feet south of Sixtieth Street. (Note 17, p. 91.)

Elevation, 33.332 feet.

B. M. 266.—*Fifty-second Street.* A ½-inch copper bolt in the top of the northeast corner of the concrete pier wall on the center line of the Manhattan Beach Division of the Long Island Railroad at Fifty-second Street, about 150 feet south of Eighteenth Avenue. The bench mark is 6 inches from the east end of the wall and 6 inches from the north end. (Note 8, p. 90.)

Elevation, 38.100 feet.

B. M. 267.—*Ocean Avenue.* A square (□) cut in the top of the low retaining wall on the south side of the Manhattan Beach Division of the Long Island Railroad, and about 8 feet west of the west line of Ocean Avenue.

Elevation, 23.245 feet.

B. M. 268.—*Fort Hamilton Parkway.* A square (ㅁ) cut in the south end of the door-sill at the apartment entrance to Borden's Milk Depot, on the westerly corner of Fort Hamilton Parkway and Sixtieth Street. The bench mark is 50 feet southwesterly from the Sixtieth Street corner of the building.

Elevation, 63.706 feet.

B. M. 269.—*Sixtieth Street.* At Borden's Milk Depot, on the westerly corner of Fort Hamilton Parkway and Sixtieth Street. The bench mark is a square (ㅁ) cut in the northwest corner of the fourth window-sill from Fort Hamilton Parkway, on the Sixtieth Street side of the building.

Elevation, 66.420 feet.

B. M. 270.—*Fifty-ninth Street.* A brass bracket in a large bluestone block in the foundation of Public School No. 105, on the northeasterly corner of Fifty-ninth Street and Tenth Avenue. The bench mark is on the Tenth Avenue side of the building, about 62 feet from Fifty-ninth Street. (Note 22, p. 91.)

Elevation, 68.208 feet.

B. M. 271.—*Fifty-eighth Street.* The center of the top of a granite city monument on the northerly corner of Fifty-eighth Street and Seventh Avenue. The top of the monument is 6 inches below the ground.

Elevation, 101.027 feet.

B. M. 272.—*Fifty-ninth Street.* A square (☐) lettered B. M., cut in the sidewalk on the southerly corner of Fifty-ninth Street and Fifth Avenue. The bench mark is 1.3 feet north of the corner of the low stone retaining wall.

Elevation, 116.957 feet.

B. M. 273.—*Sixtieth Street.* A square (ㅂ) cut in the top of the northerly corner of the brownstone door-sill at the house entrance to the three-story brick dwelling at No. 406 Sixtieth Street, on the southerly corner of Fourth Avenue. The bench mark is about 50 feet from Fourth Avenue.

Elevation, 97.720 feet.

B. M. 274.—*Third Avenue.* A cross (+) cut in the top of the 1½-inch iron anchor bolt on the southerly corner of the base of the elevated railroad column on the easterly corner of Third Avenue and Sixty-first Street. The cross is 2 feet above the sidewalk.

Elevation, 68.908 feet.

B. M. 275.—*Sixtieth Street.* A ½-inch copper bolt in the top of the limestone water-table on the westerly side of the main entrance to Public School No. 140, on the northerly side of Sixtieth Street, about 190 feet westerly from Fourth Avenue. The bench mark is 6 inches westerly from the projecting limestone pilaster on the westerly side of the entrance. (Note 18, p. 91.)

Elevation, 92.921 feet.

B. M. 276.—*Sixtieth Street.* A horizontal line on the face of a brass tablet bolt 2 inches in diameter, cemented horizontally into the limestone on the easterly side of the main entrance to Public School No. 140, on the northerly side of Sixtieth Street, about 180 feet westerly from Fourth Avenue. The bench mark is 11 inches above the ground and 1 foot easterly from the projecting limestone pilaster on the easterly side of the entrance. (Note 1, p. 89.)

Elevation, 90.637 feet.

B. M. 277.—*Fourth Avenue.* A 2-inch square (ㅂ) cut in the top of the west end of the brownstone door-sill at the Forty-seventh Street entrance to No. 4623 Fourth Avenue.

Elevation, 83.533 feet.

B. M. 278.—*Fourth Avenue.* The bottom of a square cut in the southwest corner of the bluestone coping of water-table at Engine House No. 101, on the southeasterly side of Fourth Avenue between Fifty-first Street and Fifty-second Street. The bench mark is 6 inches above the sidewalk. (Note 10, p. 90.)

Elevation, 76.933 feet.

B. M. 279.—*Fourth Avenue.* A 2-inch square (☐) cut in the top of the westerly corner of the seventh or wide landing step at the Fourth Avenue entrance to St. Michaels Church, about 25 feet southwesterly from Forty-second Street.

Elevation, 64.504 feet.

B. M. 280.—*Fourth Avenue.* A horizontal line in the end of a ⅝-inch copper bolt leaded horizontally into the granite base course at the westerly corner of Public School No. 136, on the northwesterly corner of Fourth Avenue and Forty-first Street. The bench mark faces Forty-first Street, is 5½ inches above the walk and about 65 feet westerly from Fourth Avenue. (Note 18, p. 91.)

Elevation, 51.582 feet.

B. M. 281.—*Fourth Avenue.* The highest point within a quadrant (⊾) cut in the top of the southerly corner of the base of the limestone coping on the easterly side of steps at the main entrance to Public School No. 136, on the northwesterly corner of Fourth Avenue and Forty-first Street. The bench mark is about 35 feet westerly from Fourth Avenue and 1 foot above the sidewalk.

Elevation, 52.244 feet.

B. M. 282.—*Fourth Avenue.* A square (⊡) cut in the top of the northwesterly corner of the bluestone coping on the southerly side of the second basement window from Forty-first Street, on the Fourth Avenue side of Public School No. 136, on the northwesterly corner of Forty-first Street and Fourth Avenue.

Elevation, 54.175 feet.

B. M. 283.—*Fourth Avenue.* A square (⊡) cut in the top of the curved bluestone coping at the base of iron railing in front of the easterly corner of Public School No. 82, on the southwesterly corner of Fourth Avenue and Thirty-sixth Street.

Elevation, 26.798 feet.

B. M. 284.—*Fourteenth Avenue.* The center of the top of a granite city monument on the northwest corner of Fourteenth Avenue and Sixtieth Street. The top of the monument is flush with the cement sidewalk.

Elevation, 45.290 feet.

B. M. 285.—*New Utrecht Avenue.* A square (⊡) cut in the southeast corner of the sidewalk, on the northwest corner of New Utrecht Avenue and Thirteenth Avenue. The bench mark is opposite the house entrance to No. 5418 New Utrecht Avenue.

Elevation, 55.162 feet.

B. M. 286.—*Fourteenth Avenue.* The center of the horizontal bar of a 3-inch **T** cut in the center of a large granite foundation stone on the Fifty-fourth Street side of Public School No. 103, on the easterly corner of Fourteenth Avenue and Fifty-fourth Street. The bench mark is 20 feet southeasterly from the westerly corner of the building and 1 foot above the ground.

Elevation, 59.645 feet.

B. M. 287.—*Fourteenth Avenue.* A brass bracket in the granite foundation on the Fifty-fourth Street side of Public School No. 103, on the easterly corner of Fourteenth Avenue and Fifty-fourth Street. The bench mark is 35 feet southeasterly from the westerly corner of the building. (Note 22, p. 91.)

Elevation, 59.453 feet.

B. M. 288.—*Fourteenth Avenue.* A square (□) cut in the top of the bluestone coping under the iron railing at the westerly corner of Public School No. 103, on the easterly corner of Fourteenth Avenue and Fifty-fourth Street. The bench mark is 1.8 feet from the corner of the coping.

Elevation, 58.831 feet.

B. M. 289.—*Thirteenth Avenue.* The bottom of a square cut in the southerly corner of the square foundation stone at the southerly corner of the three-story yellow brick apartment house on the northerly corner of Thirteenth Avenue and Fifty-fourth Street. (Note 10, p. 90.)

Elevation, 56.353 feet.

B. M. 290.—*Fort Hamilton Parkway.* A horizontal line in the end of a ⅝-inch copper bolt leaded horizontally into the granite base course on the southwesterly side of the 5-foot buttress at the westerly corner of Public School No. 131, on the easterly corner of Fort Hamilton Parkway and Forty-fourth Street. The bolt is 1 foot from the corner of the buttress and 14 feet from Forty-fourth Street. (Note 6, p. 90.)
Elevation, 74.706 feet.

B. M. 291.—*Fort Hamilton Parkway.* A brass bracket in the granite base course on the northwesterly side of Public School No. 131, on the easterly corner of Fort Hamilton Parkway and Forty-fourth Street. The bracket is about 40 feet from the Forty-fourth Street side of the building. (Note 22, p. 91.)
Elevation, 74.589 feet.

B. M. 292.—*Fortieth Street.* A square (⌂) cut in the top of the south corner of the lowest brownstone step at the house entrance to the two-story brick dwelling at No. 969 Fortieth Street, about 100 feet southeasterly from New Utrecht Avenue.
Elevation, 85.997 feet.

B. M. 293.—*Thirty-seventh Street.* A square (⌂) cut in the top of the southerly corner of a 8″ x 8″ granite post under the iron fence on the Thirty-seventh Street side of Greenwood Cemetery, between Seventh Avenue and Eighth Avenue. The bench mark is about 420 feet northwesterly from Eighth Avenue.
Elevation, 102.776 feet.

B. M. 294.—*Seventh Avenue.* The center of the top of a granite city monument at the corner of the fence, on the westerly corner of Seventh Avenue and Thirty-sixth Street. The top of the monument is 4 inches above the ground.
Elevation, 88.322 feet.

B. M. 295.—*Fifth Avenue.* A level triangular surface on the top of the northerly corner of the red sandstone coping at the northerly corner of the two-story brick elevated shop of the Brooklyn Rapid Transit, on the southerly corner of Fifth Avenue and Thirty-sixth Street. A corner of the curved coping above the bench mark was cut away, leaving the triangular surface, which is 3.1 feet above the ground. The letters B. M. are cut in the curved coping above the bench mark.
Elevation, 61.906 feet.

B. M. 296.—*Fifth Avenue.* A cross (+) lettered B. M., cut in the top of the westerly corner of the bluestone water-table on the Fifth Avenue side of the two-story brick elevated shop of the Brooklyn Rapid Transit, on the southerly corner of Fifth Avenue and Thirty-sixth Street. The cross is about 77 feet southwest of the corner.
Elevation, 62.775 feet.

B. M. 297.—*Foster Avenue.* A square (⌂) cut in the top of the south end of the granite basin head on the northwest corner of Foster Avenue and Ocean Parkway.
Elevation, 43.748 feet.

B. M. 298.—*Eighteenth Avenue.* A brass bracket in the granite base course on the north side of Public School No. 134, on the south side of Eighteenth Avenue between East Fourth Street and East Fifth Street. The bracket is 17.3 feet from the east side of the building. (Note 22, p. 91.)
Elevation, 49.851 feet.

B. M. 299.—*Eighteenth Avenue.* A cross (+) cut in the granite base course on the north side of Public School No. 134, on the south side of Eighteenth Avenue between East Fourth Street and East Fifth Street. The cross is 6 inches from the corner of the buttress and 14.5 feet from the east side of the building. (Note 17, p. 91.)

Elevation, 49.542 feet.

B. M. 300.—*Beverly Road.* The center of the top of a granite city monument on the southwest corner of Beverly Road and Ocean Parkway. The top of the monument is flush with the ground.

Elevation, 52.257 feet.

B. M. 301.—*Ocean Parkway.* A brass bracket on the east side of Public School No. 130, on the southwest corner of Ocean Parkway and Fort Hamilton Parkway. The bracket is about 40 feet from the south side of the building. (Note 22, p. 91.)

Elevation, 56.781 feet.

B. M. 302.—*Ocean Parkway.* A horizontal line in the end of a ¾-inch copper bolt leaded horizontally into the granite base course at the northeast corner of Public School No. 130, on the southwest corner of Ocean Parkway and Fort Hamilton Parkway. The bolt is on the north side of the building, is 6 inches from the corner, and 3½ inches above the sidewalk. (Note 6, p. 90.)

Elevation, 56.375 feet.

B. M. 303.—*Fort Hamilton Parkway.* The top of the northwest corner of a granite city monument on the southwest corner of Fort Hamilton Parkway and Thirty-sixth Street. The top of the monument is 1 inch above the ground.

Elevation, 74.557 feet.

B. M. 304.—*Parkside Avenue.* A square (⊡) cut in the top of the southwest corner of the lowest granite step of the bicycle rest on the north side of Parkside Avenue, just north of the west end of the Parade Grounds.

Elevation, 56.367 feet.

B. M. 305.—*Parkside Avenue.* A square (□) cut in the center of the cement sidewalk on the south side of Parkside Avenue, about 300 feet west of Parade Place. The bench mark is 12 feet south of the curb and is opposite the center of the Grecian Pavilion in Prospect Park.

Elevation, 60.123 feet.

B. M. 306.—*Parkside Avenue.* A cross (+) cut in the top of the southeast corner of the square base of the east column on the south side of the Grecian Pavilion in Prospect Park, just south of Parkside Avenue and about 300 feet west of Parade Place. (Note 17, p. 91.)

Elevation, 70.854 feet.

B. M. 307.—*Parkside Avenue.* The center of the top of a granite city monument on the southwest corner of Parkside Avenue and Ocean Avenue. The top of the monument is 3 inches below the sidewalk.

Elevation, 61.445 feet.

B. M. 308.—*Parkside Avenue.* A ⅞-inch copper bolt in the northeast corner of the granite base of pergola on the northwest corner of Parkside Avenue and Ocean Avenue. The bolt is 6 inches from the south end of the park wall and 2½ inches from the east edge of the base. (Note 5, p. 89.)

Elevation, 62.741 feet.

B. M. 309.—*Parkside Avenue.* A square (□) cut in the top of the southeast corner of the granite city monument on the southwest corner of Parkside Avenue and Flatbush Avenue. The top of the monument is 3 inches above the ground.

Elevation, 60.890 feet.

B. M. 310.—*Parkside Avenue.* A square (□) cut in the top of the northeast corner of the granite basin head on the northwest corner of Parkside Avenue and Flatbush Avenue.

Elevation, 59.452 feet.

B. M. 311.—*Farragut Road.* The center of the top of a granite city monument 2 feet from the corner of the curb, on the northwest corner of Farragut Road (Avenue F) and Flatbush Avenue. The top of the monument is 3 inches below the sidewalk.

Elevation, 24.387 feet.

B. M. 312.—*Farragut Road.* The center of the top of a granite city monument on the north side of Farragut Road (Avenue F) about 10 feet west of the intersection of the northerly line of Farragut Road and the westerly line of Flatbush Avenue. The top of the monument is 6 inches below the sidewalk.

Elevation, 24.586 feet.

B. M. 313.—*Rogers Avenue.* The bottom of a square cut in the top of the southwest corner of the granite water-table at the extreme southwest corner of Engine House No. 155, on the east side of Rogers Avenue about 60 feet north of Farragut Road (Avenue F). (Note 10, p. 90.)

Elevation, 26.878 feet.

B. M. 314.—*Rogers Avenue.* A ¼-inch copper bolt in the top of the granite water-table on the south side of Engine House No. 155, on the east side of Rogers Avenue about 60 feet north of Farragut Road (Avenue F). The bolt is 5 inches east of the front of the building and 2 feet above the sidewalk. (Note 7, p. 90.)

Elevation, 26.888 feet.

B. M. 315.—*Avenue I.* The center of the top of a granite city monument on the northwest corner of Avenue I and New York Avenue (East Thirty-third Street). The top of the monument is 2 inches below the ground.

Elevation, 29.023 feet.

B. M. 316.—*Ditmas Avenue.* A square (□) cut in the top of the north end of the granite basin head on the southwest corner of Ditmas Avenue and Flatbush Avenue.

Elevation, 29.902 feet.

B. M. 317.—*Snyder Avenue.* A square (□) cut in the top of the southwest corner of the lowest brownstone step at the main entrance to the Reformed Dutch Church, on the northeast corner of Snyder Avenue and Flatbush Avenue.

Elevation, 46.896 feet.

B. M. 318.—*Flatbush Avenue.* The center of the top of a granite city monument in the center of the sidewalk on the east side of Flatbush Avenue, about 305 feet south of Church Avenue. The monument is opposite the cornerstone of Erasmus Hall High School, and the top is 6 inches below the sidewalk.

Elevation, 45.900 feet.

B. M. 319.—*Flatbush Avenue.* A cross (+) cut in the small irregular space on top of the granite base course on the south side of the arched or main entrance to Erasmus Hall High School, on the east side of Flatbush Avenue between Snyder

Avenue and Church Avenue. The cross is about 300 feet south of Church Avenue, is 3.2 feet north of the north end of the inscribed cornerstone and 2.0 feet above the sidewalk. (Note 17, p. 91.)

Elevation, 48.600 feet.

B. M. 320.—*Flatbush Avenue.* The center of the top of a granite city monument on the southwest corner of Flatbush Avenue and Church Avenue. The top of the monument is 6 inches below the sidewalk.

Elevation, 48.174 feet.

B. M. 321.—*Flatbush Avenue.* A square (□) cut 8 inches from the north end of the lowest granite step at the main entrance to the Flatbush Trust Co., on the southeast corner of Flatbush Avenue and Linden Avenue.

Elevation, 53.064 feet.

B. M. 322.—*Flatbush Avenue.* A square (□) cut near the southeast corner of the lowest step of stoop at the house entrance to the Revere Apartments at No. 730 Flatbush Avenue, between Parkside Avenue and Woodruff Avenue.

Elevation, 58.916 feet.

B. M. 323.—*Utica Avenue.* A 7-16-inch copper bolt in the top of the bridge seat of the west abutment of Manhattan Beach Railroad bridge No. 104 at Utica Avenue, just south of Farragut Road (Avenue F). The bolt is 6 inches from the south edge of the south girder and 6 inches from the east face of the abutment. (Note 8, p. 90.)

Elevation, 28.470 feet.

B. M. 324.—*Utica Avenue.* A square (□) cut in the top of the southeast corner of the bridge seat of the west abutment of Manhattan Beach Railroad bridge No. 104 at Utica Avenue, just south of Farragut Road (Avenue F).

Elevation, 28.448 feet.

B. M. 325.—*East Fifty-third Street.* A square (□) cut in the top of the southeast corner of the bridge seat of the west abutment of Manhattan Beach Railroad bridge No. 102, at East Fifty-third Street (Kouwenhoven Place), just north of Farragut Road (Avenue F).

Elevation, 24.966 feet.

B. M. 326.—*Rugby.* The center of a brass bolt marking the U. S. C. & G. S. triangulation station Rugby, on the west side of East Eighty-fifth Street, between Avenue C and Avenue D. The bench mark is 4 feet below the present ground surface. (Note 25, p. 91.)

Elevation, 19.189 feet.

B. M. 327.—*Canarsie.* A square (□) cut in the top of the northwest corner of the bridge seat of the east abutment of Manhattan Beach Railroad bridge No. 88, over Rockaway Avenue, between Avenue D and Stanley Avenue.

Elevation, 27.657 feet.

B. M. 328.—*Canarsie.* A square (□) cut in the top of the southeast corner of the west abutment of Manhattan Beach Railroad bridge No. 88, over Rockaway Avenue, between Avenue D and Stanley Avenue. The bench mark is 2.5 feet above the sidewalk. Established by the Department of Docks and Ferries.

Elevation, 16.690 feet.

B. M. 329.—*Canarsie.* A ⅞-inch copper bolt in the center of a surface 2′ x 2′9″ on top of the north end of the east abutment of Manhattan Beach Railroad bridge No. 88, over Rockaway Avenue, between Avenue D and Stanley Avenue. The bolt is 2.7 feet above the sidewalk. (Note 5, p. 89.)

Elevation, 17.491 feet.

B. M. 330.—*Canarsie.* A square (□) cut in the northwest corner of the toe of the east abutment of Manhattan Beach Railroad bridge No. 88, over Rockaway Avenue between Avenue D and Stanley Avenue. Established by the Brooklyn Highway Bureau.

Elevation, 8.589 feet.

B. M. 331.—*Canarsie.* A ⅝-inch copper bolt in the top of the granite foundation in front of Engine House No. 157, on the southeast corner of Rockaway Parkway and Farragut Road (Avenue F). The bolt is 3½ inches from the face of the stone, is about 55 feet south of Farragut Road and 1.3 feet above the sidewalk. (Note 5, p. 89.)

Elevation, 22.512 feet.

B. M. 332.—*Canarsie.* The center of a brass bolt marking the U. S. C. & G. S. triangulation station Canarsie, in the northeasterly side of Canarsie Park. The bench mark is about 20 feet from the iron fence and about 203 feet from the easterly corner of the park. (Note 25, p. 91.)

Elevation, 13.573 feet.

B. M. 333.—*New Lots.* A square (□) cut in the top of the northwest corner of the bridge seat of the south abutment of Manhattan Beach Railroad bridge No. 82, over New Lots Road near Van Sinderen Avenue. The bench mark is about 105 feet east of Junius Street and was established by the Long Island Railroad.

Elevation, 36.160 feet.

B. M. 334.—*New Lots.* A square (□) cut in the top of the southwest corner of a small I-beam at the base of the elevated railroad column on the west side of the Canarsie branch of the Brooklyn Rapid Transit and on the south side of New Lots Road at Van Sinderen Avenue. The bench mark is in front of the west window of the one-story brick ticket office and 1.4 feet above the sidewalk. Established by the Brooklyn Highway Bureau.

Elevation, 25.625 feet.

B. M. 335.—*New Lots.* The center of the top of a 2-inch brass tablet bolt in the top of the southwest corner of the concrete retaining wall on the northeast corner of New Lots Road and Junius Street. The bolt is 3 feet from the extreme southwest corner of the wall. (Note 2, p. 89.)

Elevation, 37.216 feet.

B. M. 336.—*New Lots.* A square (□) cut in the top of the bluestone coping on top of the one-story brick ticket office of the Canarsie Railroad on the south side of New Lots Road at Van Sinderen Avenue. The bench mark is above the west window.

Elevation, 34.701 feet.

B. M. 337.—*New Lots.* A square (□) cut in the top of the northeast corner of the west girder of Manhattan Beach Railroad bridge No. 82, over New Lots Road between Junius Street and Van Sinderen Avenue. The bench mark is 3.3 feet above the bridge seat of the north abutment and about 150 feet east of Junius Street.

Elevation, 39.470 feet.

B. M. 338.—*New Lots.* A square (⊡) cut in the top of the northwest corner of the south abutment of concrete retaining wall on the southeast corner of New Lots Road and Junius Street. The bench mark is about 3.0 feet above the sidewalk. Established by the Department of Docks and Ferries.

Elevation, 25.702 feet.

B. M. 339.—*New Lots.* A square (⊡) cut in the top of the northeast corner of a small I-beam at the base of the elevated railroad column on the east side of the Canarsie Railroad and on the south side of New Lots Road at Van Sinderen Avenue. The bench mark is in front of the east window of the one-story brick ticket office and 1.7 feet above the sidewalk. Established by the Department of Docks and Ferries.

Elevation, 25.656 feet.

B. M. 340.—*New Lots.* A square (□) cut in the top of the concrete retaining wall on the east side of the Manhattan Beach branch of the Long Island Railroad, about 550 feet north of New Lots Road. Established by the Engineers of the Long Island Railroad, their elevation for this bench mark being 34.123 feet.

Elevation, 35.495 feet.

B. M. 341.—*East New York.* A square (⊓) cut in the concrete at the foot of the iron railing at the west end of the retaining wall on the north side of the Long Island Railroad tracks, at Atlantic Avenue and Williams Place. The bench mark is about 1.5 feet east of the prolongation of the east curb line of Williams Place.

Elevation, 67.913 feet.

B. M. 342.—*East New York.* A 2-inch square (⊡) cut in the top of the southeast corner of the large granite capstone at the east end of the retaining wall on the north side of the Long Island Railroad tracks at Atlantic Avenue and Vesta Avenue. The bench mark is 12 feet north of the center line of the railroad on Atlantic Avenue, and almost on the prolongation of the west line of Vesta Avenue.

Elevation, 69.048 feet.

B. M. 343.—*Thirty-sixth Street.* A square (⊡) cut in the top of the westerly end of the bluestone coping under the iron railing on the Thirty-sixth Street side of Public School No. 82, on the westerly corner of Thirty-sixth Street and Fourth Avenue. The bench mark is about 100 feet northwesterly from Fourth Avenue and 6 inches above the sidewalk.

Elevation, 24.373 feet.

B. M. 344.—*Fourth Avenue.* A horizontal line in the end of a ¾-inch copper bolt leaded horizontally into the granite base course on the Fourth Avenue side of Public School No. 124, between Thirteenth Street and Fourteenth Street. The bolt is 27 feet from Fourteenth Street and 6 inches above the ground. (Note 6, p. 90.)

Elevation, 51.511 feet.

B. M. 345.—*Fourth Avenue.* A brass bracket in the granite base course on the Fourth Avenue side of Public School No. 124, between Thirteenth Street and Fourteenth Street. The bracket is 3 feet south of Thirteenth Street. (Note 22, p. 91.)

Elevation, 51.550 feet.

B. M. 346.—*Fourteenth Street.* A square (⊓) cut in the top of the northwest corner of the third or largest step at the girls' entrance to the Holy Family School, on the northerly side of Fourteenth Street, about 120 feet easterly from Fourth Avenue.

Elevation, 61.554 feet.

B. M. 347.—*Prospect Park.* The top of a ¾-inch iron rod inside of a wrought-iron pipe casing in the westerly side of Prospect Park. The bench mark is 8.3 feet east of the iron fence on the east side of Coney Island Avenue, and about 10 feet south of the prolongation of the southerly line of Greenwood Avenue. (Note 21, p. 91.)

Elevation, 62.128 feet.

B. M. 348.—*Eleventh Avenue.* The top of a galvanized iron bolt in the granite base course on the westerly side of Public School No. 154, on the easterly side of Eleventh Avenue between Sherman Street and Windsor Place. The bench mark is on the south side of the north buttress, is about 35 feet from Windsor Place and 9 inches above the ground. (Note 13, p. 90.)

Elevation, 129.198 feet.

B. M. 349.—*Eleventh Avenue.* A horizontal line on the face of a brass tablet bolt 2 inches in diameter, cemented horizontally into the limestone masonry on the westerly side of Public School No. 154, on the easterly side of Eleventh Avenue between Sherman Street and Windsor Place. The bench mark is on the south side of the north buttress, is about 35 feet from Windsor Place and 5.2 feet above the ground. (Note 1, p. 89.)

Elevation, 133.587 feet.

B. M. 350.—*Prospect Park.* A ⅞-inch copper bolt in the top of the northeast corner of the second step at the base of the southerly granite column at the most westerly entrance to Prospect Park, at Fifteenth Street and Prospect Park West. (Note 5, p. 89.)

Elevation, 154.199 feet.

B. M. 351.—*Prospect Park.* The highest point within a square (□) lettered B. W. S., N. Y., cut in the top of the center of a square capstone in the wall on the westerly side of Prospect Park, about 32 feet southerly from the southerly line of Fourteenth Street. The bench mark is 3.3 feet above the sidewalk.

Elevation, 155.342 feet.

B. M. 352.—*Sixth Avenue.* A square (□) lettered B. W. S., cut in the southwest corner of the bluestone door-sill at the Sixth Avenue entrance to the grocery store at No. 571 Sixth Avenue. The bench mark is 6.3 feet from the corner of Sixteenth Street.

Elevation, 118.499 feet.

B. M. 353.—*Fifth Avenue.* A ¼-inch copper bolt in the top of the granite water-table at the southwest corner of the 144th Precinct Police Station, on the northeast corner of Sixteenth Street and Fifth Avenue. The bolt is 3.2 feet above the sidewalk. (Note 7, p. 90.)

Elevation, 92.879 feet.

P. M. 354.—*Third Avenue.* A square (□) cut in the top of the southwesterly corner of the concrete base of elevated railroad column on the northwesterly corner of Third Avenue and Forty-third Street.

Elevation, 30.760 feet.

B. M. 355.—*Second Avenue.* A square (⊓) cut in the top of the northerly corner of the first basement window-sill from Forty-fifth Street, on the westerly side of Bixby's four-story brick factory building, on the southeasterly corner of Second Avenue and Forty-fifth Street. Established by the Department of Docks and Ferries.

Elevation, 20.476 feet.

B. M. 356.—*Seventh Avenue.* A horizontal line on the face of a 2-inch brass tablet bolt cemented horizontally into the north side of the south buttress on the west side of the Manual Training High School, on the east side of Seventh Avenue from Fourth Street to Fifth Street. The bench mark is 3.8 feet from the west side of the buttress, is 2.8 feet above the sidewalk and about 54 feet from Fifth Street. (Note 1, p. 89.)

Elevation, 105.885 feet.

B. M. 357.—*Seventh Avenue.* The center of a square (⊓) cut in the top of the limestone base of the south buttress on the west side of the Manual Training High School, on the east side of Seventh Avenue from Fourth Street to Fifth Street. The bench mark is on the north side of the buttress, is 2.9 feet above the sidewalk and about 54 feet from Fifth Street.

Elevation, 106.615 feet.

B. M. 358.—*Seventh Avenue.* A ¾-inch copper bolt in the northwest corner of the lowest bluestone step at the main entrance to the Manual Training High School, on the east side of Seventh Avenue from Fourth Street to Fifth Street. (Note 5, p. 89.)

Elevation, 103.335 feet.

B. M. 359.—*Prospect Park West.* The highest point within a 2-inch square (□) cut in the center of the top of a square capstone in the park wall, on the east side of Prospect Park West, about 5 feet north of the center line of Fifth Street.

Elevation, 162.156 feet.

B. M. 360.—*Fourth Avenue.* The center of a square (□) cut in the center of the top of the sill of the southerly window on the west side of Engine House No. 139, on the east side of Fourth Avenue between Sixth Street and Seventh Street.

Elevation, 29.922 feet.

B. M. 361.—*Butler Street.* A brass bracket in the bluestone base course in front of Public School No. 133, on the northerly side of Butler Street between Fourth Avenue and Fifth Avenue. The bracket is on the west side of the 1-foot buttress 14 feet west of the main entrance. (Note 22, p. 91.)

Elevation, 33.869 feet.

B. M. 362.—*Butler Street.* A 2-inch square (ᒷ) cut in the top of the northwest corner of the limestone coping under the low iron railing on the west side of the main entrance to Public School No. 133, on the northerly side of Butler Street between Fourth Avenue and Fifth Avenue. The bench mark is 3.6 feet above the sidewalk.

Elevation, 36.476 feet.

B. M. 363.—*Third Street.* A square (□) cut in the top of the south end of the granite copingstone on the top of the south end of the concrete bulkhead on the west side of Gowanus Canal at Third Street.

Elevation, 8.071 feet.

B. M. 364.—*Smith Street.* A square (ᒷ) cut in the top of the southeast corner of the granite basin head on the southwest corner of Smith Street and Third Place.

Elevation, 43.892 feet.

B. M. 365.—*Henry Street.* The bottom of a square cut in the northeast corner of the limestone base block on the north side of the main entrance to Public School No. 142, on the west side of Henry Street at Third Place. The bench mark is 3 feet above the sidewalk. (Note 10, p. 90.)

Elevation, 39.338 feet.

B. M. 366.—*Henry Street.* A cross (+) cut in the granite base course on the north side of the main entrance to Public School No. 142, on the west side of Henry Street at Third Place. The cross is 6 inches above the sidewalk and 1 foot north of the base block on the north side of the entrance. (Note 15, p. 91.)

Elevation, 36.872 feet.

B. M. 367.—*Henry Street.* A horizontal line on the face of a 2-inch brass tablet bolt cemented horizontally into the limestone masonry of Public School No. 142, on the west side of Henry Street at Third Place. The bench mark is on the Henry Street side, is 1 foot north of the base block on the north side of the main entrance and 1.8 feet above the sidewalk. (Note 1, p. 89.)

Elevation, 38.049 feet.

B. M. 368.—*Hamilton Avenue.* A 2-inch square (□) cut in the top of the granite base of the ventilator of comfort station at Hamilton Avenue and Richards Street. The bench mark is on the south side of the ventilator and the letter B is cut alongside.

Elevation, 15.025 feet.

B. M. 369.—*Rapelyea Street.* A ⅝-inch copper bolt in the top of the northeast corner of the bluestone sill of basement window on the west side of the main entrance to the 145th Precinct Police Station, on the southeast corner of Richards Street and Rapelyea Street. The bolt is 8.5 feet from Richards Street and 9 inches above the ground. (Note 5, p. 89.)

Elevation, 15.318 feet.

B. M. 370.—*India Wharf.* The top of a ½-inch spike in the top of the backing log at the foot of India Wharf. The spike is about 35 feet from the southwesterly end of the pier.

Elevation, 7.604 feet.

B. M. 371.—*Berriman Street.* A square (⌙) cut in the top of the southwest corner of the bluestone coping under the iron railing on the west side of Public School No. 64, on the northeast corner of Berriman Street and Belmont Avenue. The bench mark is about 53 feet north of Belmont Avenue and 3 inches above the sidewalk.

Elevation, 26.606 feet.

B. M. 372.—*Lincoln Road.* A square (⌙) cut in the top of the northerly corner of the granite basin head on the southeast corner of Lincoln Road and Washington Avenue.

Elevation, 71.051 feet.

B. M. 373.—*Lincoln Road.* The top of the northeast corner of the granite city monument on the southwest corner of Lincoln Road and Nostrand Avenue. The top of the monument is 3 inches below the sidewalk.

Elevation, 55.005 feet.

B. M. 374.—*Albany Avenue.* A horizontal line on the face of a brass tablet bolt 2 inches in diameter cemented horizontally into the limestone on the north side of Public School No. 91, on the southwest corner of Albany Avenue and East New York Avenue. The bench mark is 3.5 feet from the east side of the building and 6 inches above the ground. (Note 1, p. 89.)

Elevation, 49.178 feet.

B. M. 375.—*Albany Avenue.* A cross (+) cut in the granite base course on the north side of Public School No. 91, on the southwest corner of Albany Avenue and East New York Avenue. The cross is 3.5 feet from the east side of the building and 1 inch above the ground. (Note 15, p. 91.)

Elevation, 48.791 feet.

B. M. 376.—*Albany Avenue.* A square (□) cut in the top of the northwest corner of the fourth or top step at the Albany Avenue entrance to Public School No. 91, on the southwest corner of Albany Avenue and East New York Avenue.

Elevation, 48.563 feet.

B. M. 377.—*Schenectady Avenue.* The center of the top of a granite city monument on the southwest corner of Schenectady Avenue and East New York Avenue. The top of the monument is 3 inches above the ground.

Elevation, 39.332 feet.

B. M. 378.—*Buffalo Avenue.* The top of a ¾-inch iron rod inside of a wrought-iron pipe casing in Lincoln Terrace Park. The bench mark is on the west side of Buffalo Avenue about 50 feet south of Eastern Parkway, and 4 feet west of the iron railing. (Note 21, p. 91.)

Elevation, 93.814 feet.

B. M. 379.—*Sutter Avenue.* A horizontal line in the end of a ¾-inch copper bolt set horizontally into the granite foundation on the north side of Public School No. 156, on the southeast corner of Sutter Avenue and Grafton Street. The bolt is 6 inches from the west side of the building and 6 inches above the sidewalk. (Note 6, p. 90.)

Elevation, 36.890 feet.

B. M. 380.—*East New York Avenue.* A ¾-inch copper bolt in the granite sill of the east window of the 165th Precinct Police Station, on the south side of East New York Avenue, between Thatford Avenue and Rockaway Avenue. The bolt is 7.8 feet from the east side of the building and 3.5 feet above the sidewalk. (Note 5, p. 89.)

Elevation, 57.531 feet.

B. M. 381.—*The Plaza.* A horizontal line on the face of a brass tablet bolt 2 inches in diameter on the west side of the Memorial Arch or Soldiers and Sailors' monument at the northerly entrance to Prospect Park. The bench mark is on the south side of the buttress, is 11 feet south of the door and 3.5 feet above ground. (Note 1, p. 89.)

Elevation, 146.292 feet.

B. M. 382.—*The Plaza.* A cross (+) cut in the top of the south end of the granite sill of the east door on the east side of the Memorial Arch or Soldiers and Sailors' monument, at the northerly entrance to Prospect Park. The cross is 2 inches from the south end of the door-sill. (Note 17, p. 91.)

Elevation, 143.151 feet.

B. M. 383.—*Flatbush Avenue.* A square (⊡) cut in the top of the southeast corner of the lowest step of stoop at the house entrance to No. 534 Flatbush Avenue, just north of Lincoln Road.

Elevation, 72.442 feet.

B. M. 384.—*Prospect Park.* A 1-inch cross (+) cut in the top of the northeast corner of the second or top step at the entrance to the southerly police booth at the Willink entrance to Prospect Park, at Flatbush Avenue and Malbone Street. The cross is lettered B. E. B. M. (Note 17, p. 91.)

Elevation, 89.691 feet.

B. M. 385.—*Prospect Park.* A ⅞-inch copper bolt in the top of the north end of a granite slab near the south side of the driveway at the Willink entrance to Prospect Park, at Flatbush Avenue and Malbone Street. The bolt is 4.4 feet south of the granite column on the south side of the driveway and 1 foot east of the park wall. (Note 5, p. 89.)

Elevation, 89.446 feet.

B. M. 386.—*Prospect Park.* A square (⌗) cut on the northeast corner of the granite slab on the north side of the Willink entrance to Prospect Park, at Flatbush Avenue and Malbone Street. The bench mark is 3.7 feet south of the north police booth and 3 inches above the sidewalk.

Elevation, 90.004 feet.

B. M. 387.—*Prospect Park.* A ⅞-inch copper bolt near the northeasterly edge of the granite base of the small stone shelter, in the northeasterly corner of Prospect Park at The Plaza. The bolt is about 20 feet east of Flatbush Avenue. (Note 5, p. 89.)

Elevation, 147.964 feet.

B. M. 388.—*Flatbush Avenue.* The center of the top of a granite city monument on the southeast corner of Flatbush Avenue and Eastern Parkway. The monument is on Flatbush Avenue opposite the north end of the east wall of Prospect Park.

Elevation, 144.263 feet.

B. M. 389.—*Flatbush Avenue.* A square (□) cut in the east curb of Flatbush Avenue, opposite the end of the wall at the north end of Prospect Park.

Elevation, 144.200 feet.

B. M. 390.—*Vanderbilt Avenue.* A 1½-inch square (⌗) cut in the top of the northwest corner of the bluestone coping at the corner of the high iron railing around Public School No. 9, on the northeast corner of Vanderbilt Avenue and Sterling Place. The bench mark is on Vanderbilt Avenue and about 1 foot north of the south side of the school.

Elevation, 121.519 feet.

B. M. 391.—*Eastern Parkway.* The top of a ¾-inch iron rod inside of a wrought-iron pipe casing, on the south side of Eastern Parkway about 30 feet west of the center line of Underhill Avenue. The bench mark is about 19 feet south of the curb. (Note 21, p. 91.)

Elevation, 151.368 feet.

B. M. 392.—*Eastern Parkway.* A square (⌗) cut in the top of the northeast corner of the lowest granite step at the entrance to Prospect Water Tower, on the south side of Eastern Parkway opposite Underhill Avenue.

Elevation, 153.082 feet.

B. M. 393.—*Eastern Parkway.* A ⅞-inch copper bolt in the top of the west end of the third step at the main entrance to the Museum of The Brooklyn Institute of Arts and Sciences, on the south side of Eastern Parkway, just south of Washington Avenue. (Note 5, p. 89.)

Elevation, 163.438 feet.

B. M. 394.—*Eastern Parkway.* A ¾-inch copper bolt in the southwest corner of the granite base block of the Slocum Monument, at the intersection of Eastern Parkway and Bedford Avenue. (Note 5, p. 89.)

Elevation, 147.986 feet.

B. M. 395.—*Prospect Place.* A quadrant (⊾) cut in the top of the corner of the 4-inch wide water-table at the northwest corner of the piazza of the Mount Prospect Pumping Station, on the southeast corner of Prospect Place and Underhill Avenue. The bench mark is about 50 feet east of the east curb of Underhill Avenue and 2.7 feet above the walk. Established by the Division of Water Supply of the City of Brooklyn.

Elevation, 123.432 feet.

B. M. 396.—*Grant Square.* A 2-inch square (⊓) cut in the top of the center of the west edge of the wide granite base of the Grant monument in Grant Square, at the intersection of Bedford Avenue and Rogers Avenue, just south of Dean Street. The bench mark is 2.1 feet above the street.

Elevation, 89.368 feet.

B. M. 397.—*Boerum Place.* Coast and Geodetic Survey B. M. C, on the east side of the Kings County Hall of Records, on the southwest corner of Boerum Place and Fulton Street. The bench mark is the top of the inclined water-table, 5.67 feet south of the extreme northeast corner. The stone above the point is lettered C, U. S. C. S., and an arrow on the water-table points to the bench mark. Established in May, 1887.

Elevation, 55.969 feet.

B. M. 398.—*Boerum Place.* The bottom of a square cut in the top of the extreme northeast corner of the granite water-table on the Kings County Hall of Records, on the southwest corner of Boerum Place and Fulton Street. The corner of the inclined water-table was cut away, leaving a 2-inch level square space which is the bench mark. It is 1.25 feet above the sidewalk and lettered B. M. (Note 10, p. 90.)

Elevation, 55.913 feet.

B. M. 399.—*Boerum Place.* The bottom of a square cut in the top of the extreme southeast corner of the granite water-table on the Kings County Hall of Records, on the northwest corner of Boerum Place and Livingston Street. The bench mark is 2.65 feet above the sidewalk. (Note 10, p. 90.)

Elevation, 55.884 feet.

B. M. 400.—*Boerum Place.* A small square (⊓) cut in the edge of the granite sill of the most southerly basement window on the Boerum Place side of the Kings County Hall of Records, between Livingston Street and Fulton Street. The bench mark is 10.5 feet north of the Livingston Street corner of the building.

Elevation, 54.058 feet.

B. M. 401.—*Livingston Street.* A quadrant (⊾) cut in the top of the west end of the granite basin head on the northwest corner of Livingston Street and Boerum Place.

Elevation, 51.570 feet.

B. M. 402.—*Nevins Street.* A square (⊔) cut in the top of the west end of the granite basin head on the northwest corner of Nevins Street and State Street.

Elevation, 40.987 feet.

B. M. 403.—*Bond Street.* A square (⊔) cut in the top of the east end of the granite basin head on the northeast corner of Bond Street and State Street.

Elevation, 43.488 feet.

B. M. 404.—*Bond Street.* A square (⊔) cut in the top of the west end of the granite basin head on the northwest corner of Bond Street and State Street.

Elevation, 43.512 feet.

B. M. 405.—*Pacific Street.* The top of a galvanized iron bolt in the granite base course on the north side of Public School No. 47, on the south side of Pacific Street between Third Avenue and Nevins Street. The bench mark is 18 feet east of the main entrance and 9 inches above the area. (Note 13, p. 90.)

Elevation, 39.772 feet.

B. M. 406.—*Pacific Street.* A square (⊓) cut in the top of the limestone masonry on the north side of Public School No. 47, on the south side of Pacific Street, between Third Avenue and Nevins Street. The bench mark is on the west edge of the 1-foot wide buttress, is 18 feet east of the main entrance and about 2.5 feet above the area.

Elevation, 41.469 feet.

B. M. 407.—*Pacific Street.* A square (⊔) cut in the top of the northwest corner of the bluestone coping at the west corner of the iron railing in front of Public School No. 47, on the south side of Pacific Street between Third Avenue and Nevins Street. The bench mark is 1 foot above the sidewalk and on line with the east line of No. 466 Pacific Street.

Elevation, 38.568 feet.

B. M. 408.—*Dean Street.* The bottom of a square cut in the top of the 4-inch wide circular water-table of the four-story brick factory building on the southeast corner of Dean Street and Third Avenue. The bench mark is 1.25 feet above the sidewalk. (Note 10, p. 90.)

Elevation, 32.400 feet.

B. M. 409.—*Flatbush Avenue.* A ¼-inch copper bolt in the top of the northwest corner of the granite foundation at the southwest corner of the Atlantic Avenue Depot of the Long Island Railroad, on Flatbush Avenue. The bolt is about 50 feet north of Atlantic Avenue and 3.7 feet above the sidewalk. The letters B. M. are cut in the stone below the bolt. (Note 7, p. 90.)

Elevation, 52.332 feet.

B. M. 410.—*Atlantic Avenue.* The top of an iron bolt in the top of the concrete slab on the southwest corner of Atlantic Avenue and Carlton Avenue. The bolt is 1 foot from the north end of the slab and 1.5 feet above the sidewalk.

Elevation, 66.389 feet.

B. M. 411.—*Atlantic Avenue.* A quadrant (∟) cut in the top of the south corner of the granite basin head on the southwest corner of Atlantic Avenue and Underhill Avenue.

Elevation, 80.185 feet.

B. M. 412.—*Atlantic Avenue.* A square (⊔) cut in the top of the south end of the granite basin head on the southwest corner of Atlantic Avenue and Classon Avenue.

Elevation, 83.677 feet.

B. M. 413.—*Atlantic Avenue.* A square (⊔) cut in the top of the east end of the granite basin head on the southeast corner of Bedford Avenue and Atlantic Avenue.

Elevation, 78.074 feet.

B. M. 414.—*Atlantic Avenue.* A quadrant (∟) cut in the top of the north end of the granite basin head on the northwest corner of Atlantic Avenue and New York Avenue.

Elevation, 63.861 feet.

B. M. 415.—*New York Avenue.* A horizontal line in the end of a 5/8-inch copper bolt set horizontally into the center of the south face of a granite block in the first buttress south of the main entrance to Public School No. 93, on the southeast corner of New York Avenue and Herkimer Street. The bolt is 11 feet south of the main entrance and 1.1 feet above the ground. (Note 18, p. 91.)

Elevation, 61.976 feet.

B. M. 416.—*New York Avenue.* The top of a galvanized iron bolt in the granite foundation on the west side of Public School No. 93, on the southeast corner of New York Avenue and Herkimer Street. The bolt is 2 feet south of the stairway at the main entrance and 1 foot above the ground. (Note 13, p. 90.)

Elevation, 62.107 feet.

B. M. 417.—*Atlantic Avenue.* A 5/8-inch copper bolt in the top of the east end of the granite basin head on the northeast corner of Atlantic Avenue and Brooklyn Avenue.

Elevation, 60.117 feet.

B. M. 418.—*Atlantic Avenue.* A 7-16-inch copper bolt in the top of the top course of ashlar masonry on the south side of the 152d Precinct Police Station, on the northwest corner of Atlantic Avenue and Schenectady Avenue. The bolt is 52 feet west of Schenectady Avenue.

Elevation, 66.039 feet.

B. M. 419.—*Atlantic Avenue.* A square (⊔) cut in the top of the north end of the granite basin head on the northeast corner of Atlantic Avenue and Suydam Place.

Elevation, 62.878 feet.

B. M. 420.—*Atlantic Avenue.* A quadrant (⊾) cut in the top of the southeast corner of the granite basin head on the northeast corner of Atlantic Avenue and Ralph Avenue.

Elevation, 75.907 feet.

B. M. 421.—*Herkimer Street.* A 5/8-inch copper bolt in the top of the granite coping under the iron railing at the east side of the main entrance to Public School No. 155, on the north side of Herkimer Street between Olive Place and Eastern Parkway. The bolt is 6 inches from the face of the school, and 6 inches from the east edge of the coping. (Note 5, p. 89.)

Elevation, 94.472 feet.

B. M. 422.—*Herkimer Street.* A cross (+) cut in the top of the granite foundation at the southeast corner of Public School No. 155, on the north side of Herkimer Street between Olive Place and Eastern Parkway. The cross is 5 feet above the sidewalk. (Note 17, p. 91.)

Elevation, 93.963 feet.

B. M. 423.—*Vermont Street.* A cross (+) cut in the top of the southeast corner of the fixed rim or cover flange on top of the fire hydrant on the east side of Vermont Street, about 100 feet north of New Lots Road. The cross is on the south side of the flange at the hinge.

Elevation, 26.154 feet.

B. M. 424.—*New Lots Road.* A brass bracket on the south side of Public School No. 72, on the north side of New Lots Road between Schenck Avenue and Barbey Street. The bracket is 13.5 feet from the east side of the building and 7 inches above the area. (Note 22, p. 91.)

Elevation, 21.699 feet.

B. M. 425.—*New Lots Road.* The center of the horizontal bar of a **T** cut in the copingstone on the south side of Public School No. 72, on the north side of New Lots Road between Schenck Avenue and Barbey Street. The bench mark is 13.5 feet from the east side of the building and 4.9 feet above the area.

Elevation, 26.038 feet.

B. M. 426.—*New Lots Road.* A square (□) cut in the top of the lowest step of stoop at the main entrance to Public School No. 72, on the north side of New Lots Road between Schenck Avenue and Barbey Street. The bench mark is 1½ inches from the south edge of the step and 1.25 feet from the west end. It is about 70 feet east of Schenck Avenue. Established by the Department of Docks and Ferries.

Elevation, 21.609 feet.

B. M. 427.—*New Lots Road.* A rude cross cut on top of the west end of the lowest step of stoop at the main entrance to Public School No. 72, on the north side of New Lots Road between Schenck Avenue and Barbey Street. The cross is 1 inch west of the westerly iron post. Established by the Department of Docks and Ferries.

Elevation, 21.630 feet.

B. M. 428.—*New Lots Road.* A square (ㅂ) cut in the top of the northwest corner of the granite basin head on the northeast corner of New Lots Road and Schenck Avenue.

Elevation, 22.052 feet.

B. M. 429.—*Ashford Street.* The top of a galvanized iron bolt in the granite foundation on the north side of the main entrance to Public School No. 158, on the west side of Ashford Street, between Belmont Avenue and Pitkin Avenue. The bolt is 1.5 feet above the area. (Note 13, p. 90.)

Elevation, 33.432 feet.

B. M. 430.—*Ashford Street.* The center of the horizontal bar of a **T** cut in the granite foundation on the north side of the main entrance to Public School No. 158, on the west side of Ashford Street, between Belmont Avenue and Pitkin Avenue. The bench mark is 2 feet above the area.

Elevation, 33.957 feet.

B. M. 431.—*Atlantic Avenue.* A square (□) cut in the top of the granite base block supporting two granite columns at the northwest corner of the East New York Savings Bank, on the southeast corner of Atlantic Avenue and Pennsylvania Avenue. The bench mark is between the two columns on the Atlantic Avenue side.

Elevation, 58.990 feet.

B. M. 432.—*Berriman Street.* A square (ㅂ) cut in the top of the northerly corner of the granite basin head on the northeast corner of Berriman Street and Pitkin Avenue.

Elevation, 29.503 feet.

B. M. 433.—*Berriman Street.* A ⅝-inch copper bolt in the top of the northeast corner of the limestone plinth of limestone column on the north side of the main entrance to Public School No. 64, on the east side of Berriman Street between Belmont Avenue and Pitkin Avenue. The bolt is about 135 feet north of Belmont Avenue. (Note 5, p. 89.)

Elevation, 28.388 feet.

B. M. 434.—*Belmont Avenue.* A brass bracket in the bluestone base course on the south side of Public School No. 64, on the north side of Belmont Avenue between Berriman Street and Atkins Avenue. The bench mark is about 3.5 feet from the southeast corner of the building and about 30 feet from Atkins Avenue. (Note 22, p. 91.)

Elevation, 24.515 feet.

B. M. 435.—*Crescent Street.* A cross, lettered B + M, cut in the granite base course on the west side of Public School No. 159, on the northeast corner of Crescent Street and Pitkin Avenue. The cross is 7 inches from the south side of the building and 4 inches above the ground.

Elevation, 27.900 feet.

B. M. 436.—*Crescent Street.* The center of a V-shaped mark, cut in the center of the top of the south edge of the sill of the most westerly window on the south side of Public School No. 159, on the northeast corner of Crescent Street and Pitkin Avenue. Established by the Brooklyn Sewer Bureau.

Elevation, 29.386 feet.

B. M. 437.—*Crescent Street.* The top of a galvanized iron bolt in the granite base course on the south side of Public School No. 159, on the northeast corner of Crescent Street and Pitkin Avenue. The bench mark is on the east side of a small buttress about 34 feet from the west side of the building and 10 inches above the ground. (Note 13, p. 90.)

Elevation, 27.962 feet.

B. M. 438.—*Crescent Street.* The center of the top of a granite city monument on the southeast corner of Crescent Street and Liberty Avenue. The top of the monument is 2 inches below the sidewalk.

Elevation, 38.023 feet.

B. M. 439.—*Sheridan Street.* The center of the top of a granite city monument on the southeast corner of Sheridan Street and Liberty Avenue. The top of the monument is 1 inch above the sidewalk.

Elevation, 31.593 feet.

B. M. 440.—*Grant Avenue.* A square (⊔) cut in the top of the northwest corner of the wide concrete step or platform in front of store, on the southeast corner of Grant Avenue and Liberty Avenue. The bench mark is on line with the east line of Grant Avenue.

Elevation, 34.316 feet.

B. M. 441.—*Grant Avenue.* A square (□) cut in the top of the northeast corner of the bluestone sill of the side door entrance to the store on the northwest corner of Grant Avenue and Liberty Avenue. The bench mark is about 19 feet north of Liberty Avenue. Established by the Queens Topographical Bureau.

Elevation, 33.835 feet.

B. M. 442.—*Court Street.* A square (⊔) cut in the top of the southeast corner of the granite pedestal of column on the south side of the main entrance to the old Dime Savings Bank, now the U. S. Title Insurance Co., on the southwest corner of Court Street and Remsen Street. The bench mark is about 38 feet south of Remsen Street and 2 feet above the sidewalk. Established by the Public Service Commission.

Elevation, 68.315 feet.

B. M. 443.—*Court Street.* A square (⊑) cut in the southeast corner of a monolithic flagstone on the southwest corner of Court Street and Remsen Street. The bench mark is 13.8 feet from the Court Street curb and 12.9 feet from the Remsen Street curb.

Elevation, 67.236 feet.

B. M. 444.—*Borough Hall.* A square (⊓) cut in the top of the south end of the sill of the south window on the west side of the Brooklyn Borough Hall. The bench mark is 7 feet from the south side of the building and 2.3 feet above the ground.

Elevation, 66.712 feet.

B. M. 445.—*Court Street.* A 2-inch square (□) cut in the top of the north end of the lowest step at the Court Street entrance to the Mechanics Bank, on the northwest corner of Court Street and Montague Street. The bench mark is 12 feet from the south side of the building and 6 inches above the sidewalk.

Elevation, 71.776 feet.

B. M. 446.—*Fulton Street.* A ½-inch copper bolt in the top of the northwest corner of the limestone column base on the east side of the entrance to No. 338 Fulton Street, just west of Court Street. The bolt is in the base of the third column from the west end of the building on Fulton Street, and about 2 feet above the sidewalk. (Note 8, p. 90.)

Elevation, 75.156 feet.

B. M. 447.—*Clinton Street.* A ⅝-inch copper bolt in the top of the northeast corner of the granite base of the Hamilton Monument in front of the Hamilton Club, on the southwest corner of Clinton Street and Remsen Street. (Note 8, p. 90.)

Elevation, 78.223 feet.

B. M. 448.—*Prospect Street.*. A 2½-inch square (□) cut in the top of a large base block of the Brooklyn Bridge, on the northeast corner of Prospect Street and Main Street. The bench mark is on the easterly line of Main Street about 40 feet north of the north line of Prospect Street, and 2.25 feet above the sidewalk. The letters B. M. are cut in the base of the pyramid-shaped block above the bench mark.

Elevation, 47.212 feet.

B. M. 449.—*Brooklyn Bridge.* A square (⊓) cut in the top of the anchorage on the Brooklyn side of the Brooklyn Bridge. The bench mark is on the westerly edge of the anchorage at the north roadway, between Front Street and Mercein Street, and near post No. 122. It is 8 feet west of the fender post. Established by the Department of Bridges.

Elevation, 90.837 feet.

B. M. 450.—*Brooklyn Bridge.* A square (⊑) cut in the top of the northeast corner of the granite base of elevated railroad column on the west side of the north roadway of the Brooklyn Bridge and about 27 feet north of Sand Street. Established by the Department of Bridges.

Elevation, 66.935 feet.

B. M. 451.—*Bridge Street.* A square (⊓) cut in the top of the southwest corner of the bluestone door-sill at the house entrance to tenement at No. 137 Bridge Street, about 55 feet north of Sand Street.

Elevation, 50.501 feet.

B. M. 452.—*Hudson Avenue.* Coast and Geodetic Survey B. M. F, about 200 feet east of the foot of Hudson Avenue. The bench mark is on the north side of the old stone warehouse, on the east face of the fourth buttress from the northeast corner, and consists of a rude cross (+) cut in the fourth course of stone above the foundation. Established in May, 1887.

Elevation, 10.189 feet.

B. M. 453.—*Hudson Avenue.* A square (⊔) cut in the top of the south end of the granite basin head on the east side of Hudson Avenue, about 105 feet north of the north line of Marshall Street.

Elevation, 8.516 feet.

B. M. 454.—*Hudson Avenue.* The bottom of a square cut in the top of the corner of the brownstone water-table at the corner of the four-story brick tenement on the southeast corner of Hudson Avenue and John Street. (Note 10, p. 90.)

Elevation, 21.553 feet.

B. M. 455.—*Brooklyn Navy Yard.* A 2-inch square (□) cut in the top of the west end of the second granite step at the Sand Street entrance to the one-story stone and brick guard house in the Brooklyn Navy Yard, on the north side of the Sand Street entrance. The bench mark is about 16 feet east of the gate.

Elevation, 16.864 feet.

B. M. 456.—*Brooklyn Navy Yard.* A square (□) cut in the southwest corner of Dry Dock No. 1, at the foot of Third Street, in the Brooklyn Navy Yard. The bench mark is on the massive copingstone and 5 inches from the inner corner of the dry dock. It is about 13 feet west of the center line of Third Street.

Elevation, 6.677 feet.

B. M. 457.—*City Park.* A ¾-inch copper bolt in a concrete city monument in the westerly side of City Park. The bench mark is 10 feet east of the iron railing on the east side of Navy Street, and 65.5 feet south of the iron railing on the south side of Flushing Avenue. (Note 23, p. 91.)

Elevation, 13.253 feet.

B. M. 458.—*City Park.* A square (□) cut in the top of the westerly end of the large granite gate-sill at the entrance to City Park, on the southeast corner of Flushing Avenue and Navy Street.

Elevation, 12.739 feet.

B. M. 459.—*Cumberland Street.* A cross (+) cut in the top of the brownstone coping at the corner of the iron railing on the northwest corner of Cumberland Street and Lafayette Avenue. The cross is about 1 inch above the sidewalk.

Elevation, 73.618 feet.

B. M. 460.—*Cumberland Street.* A square (□) cut in the top of the southwest corner of the iron base of elevated railroad column on the southwest corner of Cumberland Street and Myrtle Avenue. The square is 3 inches above the sidewalk.

Elevation, 56.652 feet.

B. M. 461.—*Grand Avenue.* A square (⊔) cut in the top of the southwest corner of the bluestone coping on the northeast corner of Grand Avenue and Willoughby Avenue.

Elevation, 55.804 feet.

B. M. 462.—*Taaffe Place.* A square (⊡) cut in the southwest corner of a section of the brownstone coping on the northeast corner of Taaffe Place and Willoughby Avenue. The point is on the copingstone on the north side of the curved section.

Elevation, 58.838 feet.

B. M. 463.—*Rodney Street.* The center of the top of a stone hitching post 1.6 feet high, on the northwest corner of Rodney Street and Bedford Avenue. The post is 1.5 feet west of the Bedford Avenue curb.

Elevation, 48.297 feet.

B. M. 464.—*Lawrence Street.* A square (□) cut near the north end of the granite basin head on the northeast corner of Lawrence Street and Willoughby Street.

Elevation, 44.680 feet.

B. M. 465.—*St. Edwards Street.* A ¼-inch copper bolt in the top of the southeast corner of the monolithic granite door-sill at the main entrance to the High Pressure Station on the northwest corner of St. Edwards Street and Willoughby Street. The bolt is about 70 feet north of Willoughby Street. (Note 7, p. 90.)

Elevation, 46.974 feet.

B. M. 466.—*Flushing Avenue.* A square (⊡) cut in the top of the south end of the iron basin head on the southwest corner of Flushing Avenue and Cumberland Street.

Elevation, 13.290 feet.

B. M. 467.—*Taaffe Place.* A square (□) cut in the top of the 6-inch wide coping of the granite foundation at the southwest corner of the Department of Street Cleaning building, on the northeast corner of Taaffe Place and Little Nassau Street. The bench mark is 3.7 feet above the sidewalk and 6½ inches from the corner of the stone.

Elevation, 24.359 feet.

B. M. 468.—*Wythe Avenue.* A quadrant (⊾) cut in the top of the southwest corner of the granite basin head on the northwest corner of Wythe Avenue and South Eighth Street.

Elevation, 29.992 feet.

B. M. 469.—*South Sixth Street.* A ¾-inch copper bolt leaded horizontally into a granite block forming part of the anchorage on the south side of the Williamsburg Bridge. The bolt faces South Sixth Street and is just east of the westerly line of Durham Place. It is 21 feet west of the southeast corner of the anchorage and 1.1 feet above the ground. (Note 6, p. 90.)

Elevation, 30.123 feet.

B. M. 470.—*South Sixth Street.* A square (⊡) cut in the top of the southeast corner of the 1-foot high granite coping on the northwest corner of South Sixth Street and Wythe Avenue. The bench mark is at the east end of the iron railing and about 16 feet from the west curb of Wythe Avenue.

Elevation, 36.003 feet.

B. M. 471.—*South Sixth Street.* A cross (+) cut in the south face of a granite block forming part of the anchorage on the south side of the Williamsburg Bridge. The cross is just east of the westerly line of Durham Place, is 21 feet west of the southeast corner of the anchorage and 1.9 feet above the ground. (Note 15, p. 91.)

Elevation, 30.821 feet.

B. M. 472.—*Fort Greene Park.* The top of a ¾-inch iron rod inside of a wrought-iron pipe casing in the southeasterly corner of Fort Greene Park. The bench mark is 5 feet west of the center of the park wall on Cumberland Street and about 46 feet from the line of the park wall on De Kalb Avenue. (Note 21, p. 91.)

Elevation, 70.382 feet.

B. M. 473.—*Eastern Parkway.* A 2-inch square (□) cut in the top of the south end of the bluestone curb on the easterly side of Eastern Parkway, just north of the Manhattan Beach Division of the Long Island Railroad. The bench mark is between Bushwick Avenue and Broadway.

Elevation, 104.667 feet.

B. M. 474.—*Bushwick Avenue.* A square (⊡) cut in the top of the northeast corner of the granite basin head on the southwest corner of Bushwick Avenue and Granite Street.

Elevation, 90.360 feet.

B. M. 475.—*Bushwick Avenue.* A square (□) cut in the top of the center of the granite basin head on the northwest corner of Bushwick Avenue and Schaeffer Street.

Elevation, 53.103 feet.

B. M. 476.—*Bushwick Avenue.* A square (□) cut in the top of the brownstone balustrade on the northwesterly side of the Bushwick Avenue entrance to Public School No. 56, on the corner of Madison Street. The bench mark is near the westerly corner of the stone at the end of the low iron railing and about 4 feet above the ground. It is about 30 feet from Madison Street.

Elevation, 55.265 feet.

B. M. 477.—*Bushwick Avenue.* A cross (+), lettered B. M., cut in the top of the bluestone coping at the easterly corner of the retaining wall, on the northwesterly corner of Bushwick Avenue and Willoughby Avenue. (Note 17, p. 91.)

Elevation, 70.660 feet.

B. M. 478.—*Troutman Street.* A square (□) cut in the bluestone coping at the base of the iron railing in front of Public School No. 53, on the southeasterly side of Troutman Street, between Central Avenue and Hamburg Avenue. The bench mark is about 12.5 feet easterly from the westerly side of the building and about 186 feet easterly from Central Avenue.

Elevation, 62.895 feet.

B. M. 479.—*Willoughby Avenue.* A square (⊡) cut in the top of the granite city monument on the easterly corner of Willoughby Avenue and Irving Avenue. The monument is at the corner of the iron fence surrounding Public School No. 123, and flush with the sidewalk.

Elevation, 38.992 feet.

B. M. 480.—*Willoughby Avenue.* A brass bracket bolted to the limestone masonry on the northwesterly side of Public School No. 123, on the easterly corner of Willoughby Avenue and Irving Avenue. The bracket is on the Willoughby Avenue side of the building, is about 225 feet from Irving Avenue and 2.7 feet from the easterly side of the building. (Note 22, p. 91.)

Elevation, 39.007 feet.

B. M. 481.—*Bushwick Avenue.* This is the center of the top of a 2′ x 2′ x 3′ limestone post at the northwesterly end of the low red brick retaining wall on the Bushwick Avenue side of the Brooklyn Public Library, on the southerly corner of

Bushwick Avenue and De Kalb Avenue. The bench mark is about 18 feet southerly from the southerly line of De Kalb Avenue, is on the second limestone post and about 18 feet from the Bushwick Avenue curb. Established by the Brooklyn Sewer Bureau.

Elevation, 59.354 feet.

B. M. 482.—*Vernon Avenue.* A 7-16-inch copper bolt in the top of the south-easterly granite block in the center of the small public park at the intersection of Vernon Avenue, Stuyvesant Avenue and Broadway. The bolt is near the south edge of the level surface and 1.25 feet from the outer edge of the stone. (Note 14, p. 91.)

Elevation, 70.207 feet.

B. M. 483.—*Vernon Avenue.* A cross (+) cut in the top of the southwest corner of the iron base of the elevated railroad column nearest to the northwest corner of Vernon Avenue and Broadway. The cross is on the westerly side of Broadway and almost on line with the northerly line of Vernon Avenue. It is 2 inches above the sidewalk. (Note 17, p. 91.)

Elevation, 67.634 feet.

B. M. 484.—*Vernon Avenue.* A square (⊡) cut near the east end of the lowest step of stoop at the house entrance to No. 390 Vernon Avenue, about 120 feet west of Stuyvesant Avenue.

Elevation, 72.463 feet.

B. M. 485.—*Vernon Avenue.* This bench mark is on the cornerstone of the Prudential Savings Bank, on the southeast corner of Vernon Avenue and Stuyvesant Avenue. The bench mark is the center of the horizontal bar in the letter A, in the inscription, A. D. 1909, on the cornerstone, which is in the northwest corner of the building.

Elevation, 72.459 feet.

B. M. 486.—*Flushing Avenue.* A square (⊡) cut in the top of the north end of the granite basin head on the northeast corner of Flushing Avenue and Harrison Avenue.

Elevation, 14.002 feet.

B. M. 487.—*Sumner Avenue.* A square (□) cut in the top of the southwest corner of the granite base of the elevated railroad column on the northeast corner of Summer Avenue and Myrtle Avenue.

Elevation, 64.270 feet.

B. M. 488.—*South Ninth Street.* A ½-inch copper bolt in the top of the south-easterly corner of the large granite base block supporting limestone column and lamppost on the easterly side of the main entrance to the Williamsburg Y. M. C. A., on the southeast corner of South Ninth Street and Marcy Avenue. The bolt is 4.4 feet above the sidewalk and 5.5 inches from the base of the column. (Note 5, p. 89.)

Elevation, 50.256 feet.

B. M. 489.—*Driggs Avenue.* A square (□) cut in the top of the southeast corner of the granite base of the elevated railroad column on the northwest corner of Driggs Avenue and Broadway. The bench mark is on the west curb line of Driggs Avenue and 9 inches above the sidewalk.

Elevation, 49.312 feet.

B. M. 490.—*Bedford Avenue.* A 2-inch square (□) cut in the top of the horizontal surface of the granite block at the base of steel column under the Williamsburg Bridge, and on the southwest corner of Bedford Avenue and South Fifth Street. The bench mark is on the south side of the column, which is under the north edge of the north roadway of the bridge. Established by the Department of Bridges.
Elevation, 51.526 feet.

B. M. 491.—*Bedford Avenue.* A square (⊓), lettered B. W. S., N. Y., cut in the top of the west edge of the granite water-table of the Williamsburg Bridge, on the east side of Bedford Avenue and 1 foot south of South Fifth Street. The bench mark is 9 inches above the sidewalk.

Elevation, 52.320 feet.

B. M. 492.—*Wythe Avenue.* A square (□), lettered B. M., cut in the top of the bluestone coping on the west side of the Wythe Avenue Station of the Brooklyn Union Gas Company, on the east side of Wythe Avenue between North Twelfth Street and North Thirteenth Street. The bench mark is 5 feet from the north end of the building, is about 50 feet south of North Thirteenth Street and 3.5 feet above the ground.

Elevation, 18.281 feet.

B. M. 493.—*Kent Avenue.* A quadrant (⊾) cut in the west end of the south abutment of old highway bridge on Kent Avenue at North Fourteenth Street.
Elevation, 12.077 feet.

B. M. 494.—*Franklin Street.* A quadrant (⊾) cut in the top of the east end of the granite basin head on the southeast corner of Franklin Street and Oak Street.
Elevation, 14.363 feet.

B. M. 495.—*Franklin Street.* A square (⊓) cut in the top of the east end of the granite basin head on the northeast corner of Franklin Street and Oak Street.
Elevation, 14.302 feet.

B. M. 496.—*South Sixth Street.* A small cross (+), lettered U. S., cut in the west face of the second course of granite in the south side of the Brooklyn anchorage of the Williamsburg Bridge, at South Sixth Street and Durham Place. The cross is about 42 feet west of the east side of the anchorage, is 4.9 feet north of the south side, and 4 feet above the ground. (Note 15, p. 91.)

Elevation, 32.643 feet.

B. M. 497.—*South Sixth Street.* On the north side of the five-story brick factory building, on the southeast corner of South Sixth Street and Kent Avenue. The bench mark is a 2-inch square (⊓), lettered B. M., cut in the top of the northwest corner of the sill of the fourth window east of Kent Avenue. The bench mark is 3.8 feet above the ground and about 40 feet east of Kent Avenue. Established by the Department of Bridges.

Elevation, 26.975 feet.

B. M. 498.—*South Sixth Street.* On the north side of the two-story brick factory building on the southeast corner of South Sixth Street and Durham Place. The bench mark is the highest point within a 2-inch square (□) cut in the sill of the second window east of Durham Place. The bench mark is 11.75 feet east of Durham Place, is 8 inches from the east end of the sill and 3.9 feet above the ground. Established by the Department of Bridges.

Elevation, 33.279 feet.

B. M. 499.—*Greenpoint Avenue.* A ¼-inch copper bolt in the top of the large granite block on the west side of the wide entrance to No. 124 Greenpoint Avenue, between Manhattan Avenue and Franklin Street. The building is a two-story limestone structure, occupied by Hook and Ladder Company No. 106. The bolt is 6 feet east of the west side of the building and 4.4 feet above the sidewalk. (Note 7, p. 90.)

Elevation, 35.297 feet.

B. M. 500.—*Dupont Street.* The center of the surface within a square (□) cut in the top of the northeast corner of the bluestone door-sill at the main entrance to Public School No. 31, on the south side of Dupont Street, between Franklin Street and Manhattan Avenue.

Elevation, 19.072 feet.

B. M. 501.—*Dupont Street.* A brass bracket in the bluestone base course on the north side of Public School No. 31, on the south side of Dupont Street, between Franklin Street and Manhattan Avenue. The bracket is about 15 feet west of the main entrance. (Note 22, p. 91.)

Elevation, 18.758 feet.

B. M. 502.—*Dupont Street.* A square (ㅂ) cut in the top of the southwest corner of the lowest bluestone step at the house entrance to the three-story frame tenement at No. 131 Dupont Street, about 110 feet east of Manhattan Avenue. Established by the Public Service Commission.

Elevation, 16.391 feet.

B. M. 503.—*Manhattan Avenue.* A square (ㅂ) cut in the top of the southeast corner of the granite copingstone on the east side of the approach to the highway bridge (Jackson Avenue Bridge) over Newtown Creek at Manhattan Avenue. The bench mark is at the south end of the iron railing and almost exactly on line with the north line of Ash Street.

Elevation, 18.353 feet.

B. M. 504.—*Manhattan Avenue.* A 2-inch square (□) cut in the top of the granite coping of the bulkhead on the south side of Newtown Creek at Manhattan Avenue. The bench mark is 5.5 feet west of the east line of Manhattan Avenue.

Elevation, 8.419 feet.

# STANDARD ELEVATIONS OF BENCH MARKS
## IN THE
## BOROUGH OF QUEENS
## AND IN
## NASSAU COUNTY.

| No. of Bench Mark. | LOCALITY. | Elevation Above Mean Sea Level. | |
|---|---|---|---|
| | | Meters. | Feet. |
| 505 | Belle Harbor | 3.1081 | 10.197 |
| 506 | " " | 2.5107 | 8.237 |
| 507 | " " | 2.3961 | 7.861 |
| 508 | Rockaway Park | 2.1087 | 6.918 |
| 509 | " " | 4.5039 | 14.777 |
| 510 | " " | 3.2635 | 10.707 |
| 511 | " " | 3.2678 | 10.721 |
| 512 | Seaside | 2.8551 | 9.367 |
| 513 | Hollands | 2.4587 | 8.067 |
| 514 | " | 2.4027 | 7.883 |
| 515 | Hammels | 2.9925 | 9.818 |
| 516 | Arverne | 2.8690 | 9.413 |
| 517 | " | 2.2233 | 7.294 |
| 518 | " | 2.7500 | 9.022 |
| 519 | " | 2.8417 | 9.323 |
| 520 | " | 2.0780 | 6.818 |
| 521 | Edgemere | 2.0493 | 6.723 |
| 522 | " | 1.8343 | 6.018 |
| 523 | " | 2.1708 | 7.122 |
| 524 | " | 2.9354 | 9.631 |
| 525 | " | 2.2588 | 7.411 |
| 526 | Far Rockaway | 8.6711 | 28.449 |
| 527 | " " | 8.8227 | 28.946 |
| 528 | " " | 9.9714 | 32.715 |
| 529 | " " | 7.9895 | 26.212 |
| 530 | " " | 9.6433 | 31.638 |
| 531 | " " | 7.3406 | 24.083 |
| 532 | " " | 8.8593 | 29.066 |
| 533 | Seaside | 2.3260 | 7.631 |
| 534 | Lawrence, Nassau County | 9.1167 | 29.910 |
| 535 | " " " | 7.9824 | 26.189 |
| 536 | " " " | 7.9712 | 26.152 |
| 537 | " " " | 7.8348 | 25.705 |
| 538 | " " " | 8.3952 | 27.543 |
| 539 | Cedarhurst, Nassau County | 5.5493 | 18.206 |
| 540 | " " " | 5.2549 | 17.241 |
| 541 | " " " | 4.4319 | 14.540 |
| 542 | " " " | 2.0992 | 6.887 |
| 543 | Bay Head | 2.1713 | 7.124 |
| 544 | " " | 1.4285 | 4.687 |
| 545 | " " | 1.6033 | 5.260 |
| 546 | " " | 4.5846 | 15.041 |
| 547 | Springfield | 1.9453 | 6.382 |
| 548 | " | 1.7632 | 5.785 |
| 549 | " | 1.1706 | 3.841 |
| 550 | " | 3.6179 | 11.870 |

| No. of Bench Mark. | Locality. | Elevation Above Mean Sea Level. | |
|---|---|---|---|
| | | Meters. | Feet. |
| 551 | Springfield | 3.7570 | 12.326 |
| 552 | " | 3.1005 | 10.172 |
| 553 | " | 6.2264 | 20.428 |
| 554 | " | 3.8400 | 12.598 |
| 555 | " | 4.6500 | 15.256 |
| 556 | " | 5.6853 | 18.653 |
| 557 | " | 7.7360 | 25.381 |
| 558 | " | 7.7350 | 25.377 |
| 559 | Rosedale Terrace | 6.5276 | 21.416 |
| 560 | " " | 6.6139 | 21.699 |
| 561 | Rosedale | 5.8376 | 19.152 |
| 562 | St. Albans Heights | 15.0181 | 49.272 |
| 563 | Springfield Boulevard | 10.1475 | 33.292 |
| 564 | " " | 12.6070 | 41.362 |
| 565 | Rockaway Road | 2.0999 | 6.889 |
| 566 | Conduit | 5.5016 | 18.050 |
| 567 | Three Mile Mill Road | 6.3163 | 20.723 |
| 568 | Rockaway Boulevard | 4.7403 | 15.552 |
| 569 | " " | 10.9099 | 35.794 |
| 570 | " " | 11.4973 | 37.721 |
| 571 | " " | 6.1794 | 20.274 |
| 572 | Ozone Park | 11.6724 | 38.295 |
| 573 | " " | 11.7337 | 38.496 |
| 574 | " " | 11.1879 | 36.706 |
| 575 | Woodhaven | 7.6164 | 24.988 |
| 576 | " | 8.2266 | 26.990 |
| 577 | Aqueduct | 7.2741 | 23.865 |
| 578 | " | 6.8070 | 22.333 |
| 579 | Ozone Park | 12.6999 | 41.666 |
| 580 | Woodhaven | 12.9264 | 42.409 |
| 581 | " | 13.4108 | 43.999 |
| 582 | Jamaica Avenue | 29.5889 | 97.076 |
| 583 | " " | 26.4547 | 86.794 |
| 584 | " " | 24.8616 | 81.567 |
| 585 | Little Neck Road | 32.7538 | 107.460 |
| 586 | " " " | 35.1731 | 115.397 |
| 587 | Jamaica Avenue | 24.2110 | 79.432 |
| 588 | " " | 24.3644 | 79.936 |
| 589 | Springfield Boulevard | 22.2055 | 72.853 |
| 590 | " " | 21.3660 | 70.098 |
| 591 | Jamaica Avenue | 22.1266 | 72.594 |
| 592 | Braddock Avenue (Rocky Hill Road) | 47.1912 | 154.827 |
| 593 | Madison Avenue | 26.5498 | 87.104 |
| 594 | " " | 25.9980 | 85.295 |
| 595 | " " | 26.3493 | 86.448 |
| 596 | Palatina Avenue | 18.8826 | 61.951 |
| 597 | Jamaica | 16.7733 | 55.030 |
| 598 | " | 18.5404 | 60.828 |
| 599 | " | 18.9144 | 62.055 |
| 600 | " | 18.2893 | 60.004 |
| 601 | " | 18.3064 | 60.060 |
| 602 | " | 18.1685 | 59.608 |
| 603 | " | 30.8003 | 101.051 |
| 604 | " | 30.9185 | 101.438 |
| 605 | " | 14.2578 | 46.777 |
| 606 | " | 20.3642 | 66.812 |
| 607 | " | 20.2147 | 66.321 |
| 608 | Richmond Hill | 19.5046 | 63.991 |
| 609 | " " | 19.3284 | 63.413 |

| No. of Bench Mark. | Locality. | Elevation Above Mean Sea Level. | |
|---|---|---|---|
| | | Meters. | Feet. |
| 610 | Richmond Hill | 18.4704 | 60.598 |
| 611 | " " | 14.0307 | 46.032 |
| 612 | Glendale | 39.9509 | 131.072 |
| 613 | " | 26.1606 | 85.829 |
| 614 | " | 29.2769 | 96.053 |
| 615 | " | 28.7514 | 94.329 |
| 616 | " | 24.6205 | 80.776 |
| 617 | " | 24.6164 | 80.762 |
| 618 | " | 30.0509 | 98.592 |
| 619 | " | 28.5148 | 93.552 |
| 620 | " | 29.8592 | 97.963 |
| 621 | " | 28.0637 | 92.072 |
| 622 | " | 26.6042 | 87.284 |
| 623 | Creek Street | 7.8174 | 25.648 |
| 624 | " " | 9.0764 | 29.778 |
| 625 | " " | 9.5226 | 31.242 |
| 626 | Laurel Hill | 8.1136 | 26.620 |
| 627 | " " | 4.2695 | 14.008 |
| 628 | " " | 4.4053 | 14.453 |
| 629 | " " | 4.4040 | 14.449 |
| 630 | " " | 4.4017 | 14.441 |
| 631 | " " | 4.2279 | 13.871 |
| 632 | Greenpoint Avenue | 5.1221 | 16.805 |
| 633 | " " | 11.2453 | 36.894 |
| 634 | " " | 15.8698 | 52.066 |
| 635 | " " | 16.3533 | 53.652 |
| 636 | Borden Avenue | 5.0853 | 16.684 |
| 637 | " " | 5.0456 | 16.554 |
| 638 | " " | 5.1603 | 16.930 |
| 639 | " " | 5.0374 | 16.527 |
| 640 | " " | 5.0370 | 16.526 |
| 641 | " " | 5.0673 | 16.625 |
| 642 | Vernon Avenue | 2.5453 | 8.351 |
| 643 | " " | 2.8200 | 9.252 |
| 644 | " " | 4.0281 | 13.216 |
| 645 | Borden Avenue | 3.8973 | 12.768 |
| 646 | " " | 2.4023 | 7.882 |
| 647 | Maspeth | 19.4437 | 63.792 |
| 648 | " | 26.7312 | 87.701 |
| 649 | Newtown | 17.1993 | 56.428 |
| 650 | " | 15.8343 | 51.950 |
| 651 | " | 15.6090 | 51.211 |
| 652 | Elmhurst | 11.2727 | 36.984 |
| 653 | " | 21.1281 | 69.318 |
| 654 | Queensboro Bridge | 6.2262 | 20.427 |
| 655 | " " | 6.3499 | 20.833 |
| 656 | " " | 8.8750 | 29.117 |
| 657 | Jackson Avenue | 11.5348 | 37.844 |
| 658 | Webster Avenue | 10.2266 | 33.552 |
| 659 | Prospect Street | 14.1605 | 46.458 |
| 660 | Jackson Avenue | 14.2100 | 46.621 |
| 661 | " " | 14.5407 | 47.706 |
| 662 | Astoria | 15.4646 | 50.737 |
| 663 | " | 5.0803 | 16.668 |
| 664 | " | 2.9807 | 9.779 |
| 665 | " | 8.9722 | 29.436 |
| 666 | " | 10.0645 | 33.020 |
| 667 | " | 9.8773 | 32.406 |
| 668 | " | 9.7625 | 32.029 |

| No. of Bench Mark. | LOCALITY. | Elevation Above Mean Sea Level. Meters. | Feet. |
|---|---|---|---|
| 669 | Astoria | 4.0426 | 13.263 |
| 670 | " | 2.4925 | 8.177 |
| 671 | " | 18.9806 | 62.272 |
| 672 | " | 20.4731 | 67.169 |
| 673 | " | 20.3751 | 66.847 |
| 674 | " | 13.8521 | 45.446 |
| 675 | " | 2.0611 | 6.762 |
| 676 | " | 5.0866 | 16.688 |
| 677 | Jackson Avenue | 12.3138 | 40.400 |
| 678 | Trains Meadow Road | 14.6578 | 48.090 |
| 679 | Jackson Avenue | 22.0772 | 72.432 |
| 680 | " " | 22.8458 | 74.953 |
| 681 | " " | 20.0799 | 65.879 |
| 682 | " " | 8.7378 | 28.667 |
| 683 | Flushing Bridge | 5.1150 | 16.782 |
| 684 | " " | 5.1254 | 16.816 |
| 685 | Flushing | 7.3479 | 24.107 |
| 686 | " | 11.7852 | 38.665 |
| 687 | " | 11.8372 | 38.836 |
| 688 | " | 15.1964 | 49.857 |
| 689 | " | 22.1629 | 72.713 |
| 690 | " | 22.6476 | 74.303 |
| 691 | " | 22.1926 | 72.810 |
| 692 | Kissena Road (Jamaica Road) | 22.5276 | 73.909 |
| 693 | " " " " | 22.8816 | 75.071 |
| 694 | " " " " | 35.8481 | 117.612 |
| 695 | " " " " | 26.0289 | 85.397 |
| 696 | " " " " | 4.2659 | 13.996 |
| 697 | " " " " | 15.1296 | 49.638 |
| 698 | Beechhurst Avenue (Whitestone Road) | 20.0368 | 65.737 |
| 699 | Bayside | 24.7880 | 81.325 |
| 700 | " | 33.5218 | 109.979 |
| 701 | " | 4.8843 | 16.025 |
| 702 | " | 22.4467 | 73.644 |
| 703 | " | 21.3513 | 70.050 |
| 704 | " | 20.5870 | 67.543 |
| 705 | Springfield Boulevard (Rocky Hill Road) | 37.1727 | 121.957 |
| 706 | Douglaston | 12.3417 | 40.491 |
| 707 | " | 13.5809 | 44.557 |
| 708 | Little Neck | 10.2984 | 33.787 |
| 709 | " " | 11.1981 | 36.739 |
| 710 | " " | 25.3355 | 83.122 |
| 711 | " " | 56.4318 | 185.143 |
| 712 | " " | 48.1206 | 157.876 |
| 713 | " " | 56.0251 | 183.809 |
| 714 | " " | 81.2258 | 266.488 |
| 715 | " " | 81.2173 | 266.460 |
| 716 | " " | 72.2519 | 237.046 |
| 717 | Flushing | 2.6730 | 8.770 |
| 718 | College Point | 13.2987 | 43.631 |
| 719 | " " | 13.6040 | 44.632 |
| 720 | " " | 2.9225 | 9.588 |
| 721 | " " | 3.1169 | 10.226 |
| 722 | " " | 2.9522 | 9.686 |
| 723 | " " | 6.6033 | 21.664 |
| 724 | " " | 12.8080 | 42.021 |
| 725 | " " | 17.2510 | 56.598 |
| 726 | " " | 17.2336 | 56.541 |
| 727 | " " | 12.7930 | 41.972 |

| No. of Bench Mark. | Locality. | Elevation Above Mean Sea Level. | |
|---|---|---|---|
| | | Meters. | Feet. |
| 728 | College Point | 17.5109 | 57.450 |
| 729 | Whitestone | 15.6630 | 51.388 |
| 730 | " | 15.7820 | 51.778 |
| 731 | " | 3.6050 | 11.827 |
| 732 | Willets Point | 3.5630 | 11.689 |
| 733 | " " | 4.0965 | 13.440 |
| 734 | " " | 4.7536 | 15.596 |
| 735 | " " | 4.7547 | 15.599 |
| 736 | " " | 14.9329 | 48.992 |
| 737 | " " | 3.3544 | 11.005 |

# DESCRIPTIONS OF BENCH MARKS

IN THE

# BOROUGH OF QUEENS

AND IN

# NASSAU COUNTY.

B. M. 505.—*Belle Harbor.* A horizontal line on the face of a brass tablet bolt 2 inches in diameter set horizontally into the northeasterly brick buttress of the tower of the St. Francis de Sales Roman Catholic Church, on the northwest corner of Beach One Hundred and Twenty-ninth Street (Chester Avenue) and Rockaway Beach Boulevard (Washington Avenue). The bolt is 1.5 feet above the ground. (Note 1, p. 89.)

Elevation, 10.197 feet.

B. M. 506.—*Belle Harbor.* A square (ㅁ) cut in the top of the northwest corner of the west sill at the gate entrance to Belle Harbor, on the northwest corner of Beach One Hundred and Twenty-sixth Street (Pelham Avenue) and Rockaway Beach Boulevard (Washington Avenue).

Elevation, 8.237 feet.

B. M. 507.—*Belle Harbor.* A square (ㅁ), lettered U. S., cut in the top of the north edge of the cement sidewalk on the northwest corner of Beach One Hundred and Twenty-eighth Street (Winthrop Avenue) and Rockaway Beach Boulevard (Washington Avenue). The bench mark is about 3 feet west of the west curb. (Note 16, p, 91.)

Elevation, 7.861 feet.

B. M. 508.—*Rockaway Park.* A square (ㅁ), lettered U. S., cut in the top of the east end of the cement sidewalk on the northwest corner of Beach One Hundred and Seventeenth Street (Sixth Avenue) and Rockaway Beach Boulevard (Washington Avenue). The bench mark is 6 inches from the west curb. (Note 16, p. 91.)

Elevation, 6.918 feet.

B. M. 509.—*Rockaway Park.* A horizontal line in the end of a ⅝-inch copper bolt leaded horizontally into the red sandstone at the northeast corner of Public School No. 43, at Rockaway Beach Boulevard (Washington Avenue) and Beach One Hundred and Tenth Street (Eastern Avenue). The bolt is on the north side of the building and about 64 feet south of Rockaway Beach Boulevard. It is 4.5 feet above the ground and 1.25 feet from the east end of the building. (Note 6, p. 90.)

Elevation, 14.777 feet.

B. M. 510.—*Rockaway Park.* A square (ㅁ) cut in the top of the east edge of the curved portion of the first cement step at the main entrance to Public School No. 43, at Rockaway Beach Boulevard (Washington Avenue) and Beach One Hundred and Tenth Street (Eastern Avenue). The bench mark is 2 feet from the south end of the step. Established by the Queens Topographical Bureau.

Elevation, 10.707 feet.

B. M. 511.—*Rockaway Park.* A square (□) cut in the top of the first cement step at the main entrance to Public School No. 43, at Rockaway Beach Boulevard (Washington Avenue) and Beach One Hundred and Tenth Street (Eastern Avenue). The bench mark is about 6 feet from the south end of the step and 8 inches from the east edge.

Elevation, 10.721 feet.

B. M. 512.—*Seaside.* A square (□) cut in the top of the curved curb on the northeast corner of Beach One Hundred and First Street (Hollywood Avenue) and Rockaway Beach Boulevard.

Elevation, 9.367 feet.

B. M. 513.—*Hollands.* A square (□), lettered U. S., cut in the top of the east end of the first cement step on the north side of the east wing of Public School No. 44, on the southeast corner of Beach Ninety-fourth Street (Academy Avenue) and Rockaway Beach Boulevard. The bench mark is 1 foot from the north edge of the step and 4 inches above the walk. (Note 16, p. 91.)

Elevation, 8.067 feet.

B. M. 514.—*Hollands.* A square (⊓) cut in the stone base between two pilasters on the west side of the Boulevard entrance to Public School No. 44, on the southeast corner of Beach Ninety-fourth Street (Academy Avenue) and Rockaway Beach Boulevard. The bench mark is about 32 feet easterly from the west side of the building and 4½ inches above the walk.

Elevation, 7.883 feet.

B. M. 515.—*Hammels.* A horizontal line in the end of a ¾-inch copper bolt leaded horizontally into the granite foundation at the northeast corner of Sub. Sta. Power House No. 5 of the Long Island Railroad, at Hammels Station. The bolt is in the east side of the building, is 7 inches from the north side and 1.5 feet above the ground. (Note 6, p. 90.)

Elevation, 9.818 feet.

B. M. 516.—*Arverne.* A square (⊓) cut near the west end of the first step at the main or south entrance to Public School No. 42, on the north side of Rockaway Beach Boulevard between Beach Sixty-eighth Street (Jessica Avenue) and Beach Sixty-seventh Street (Vernam Avenue). The bench mark is 8 inches from the west end of the step. Established by the Queens Topographical Bureau.

Elevation, 9.413 feet.

B. M. 517.—*Arverne.* A horizontal line in the end of a ⅝-inch copper bolt leaded horizontally into the south face of a granite block near the southeast corner of the two-story stone building on the northwest corner of Beach Sixty-sixth Street (Gaston Avenue) and the Long Island Railroad. The bolt is 1.7 feet above the ground, is 15.3 feet from the north side of the building, and on the northeasterly side of the office entrance at the southeast corner of the building. (Note 6, p. 90.)

Elevation, 7.294 feet.

B. M. 518.—*Arverne.* A square (⊓), lettered U. S., cut in the top of the coping-stone at the west side of the cellar entrance to Public School No. 42, on the north side of Rockaway Beach Boulevard, between Beach Sixty-eighth Street (Jessica Avenue) and Beach Sixty-seventh Street (Vernam Avenue). The bench mark is on the south side of the building east of the main entrance. (Note 16, p. 91.)

Elevation, 9.022 feet.

B. M. 519.—*Arverne.* A square (⊓) cut in the top of the west end of the sill of cellar window on the south side of Public School No. 42, on the north side of Rockaway Beach Boulevard, between Beach Sixty-eighth Street (Jessica Avenue) and Beach Sixty-seventh Street (Vernam Avenue). The bench mark is 4.3 feet from the west side of the building.

Elevation, 9.323 feet.

B. M. 520.—*Arverne.* A square (⊓), lettered B. E., cut in the top of the west end of the bluestone door-sill on the south side of the Beach Sixtieth Street (Straiton Avenue) station of the Long Island Railroad at Arverne.

Elevation, 6.818 feet.

B. M. 521.—*Edgemere.* A square (□) cut in the concrete at the base of flag pole on the east side of the Half Way House, on the southwest corner of Beach Forty-fourth Street (Frank Avenue) and Rockaway Beach Boulevard. The bench mark is at the west side of the flag pole.

Elevation, 6.723 feet.

B. M. 522.—*Edgemere.* The center of the top of a granite city monument on the north side of Rockaway Beach Boulevard, about 275 feet east of Beach Forty-fourth Street (Frank Avenue).

Elevation, 6.018 feet.

B. M. 523.—*Edgemere.* A ½-inch copper bolt in the east end of the bluestone step on the north side of the Long Island Railroad station. The bolt is 3 feet from the east side of the building. (Note 8, p. 90.)

Elevation, 7.122 feet.

B. M. 524.—*Edgemere.* The surface between two parallel lines (| |) cut in the top of the center of the bluestone sill of the double window of the telegraph office on the north side of the Long Island Railroad station.

Elevation, 9.631 feet.

B. M. 525.—*Edgemere.* A square (⊔), lettered U. S., cut in the top of the west end of the bluestone door-sill at the west entrance on the north side of the Long Island Railroad station. (Note 16, p. 91.)

Elevation, 7.411 feet.

B. M. 526.—*Far Rockaway.* A 2-inch square (□) cut in the top of the west end of the bluestone sill of the east door on the south side of the Long Island Railroad station.

Elevation, 28.449 feet.

B. M. 527.—*Far Rockaway.* A square (⊔) cut in the top of the westerly corner of the third bluestone step at the main entrance to the Public Library, on the southeasterly corner of Far Rockaway Boulevard (Central Avenue) and Mott Avenue. Established by the Queens Topographical Bureau.

Elevation, 28.946 feet.

B. M. 528.—*Far Rockaway.* A horizontal line in the end of a ¾-inch copper bolt set horizontally into the brick on the Far Rockaway Boulevard (Central Avenue) side of the Public Library building on the southeasterly corner of Mott Avenue. The bolt is 1.15 feet from the Mott Avenue side and 5 feet above the ground. (Note 6, p. 90.)

Elevation, 32.715 feet.

B. M. 529.—*Far Rockaway.* A square (⌂) cut in the top of the southwest corner of the wide bluestone door-sill at the house entrance to the Horton Building (four-story brick) at No. 337 Far Rockaway Boulevard (Central Avenue), about 127 feet west of Dinsmore Street (Cleveland Avenue).

Elevation, 26.212 feet.

B. M. 530.—*Far Rockaway.* A cross, lettered B + M, cut in the fifth course of brick above the walk, at the northwest corner of Public School No. 39, on the south side of State Street, between Dinsmore Street (Cleveland Avenue) and Bayport Place (Nostrand Avenue). The cross faces State Street and is 7 inches from the corner of the building. (Note 15, p. 91.)

Elevation, 31.638 feet.

B. M. 531.—*Far Rockaway.* The center of the top of a granite city monument on the southeast corner of Far Rockaway Boulevard (Central Avenue) and Bayport Place (Nostrand Avenue). The top of the monument is 1½ inches below the sidewalk.

Elevation, 24.083 feet.

B. M. 532.—*Far Rockaway.* A square (⌂) cut in the top of the northwest corner of the bluestone coping on the north balustrade of house at No. 1 Beach Ninth Street (Oak Street), on the southeast corner of Far Rockaway Boulevard (Central Avenue). The bench mark is 3.6 feet above the sidewalk.

Elevation, 29.066 feet.

B. M. 533.—*Seaside.* A square (□) cut in the top of the backing log above the tide staff on the east side of the dock at the foot of Beach One Hundred and Second Street (Henry Street), in Jamaica Bay. The tide-staff was established on April 30, 1901, by the Department of Docks and Ferries.

Elevation, 7.631 feet.

B. M. 534.—*Lawrence, Nassau County.* A cross, lettered B + M, cut in the south face of the bluestone sill of the east basement window, on the south side of the two-story and attic frame building known as the Lawrence Academy, about 150 feet north of Central Avenue and about 250 feet east of Winchester Place. The cross is 6 feet from the east side of the building and 8 inches above the ground.

Elevation, 29.910 feet.

B. M. 535.—*Lawrence, Nassau County.* A horizontal line on the face of a brass tablet bolt 2 inches in diameter, set horizontally into the limestone at the northwest corner of the four-story red brick building on the southeast corner of Rockaway Turnpike and Central Avenue. The bench mark faces Central Avenue, is 2 feet from the west side of the building and 6 inches above the sidewalk. (Note 1, p. 89.)

Elevation, 26.189 feet.

B. M. 536.—*Lawrence, Nassau County.* A ⅝-inch brass bolt set vertically in the east end of the bluestone door-sill at the east store entrance to the four-story red brick building on the southeast corner of Rockaway Turnpike and Central Avenue. The bolt is on the Central Avenue side and about 45 feet east of Rockaway Turnpike. Established by the Town of Lawrence.

Elevation, 26.152 feet.

B. M. 537.—*Lawrence, Nassau County.* A ⅝-inch brass bolt set vertically in the northeast corner of the bluestone door-sill at the entrance to the Bank of Lawrence, on the south side of Central Avenue just south of Washington Avenue. The building is two-story and attic and built of stone. The bench mark is about 16 feet west of Washington Avenue. Established by the Town of Lawrence.

Elevation, 25.705 feet.

B. M. 538.—*Lawrence, Nassau County.* A cross (+) cut in the top of a section of steel rail driven into the ground on the northeast corner of Rockaway Turnpike and the Far Rockaway Branch of the Long Island Railroad.

Elevation, 27.543 feet.

B. M. 539.—*Cedarhurst, Nassau County.* The center of a small bluestone monument on the northeast corner of Rockaway Turnpike and Pacific Street. The top of the monument is 4 inches above the ground.

Elevation, 18.206 feet.

B. M. 540.—*Cedarhurst, Nassau County.* A cross, lettered B + M, cut in the brick foundation at the southwest corner of the two-story and attic frame house on the south side of West Broadway, about 400 feet east of Rockaway Turnpike. The cross is on the west side of the house, is on the fifth course of brick above the cement walk and 4 inches from the corner. (Note 15, p. 91.)

Elevation, 17.241 feet.

B. M. 541.—*Cedarhurst, Nassau County.* A cross (+) cut in the head of a railroad spike driven vertically in a flat surface on the exposed root on the northerly side of a large willow tree on the westerly side of Rockaway Turnpike, about 250 feet south of Rugby Road. The tree containing the bench mark is about 50 feet south of the old toll gate house.

Elevation, 14.540 feet.

B. M. 542.—*Cedarhurst, Nassau County.* The head of a 20d. copper nail driven vertically in the top of the north end of the string-piece on the east side of the highway bridge on Rockaway Turnpike over Motts Creek. This is the first creek crossing Rockaway Turnpike north of the Far Rockaway Branch of the Long Island Railroad.

Elevation, 6.887 feet.

B. M. 543.—*Bay Head.* The head of a 20d. copper nail driven vertically in the top of the north end of the string-piece on the west side of the highway bridge on Rockaway Road over Hook Creek. This is the second creek north of the Far Rockaway Branch of the Long Island Railroad and is about 100 feet south of Third Street, in Meadowmere Park.

Elevation, 7.124 feet.

B. M. 544.—*Bay Head.* The head of a 20d. copper nail driven vertically in the top of the south edge of a pile level with the roadway and on the east edge of Rockaway Road about 22 feet north of the angle point, between the Bay Head Yacht Club and the Bay Head Hotel. The bench mark is 4 feet south of New York Telephone Company pole No. 327.

Elevation, 4.687 feet.

B. M. 545.—*Bay Head.* The center of the top of a granite city monument set by the Queens Topographical Bureau. It is on the easterly side of Rockaway Road at the angle point, between the Bay Head Yacht Club and the Bay Head Hotel. The top of the monument is flush with the ground.

Elevation, 5.260 feet.

B. M. 546.—*Bay Head.* Queens Topographical Bureau B. M. No. 520. "The underside of north end of cap over saloon door in G. E. Warner's Hotel, on the west side of Rockaway Turnpike, north side of Hook Creek." The bench mark is on the south side of the hotel (now occupied by Wm. Schlott) and on the north end of the cap. The hotel is on the north side of the canal just east of Hook Creek, about 600 feet north of the angle point in Rockaway Road.

Elevation, 15.041 feet.

B. M. 547.—*Springfield.* A horizontal line in the end of a ½-inch copper bolt leaded horizontally into the brick foundation at the northerly corner of the two-story frame house occupied by John Dooley, on the southwesterly side of Rockaway Road about 225 feet southeasterly from Springfield Road. The bolt faces Rockaway Road, is on the sixth course of brick above the ground and 6 inches from the corner. (Note 18, p. 91.)

Elevation, 6.382 feet.

B. M. 548.—*Springfield.* The head of a railroad spike driven vertically in the exposed root on the west side of a large willow tree on the southeast corner of Springfield Road and Rockaway Turnpike.

Elevation, 5.785 feet.

B. M. 549.—*Springfield.* The center of a brass bolt marking the U. S. C. & G. S. triangulation station Jaeger, on the southwesterly side of Rockaway Road at New York Avenue. The bolt is exposed, is about 10 feet west of the center line of New York Avenue and about 25 feet southerly from Rockaway Road. (Note 25, p. 91.)

Elevation, 3.841 feet.

B. M. 550.—*Springfield.* A cross, lettered + B. M., cut in the east face of the brick foundation at the southeast corner of the two-story and attic frame house near the southwest corner of New York Avenue and Cherry Avenue. The bench mark is about 130 feet south of Cherry Avenue, is 6 inches above the ground and 2 inches from the corner of the house. (Note 15, p. 91.)

Elevation, 11.870 feet.

B. M. 551.—*Springfield.* A quadrant (⦣) cut in the top of the east end of the bluestone coping of the old pump well just south of the Springfield pumping station, on the west side of Springfield Road, north of Cherry Avenue and about 1,000 feet south of the conduit.

Elevation, 12.326 feet.

B. M. 552.—*Springfield.* A ⅞-inch brass bolt set vertically in the top of the northwest corner of the brick vault or suction gate chamber about 59 feet south of the Springfield pumping station, on the west side of Springfield Road, north of Cherry Avenue and about 1,000 feet south of the conduit. The bolt is on the west side of the road and about 2 feet below grade. Established by the Brooklyn Bureau of Water Supply.

Elevation, 10.172 feet.

B. M. 553.—*Springfield.* Queens Topographical Bureau B. M. "The underside of pilaster cap on the south side of the pilaster on north side of main entrance to pumping station on Springfield Road, north of Cherry Avenue." A square (⊡) was cut in the stone at the bench mark.

Elevation, 20.428 feet.

B. M. 554.—*Springfield.* A horizontal line in the end of a ⅝-inch copper bolt set horizontally into the limestone base of the yellow brick pilaster on the north side of the main entrance to the Springfield pumping station, on the west side of Springfield Road north of Cherry Avenue and about 1,000 feet south of the conduit. The bolt faces the east, is 12 feet from the north side of the building and 3 inches above the ground. The letters B. M. are cut in the stone. (Note 6, p. 90.)

Elevation, 12.598 feet.

B. M. 555.—*Springfield.* A square (□) cut in the top of the northeast corner of the granite coping of parapet wall on the north end of the culvert crossing the conduit, about 200 feet west of Springfield Road and about 60 feet south of the center line of the Freeport trolley line. The bench mark is on the north side of the conduit and at the foot of the iron railing.

Elevation, 15.256 feet.

B. M. 556.—*Springfield.* A quadrant (⊾) cut in the top of the northeast corner of the granite curb of the manhole on the conduit line, about 216 feet west of Springfield Road.

Elevation, 18.653 feet.

B. M. 557.—*Springfield.* A 2-inch cross, lettered B + M, cut in the brick foundation at the southwest corner of the First Presbyterian Church, on the northeast corner of Springfield Boulevard (Springfield Road) and Carson Street (Park Avenue), just north of the Montauk Division of the Long Island Railroad. The cross faces Park Avenue, is 1 foot from the concrete block at the corner of the church and 1 foot above the ground. (Note 15, p. 91.)

Elevation, 25.381 feet.

B. M. 558.—*Springfield.* A cross (+) cut in the top of the west nut on the fixed rim of the fire hydrant on the southwest corner of Springfield Boulevard (Springfield Road) and Merrick Road. The bench mark is 1.4 feet above the ground.

Elevation, 25.377 feet.

B. M. 559.—*Rosedale Terrace.* A square (□) cut in the top of the west end of the copingstone on the south end of the culvert connecting the storage reservoirs on both sides of Merrick Road, just west of Brookville Boulevard (Foster's Meadow Road). The bench mark is on the south side of Merrick Road, is south of the board fence and about 155 feet west of Brookville Boulevard.

Elevation, 21.416 feet.

B. M. 560.—*Rosedale Terrace.* A square (□) cut in the top of a granite city monument set in concrete 3′ x 3′ x 4′ on the southeast corner of Merrick Road and Brookville Boulevard (Foster's Meadow Road). This monument was set by the Queens Topographical Bureau. It is 41.80 feet from the corner of the fence on the southwest corner of both roads and 59.75 feet from the corner of the fence on the northwest corner. The top of the monument is 1.2 feet below the ground.

Elevation, 21.699 feet.

B. M. 561.—*Rosedale.* A quadrant (⊾) cut in the top of the southeast corner of the granite curb of manhole on the conduit line, about 310 feet east of Rosedale Boulevard (Rosedale Avenue) and about 25 feet north of the Montauk Division of the Long Island Railroad.

Elevation, 19.152 feet.

B. M. 562.—*St. Albans Heights.* The center of a brass bolt marking the U. S. C. & G. S. triangulation station Lang. The bench mark is in the rear of a vacant lot, is about 100 feet east of Two Hundred and First Street (Bank Street) and about 175 feet north of One Hundred and Twentieth Avenue (Banks Avenue). (Note 25, p. 91.)

Elevation, 49.272 feet.

B. M. 563.—*Springfield Boulevard.* The head of a railroad spike driven horizontally into the west side of an electric light pole on the west side of Springfield Boulevard (Springfield Road). The bench mark is on the southwest corner of a narrow road running from Springfield Boulevard to One Hundred and Twentieth Avenue (Banks Avenue), is just north of the northwest corner of the land owned by the Montefiore Cemetery and 1 foot above the ground.

Elevation, 33.292 feet.

B. M. 564.—*Springfield Boulevard.* The top of a railroad spike driven vertically in the root of an old maple tree on the southwest corner of Springfield Boulevard (Springfield Road) and One Hundred and Seventeenth Avenue (Central Avenue).

Elevation, 41.362 feet.

B. M. 565.—*Rockaway Road.* A railroad spike driven vertically in the root on the northwesterly side of a large maple tree on the west side of Rockaway Road, about 30 feet north of Farmers Avenue.

Elevation, 6.889 feet.

B. M. 566.—*Conduit.* A quadrant (⊾) cut in the top of the southeast corner of the 5-inch wide granite manhole curb on the southerly side of the conduit line and about 500 feet west of Rockaway Road.

Elevation, 18.050 feet.

B. M. 567.—*Three Mile Mill Road.* A horizontal line in the end of a ½-inch copper bolt set horizontally into the brick foundation at the northeast corner of Public School No. 45, on the west side of Three Mile Mill Road, about 375 feet south of Rockaway Boulevard. The bolt is on the east side of the building, is 9 inches above the ground and 11½ inches from the north side of the building. The letters B. M. are cut in the brick. (Note 18, p. 91.)

Elevation, 20.723 feet.

B. M. 568.—*Rockaway Boulevard.* A square (□) cut in the top of the northwest corner of the 6-inch brownstone curb of manhole on the Baisleys Pond branch conduit, about 100 feet south of Rockaway Boulevard and about 90 feet east of Three Mile Mill Road.

Elevation, 15.552 feet.

B. M. 569.—*Rockaway Boulevard.* A horizontal line in the end of a ½-inch copper bolt set horizontally into the brick foundation near the southwest corner of the three-story frame dwelling owned by J. R. Fredericks, on the north side of Rockaway Boulevard about 150 feet east of Frost Avenue (Lincoln Avenue). The bolt is on the west side of the house, is 11 inches from the corner and 1 foot above the ground. The letters B. M. are cut in the brick above. (Note 18, p. 91.)

Elevation, 35.794 feet.

B. M. 570.—*Rockaway Boulevard.* Queens Topographical Bureau B. M. "The center of main entrance on threshold under main front door when closed, at James Fredericks' house, north side of Rockaway Road, just east of Lincoln Avenue." A brass nail was driven into the threshold at the bench mark.

Elevation, 37.721 feet.

B. M. 571.—*Rockaway Boulevard.* The center of the top of a granite city monument set in concrete 3′ x 3′ x 4′ on the northwest corner of Rockaway Boulevard and Neosho Street (Hawtree Creek Road). This monument was set by the Queens Topographical Bureau. It is 3 feet east of the end of the cement sidewalk and the top is 1 foot below the ground.

Elevation, 20.274 feet.

B. M. 572.—*Ozone Park.* A square (ⴑ) cut in the top of the northeast corner of the small I-beam on the east side of the base of the steel transmission tower on the west side of the Rockaway Beach Division of the Long Island Railroad and about 8 feet south of Rockaway Boulevard (Rockaway Plank Road). The bench mark is 1 foot above the concrete base of the tower.

Elevation, 38.295 feet.

B. M. 573.—*Ozone Park.* The intersection of a cross (+) cut in the top of the 1¾-inch anchor bolt on the east side of the base of steel transmission tower on the west side of the Rockaway Beach Division of the Long Island Railroad and about 8 feet south of Rockaway Boulevard (Rockaway Plank Road). The bench mark is 5 inches from the north end of the I-beam.

Elevation, 38.496 feet.

B. M. 574.—*Ozone Park.* A cross (+) cut in the top of the ¾-inch anchor bolt on the east side of the base of the northeasterly steel column supporting the overhead switch tower over the tracks of the Rockaway Beach Division of the Long Island Railroad at Ocean Avenue and Liberty Avenue. The bolt is about 15 feet south of Liberty Avenue and on the east side of the railroad.

Elevation, 36.706 feet.

B. M. 575.—*Woodhaven.* A square (ⴑ) cut in the top of the southwest corner of the monolithic bluestone door-sill at the main entrance to Public School No. 64, on the north side of Jerome Avenue (Broadway) between Weymouth Street and Culloden Street.

Elevation, 24.988 feet.

B. M. 576.—*Woodhaven.* A 7/16-inch copper bolt set vertically in the center of the top of the bluestone sill of double window on the south side of Public School No. 64, on the north side of Jerome Avenue (Broadway) between Weymouth Street and Culloden Street. The bench mark is about 300 feet east of Halifax Street, is about 10 feet west of the main entrance and 2.8 feet above the ground. (Note 8, p. 90.)

Elevation, 26.990 feet.

B. M. 577.—*Aqueduct.* A square (ⴑ) cut in the top of the southwest corner of the 20-inch square bluestone block under the southwest corner of the frame stable at the Aqueduct Race Track. The bench mark is about 32 feet east of the east rail of the Rockaway Beach Division of the Long Island Railroad and about 185 feet north of the north side of North Conduit Avenue (Old South Road). It is 6 inches above the ground.

Elevation, 23.865 feet.

B. M. 578.—*Aqueduct.* A square (□) cut in the top of a granite city monument set in concrete 3′ x 3′ x 4′ on the northeast corner of North Conduit Avenue (Old South Road) and the Rockaway Beach Division of the Long Island Railroad. This monument was set by the Queens Topographical Bureau. It is 7.55 feet south of the porch of the two-story and attic frame hotel on the corner, measured at a point 10.5 feet east of the southwest corner of the porch. The top of the monument is 8 inches below the ground.

Elevation, 22.333 feet.

B. M. 579.—*Ozone Park.* A square (□) cut in the top of the south end of the bluestone door-sill at the office entrance to John B. Reimer's coal office (one-story yellow brick), on the east side of Ocean Avenue about 237 feet south of Jerome Avenue (Broadway). The bench mark is 1.25 feet above the sidewalk.

Elevation, 41.666 feet.

B. M. 580.—*Woodhaven.* A horizontal line in the end of a ¾-inch copper bolt set horizontally into the granite foundation at the northeast corner of Sub. Sta. Power House No. 3 of the Long Island Railroad, on the southwest corner of Atlantic Avenue and Ocean Avenue. The bolt is 8 inches from the east side of the building and 1.5 feet above the ground. (Note 6, p. 90.)

Elevation, 42.409 feet.

B. M. 581.—*Woodhaven.* A square (□) cut in the top of the west end of the brownstone door-sill at the west or store entrance to the two-story brick house at No. 4264 Atlantic Avenue. The bench mark is on the south side of Atlantic Avenue about 75 feet west of Freedom Avenue (Union Avenue).

Elevation, 43.999 feet.

B. M. 582.—*Jamaica Avenue.* A ⅝-inch copper bolt set vertically in the top of the north abutment of railroad bridge over Jamaica Avenue (Jericho Turnpike), about 200 feet east of Little Neck Road. The bolt is on the west side of the bridge, 8 inches west of the base of the concrete bridge seat. (Note 5, p. 89.)

Elevation, 97.076 feet.

B. M. 583.—*Jamaica Avenue.* A square (⊔) cut in the top of the southeast corner of the second course of brownstone masonry above the ground, at the east end of the north abutment of railroad bridge over Jamaica Avenue (Jericho Turnpike), about 200 feet east of Little Neck Road. The bench mark is 3 feet above the ground and the letters B. M. are cut in the stone on the north side of the point.

Elevation, 86.794 feet.

B. M. 584.—*Jamaica Avenue.* A square (□) cut in the top of a granite city monument set in concrete 3′ x 3′ x 4′ on the south side of Jamaica Avenue (Jericho Turnpike) almost on line with the westerly line of Braddock Avenue (Rocky Hill Road). This monument was set by the Queens Topographical Bureau. It is 3 feet north of the picket fence and the top is 1 foot below the sidewalk.

Elevation, 81.567 feet.

B. M. 585.—*Little Neck Road.* A cross (+) cut in the west face of the foundation stone at the northwest corner of John Froehlichs' two-story and attic frame house, on the east side of Little Neck Road, about 550 feet south of the Motor Parkway. The cross is 6 inches above the ground. (Note 15, p. 91.)

Elevation, 107.460 feet.

B. M. 586.—*Little Neck Road.* A 1½-inch iron bolt in the top of the south end of the east abutment of Motor Parkway bridge over Little Neck Road. The bolt is 5.5 feet above the ground and 1 foot from the southwest corner of the abutment.

Elevation, 115.397 feet.

B. M. 587.—*Jamaica Avenue.* The top of a railroad spike driven vertically in the root on the north side of the large tree on the northeast corner of Jamaica Avenue (Jericho Turnpike) and Two Hundred and Twenty-second Avenue (Madison Avenue). The tree is on line with the east curb.

Elevation, 79.432 feet.

B. M. 588.—*Jamaica Avenue.* The top of a railroad spike driven vertically in the root on the north side of the large maple tree on the north side of Jamaica Avenue (Jericho Turnpike), about 35 feet west of Springfield Boulevard (Lincoln Avenue).

Elevation, 79.936 feet.

B. M. 589.—*Springfield Boulevard.* A square (□) cut in the top of the east end of the third stone step at the main entrance to Public School No. 34, on the west side of Springfield Boulevard (Springfield Road) just south of Hollis Avenue. The bench mark is at the base of the east balustrade on the north side of the building.

Elevation, 72.853 feet.

B. M. 590.—*Springfield Boulevard.* A horizontal line in the end of a ¾-inch copper bolt set horizontally into the limestone masonry at the northeast corner of Public School No. 34, on the west side of Springfield Boulevard (Springfield Road), just south of Hollis Avenue. The bolt is on the east side of the building, is 1.7 feet above the ground and 7 inches from the north side. (Note 6, p. 90.)

Elevation, 70.098 feet.

B. M. 591.—*Jamaica Avenue.* A square (□) cut in the top of a granite city monument set in concrete 3′ x 3′ x 4′ on the northwest corner of Jamaica Avenue (Jericho Turnpike) and Two Hundred and Twelfth Street (Queens Road). This monument was set by the Queens Topographical Bureau. It is 2 feet east of the cement sidewalk, is 8.5 feet north of the north curb and the top is 8 inches below the sidewalk.

Elevation, 72.594 feet.

B. M. 592.—*Braddock Avenue.* The surface within a semi-circle cut in the top of a large boulder in the west gutter of Braddock Avenue (Rocky Hill Road), about 75 feet north of the old Stewart Railroad. Established by the Queens Topographical Bureau.

Elevation, 154.827 feet.

B. M. 593.—*Madison Avenue.* A square (ᄇ) cut in the top of the southeast corner of the bluestone door-sill at the main entrance to Public School No. 33, on the northwest corner of Two Hundred and Twenty-second Avenue (Madison Avenue) and Carrington Avenue (Pine Street). The bench mark is 8.5 feet west of the east side of the building.

Elevation, 87.104 feet.

B. M. 594.—*Madison Avenue.* U. S. Geological Survey bench mark, on the northwest corner of Two Hundred and Twenty-second Avenue (Madison Avenue) and Carrington Avenue (Pine Street). The bench mark is the center of the top of a 3¾-inch bronze cap stamped U. S. Geological Survey B. M. 85 W. P., and screwed upon a 4-foot wrought-iron pipe set in the ground 5 inches west of the cement sidewalk on Madison Avenue and 11.5 feet north of the corner of the fence. It is 1.25 feet above the ground. Established in 1897.

Elevation, 85.295 feet.

B. M. 595.—*Madison Avenue.* A horiontal line in the end of a ¾-inch copper bolt set horizontally into the limestone masonry on the south side of Public School No. 33, on the northwest corner of Two Hundred and Twenty-second Avenue (Madison Avenue) and Carrington Avenue (Pine Street). The bolt is 1 foot east of the east side of the basement door and 6 inches above the bluestone coping at the base of the railing. (Note 6, p. 90.)

Elevation, 86.448 feet.

B. M. 596.—*Palatina Avenue.* A square (⌴) cut in the top of the northeast corner of the bluestone sill of the third basement window south of the main entrance to Public School No. 35, on the east side of One Hundred and Ninety-first Street (Palatina Avenue) about 300 feet north of Jamaica Avenue (Jericho Turnpike). The bench mark is on the west side of the building, is 1 foot above the ground, and the letters B. E. are cut alongside.

Elevation, 61.951 feet.

B. M. 597.—*Jamaica.* A square (□) cut in the top of a granite city monument set in concrete 3′ x 3′ x 4′ on the northwest corner of Homer Lee Avenue and Jamaica Avenue (Fulton Street). This monument was set by the Queens Topographical Bureau. It is 1.5 feet east of the end of the cement sidewalk and 6.3 feet south of the southeast corner of the old one-story frame house. The top of the monument is 7 inches below the ground.

Elevation, 55.030 feet.

B. M. 598.—*Jamaica.* A ⅝-inch copper bolt in a granite city monument on the southwest corner of Jamaica Avenue (Fulton Street) and Cliffside Avenue (Canal Street). The top of the monument is 5 inches below the sidewalk.

Elevation, 60.828 feet.

B. M. 599.—*Jamaica.* A square (⌴) cut in the top of the east end of the bluestone door-sill at the house entrance to No. 452 Jamaica Avenue (Fulton Street). The bench mark is about 35 feet west of Bergen Avenue (Smith Street).

Elevation, 62.055 feet.

B. M. 600.—*Jamaica.* A horizontal line in the end of a ¾-inch copper bolt set horizontally in the south face of the first course of brownstone at the southwest corner of the Jamaica Town Hall, on the northeast corner of Jamaica Avenue (Fulton Street) and Underhill Avenue (Flushing Avenue). The bolt is 5 inches above the walk and 5 inches from the corner. (Note 18, p. 91.)

Elevation, 60.004 feet.

B. M. 601.—*Jamaica.* A level rectangular surface 2″ x ½″ on top of the center of the west edge of the brownstone sill of the most southerly basement window on the west side of the Jamaica Town Hall, on the northeast corner of Jamaica Avenue (Fulton Street) and Underhill Avenue (Flushing Avenue). The bench mark is 10 feet north of the south side of the building. Established by the Queens Topographical Bureau.

Elevation, 60.060 feet.

B. M. 602.—*Jamaica.* A square (⌴) cut in the top of the east edge of the brownstone coping at the west end of the area on the south side of the Jamaica Town Hall, on the northeast corner of Jamaica Avenue (Fulton Street) and Underhill Avenue (Flushing Avenue). The bench mark is 2 feet from the west side of the building and 8 inches from the south side.

Elevation, 59.608 feet

B. M. 603.—*Jamaica.* A ¾-inch copper bolt set vertically in the top of the west end of the lowest step of the easterly stairway on the south side of the State Normal School, on the easterly side of Parsons Avenue (Jamaica Road), just north of Gilman Avenue (Highland Avenue). (Note 5, p. 89.)

Elevation, 101.051 feet.

B. M. 604.—*Jamaica.* The center of the surface within a square (□) cut in the top of the north end of the door-sill at the east entrance to the State Normal School, on the easterly side of Parsons Avenue (Jamaica Road), just north of Gilman Avenue (Highland Avenue). Established by the Queens Topographical Bureau.

Elevation, 101.438 feet.

B. M. 605.—*Jamaica.* A copper bolt in a granite city monument set in the lawn in King Park. The bench mark is 58.7 feet south of the south side of the two-story and attic mansion (King's Manor) or 22.2 feet south of the flag pole. It is about 18 feet east of the east line of Ezra Street (Division Street) and about 125 feet north of the iron picket fence on the north side of Jamaica Avenue (Fulton Street). (Note 24, p. 91.)

Elevation, 46.777 feet.

B. M. 606.—*Jamaica.* A square (□) cut in the top of the southeast corner of the second step at the girls' entrance to Public School No. 82, on the west side of Kaplan Street (Kaplan Avenue), between Leroy Place (Hammond Street) and Willett Street (Horton Avenue). The bench mark is near the southeast corner of the building, and faces Kaplan Street.

Elevation, 66.812 feet.

B. M. 607.—*Jamaica.* A horizontal line in the end of a ⅝-inch copper bolt set horizontally into the south face of a granite block at the southeast corner of Public School No. 82, on the west side of Kaplan Street (Kaplan Avenue), between Leroy Place (Hammond Street) and Willett Street (Horton Avenue). The bolt is 6 inches from the corner and 4 inches above the walk. (Note 6, p. 90.)

Elevation, 66.321 ft.

B. M. 608.—*Richmond Hill.* A square (□) cut in the top of the northeast corner of the limestone base block of iron column on the west side of the house entrance to No. 5058 Jamaica Avenue (three-story yellow brick). The bench mark is on the south side of Jamaica Avenue about 110 feet east of the center line of Magnolia Avenue (Ellsworth Avenue).

Elevation, 63.991 feet.

B. M. 609.—*Richmond Hill.* A square (□) cut in the top of the extreme north-east corner of the bluestone door-sill at the Jamaica Avenue entrance to the A. L. Reed Leather Co.'s factory (four-story brick) on the south side of Jamaica Avenue between Birch Street (Beech Street) and Spruce Street.

Elevation, 63.413 feet.

B. M. 610.—*Richmond Hill.* A ⅜-inch copper bolt in a granite city monument on the southwest corner of Lefferts Avenue and Jamaica Avenue. The top of the monument is 5 inches below the sidewalk.

Elevation, 60.598 feet.

B. M. 611.—*Richmond Hill.* The center of the top of a granite city monument on the southwest corner of Hatch Avenue (Napier Avenue) and Jamaica Avenue. The top of the monument is 7 inches below the sidewalk.

Elevation, 46.032 feet.

B. M. 612.—*Glendale.* A square (ᒷ) cut in the top of the southeast corner of the lowest bluestone step at the main entrance to the three-story brick apartment house at No. 3127 Myrtle Avenue, about 140 feet east of Woodhaven Avenue.*

Elevation, 131.072 feet.

B. M. 613.—*Glendale.* A square (ᒷ) cut in the top of the southeast corner of the lowest bluestone step at the entrance to the two-story and basement yellow brick house on the west side of Ford Avenue (No. 21), about 300 feet north of Myrtle Avenue.

Elevation, 85.829 feet.

B. M. 614.—*Glendale.* The center of the top of a granite city monument on the northwest corner of Proctor Street (Clinton Avenue) and Myrtle Avenue. The top of the monument is 3 inches above the sidewalk.

Elevation, 96.053 feet.

B. M. 615.—*Glendale.* A square (□) cut in the top of the southwest corner of the raised cement walk on the northeast corner of Tompkins Place and Myrtle Avenue.

Elevation, 94.329 feet.

B. M. 616.—*Glendale.* A square (□) cut in the top of the north end of the bluestone door-sill at the boys' entrance to Public School No. 68, on the northerly side of St. Felix Avenue (Bergen Avenue) between Seneca Avenue (Henry Street) and Rathjen Street.. The bench mark is on the east side of the building, about 22 feet from the southeast corner.

Elevation, 80.776 feet.

B. M. 617.—*Glendale.* A horizontal line in the end of a ⅝-inch copper bolt set horizontally into the granite foundation at the northwest corner of Public School No. 68, on the north side of St. Felix Avenue (Bergen Avenue) between Seneca Avenue (Henry Street) and Rathjen Street. The bolt is on the west side of the building, is 4½ inches from the north side, and 6½ inches above the ground. (Note 18, p. 91.)

Elevation, 80.762 feet.

B. M. 618.—*Glendale.* A square (□) cut in the top of the north end of the lowest step at the basement entrance to Public School No. 67, on the north side of Central Avenue between Tompkins Place and Olmsted Place (Webster Avenue). The bench mark is on the south side of the building.

Elevation, 98.592 feet.

B. M. 619.—*Glendale.* A cross, lettered B. + M., cut in the northerly face of the southerly abutment of the Brooklyn Rapid Transit railroad bridge over the Montauk Division of the Long Island Railroad, on the southerly side of Lutheran Cemetery. The cross is almost under the easterly girder of the bridge and about 2.9 feet above the Long Island Railroad tracks. (Note 15, p. 91.)

Elevation, 93.552 feet.

B. M. 620.—*Glendale.* A quadrant (ᒪ) cut in the top of the southwest corner of the third step at the westerly end of the northerly abutment of the Brooklyn Rapid Transit railroad bridge over the Montauk Division of the Long Island Railroad, on the southerly side of Lutheran Cemetery.

Elevation, 97.963 feet.

*Reported as having been disturbed by re-grading during 1914.

B. M. 621.—*Glendale.* A square (◻) cut in the top of the southeast corner of granite basin head on the northeast corner of Metropolitan Avenue and Fresh Pond Road. The bench mark is north of the Montauk Division of the Long Island Railroad and on the north curb line of Metropolitan Avenue.

Elevation, 92.072 feet.

B. M. 622.—*Glendale.* A square (◻) cut in the top of a granite city monument set in concrete 3′ x 3′ x 4′ on the northerly side of Woodhaven Avenue (Trotting Course Lane) about 1,000 feet north of Myrtle Avenue or about 325 feet north of the Rockaway Beach Division of the Long Island Railroad. This monument was set by the Queens Topographical Bureau. It is 4.5 feet from the board fence, measured at a point opposite the southeasterly corner of the Wicke ribbon factory. The top is 1 foot below the ground.

Elevation, 87.284 feet.

B. M. 623.—*Creek Street.* The center of the top of a granite city monument on the west side of Creek Street about 170 feet south of Grand Street. The top of the monument is flush with the ground.

Elevation, 25.648 feet.

B. M. 624.—*Creek Street.* A square (◻) cut in the top of the westerly corner of the bluestone door-sill at the girls' entrance to Public School No. 86, on the westerly side of Creek Street between Grand Street and Frederick Street. The bench mark is on the northerly side of the building about 21.7 feet from the Creek Street side.

Elevation, 29.778 feet.

B. M. 625.—*Creek Street.* A horizontal line in the end of a ⅝-inch copper bolt set horizontally into the second course of bluestone at the northeasterly corner of Public School No. 86, on the westerly side of Creek Street between Grand Street and Frederick Street. The bolt is on the Creek Street side of the building, is 3½ inches from the north side and 1.9 feet above the area. (Note 18, p. 91.)

Elevation, 31.242 feet.

B. M. 626.—*Laurel Hill.* The intersection of a cross (+) cut in the top of a large boulder on the southwest corner of Munich Street and the Montauk Division of the Long Island Railroad. The bench mark is at the northeast corner of the two-story red brick factory building of the National Enamel and Stamping Co.

Elevation, 26.620 feet.

B. M. 627.—*Laurel Hill.* A square (◻) cut in the center of the top of a large granite copingstone at the easterly end of the northerly abutment of the highway bridge over Newtown Creek, at Laurel Hill Boulevard. The bench mark is 1 foot northeasterly from the intersection of the iron railings at the northeasterly corner of the bridge. (Note 9, p. 90.)

Elevation, 14.008 feet.

B. M. 628.—*Laurel Hill.* A square (◻) cut in the top of the wide end of the granite copingstone at the westerly end of the northerly abutment of the highway bridge over Newtown Creek at Laurel Hill Boulevard. The bench mark is 6 inches northwesterly from the corner of the abutment near the westerly railing and above the westerly end of the bridge seat.

Elevation, 14.453 feet.

B. M. 629.—*Laurel Hill.* A triangle cut in the top of the granite coping on the northerly side of the highway bridge over Newtown Creek at Laurel Hill Boulevard. The bench mark is 2.1 feet easterly from the westerly side of the bridge and is just west of the center line of the westerly foot path.

Elevation, 14.449 feet.

B. M. 630.—*Laurel Hill.* A square (□) cut in the top of the granite coping on the northerly side of the highway bridge over Newtown Creek at Laurel Hill Boulevard. The bench mark is 2.5 feet easterly from the westerly side of the bridge and is just west of the center line of the westerly foot path.

Elevation, 14.441 feet.

B. M. 631.—*Laurel Hill.* The center of a granite city monument on the northerly corner of Review Avenue and Laurel Hill Boulevard. The monument is 5 feet south of the southeasterly corner of Calvary Cemetery, and 3 feet northwesterly from the northwesterly curb of Laurel Hill Boulevard. The top of the monument is 1 inch below the ground.

Elevation, 13.871 feet.

B. M. 632.—*Greenpoint Avenue.* A square (⊡) cut in the south end of the granite basin head on the southeast corner of Greenpoint Avenue and Review Avenue.

Elevation, 16.805 feet.

B. M. 633.—*Greenpoint Avenue.* The center of the top of a granite city monument on the northwesterly corner of Greenpoint Avenue and Bradley Avenue. The top of the monument is 5 inches below the sidewalk.

Elevation, 36.894 feet.

B. M. 634.—*Greenpoint Avenue.* A square (⊡) cut in the top of the iron protective plate in front of Engine House No. 259, on the northerly side of Greenpoint Avenue between Bradley Avenue and Gale Street. The bench mark is on the east side of the east entrance, about 84 feet from Gale Street. It is 3.7 feet above the sidewalk and the letters B. M. are cut in the limestone above the point.

Elevation, 52.066 feet.

B. M. 635.—*Greenpoint Avenue.* The top of a ¾-inch iron rod inside of a wrought-iron pipe casing in the northerly side of Calvary Cemetery. The bench mark is about 255 feet west of Borden Avenue, is 2 feet south of the inside of the cemetery wall and about 90 feet east of the Greenpoint Avenue entrance. (Note 21, p. 91.)

Elevation, 53.652 feet.

B. M. 636.—*Borden Avenue.* A square (□) cut in the granite copingstone on top of the easterly abutment of the highway bridge on Borden Avenue, over Dutch Kills Creek. The bench mark is on the south side of Borden Avenue about 12 feet south of the south curb, and 6 inches east of the east edge of the iron railing.

Elevation, 16.684 feet.

B. M. 637.—*Borden Avenue.* A square (⊡) cut in the top of the northeast corner of the 3.7 feet wide granite copingstone on the south end of the east abutment of the highway bridge on Borden Avenue, over Dutch Kills Creek. The bench mark is on line with the south curb of the bridge. Established by the Department of Bridges.

Elevation, 16.554 feet.

B. M. 638.—*Borden Avenue.* A cross (+) cut in the top of the southeasterly anchor bolt at the base of electric light pole on the north end of the easterly abutment of the highway bridge on Borden Avenue, over Dutch Kills Creek. The bench mark is on line with the north side of the bridge and 3 feet west of the bridge tender's house. Established by the Queens Topographical Bureau.

Elevation, 16.930 feet.

B. M. 639.—*Borden Avenue.* A square (⊡) lettered U. S., cut in the top of the north end of the east abutment of the highway bridge on Borden Avenue, over Dutch Kills Creek. The bench mark is on the southwest corner of the 3.7 feet wide copingstone, is on the west edge of the abutment and on line with the north curb. (Note 16, p. 91.)

Elevation, 16.527 feet.

B. M. 640.—*Borden Avenue.* A square (⊡) lettered U. S., cut in the top of the southeast corner of the 2.5 feet wide granite copingstone on the north end of the west abutment of the highway bridge on Borden Avenue, over Dutch Kills Creek. The bench mark is on the east edge of the west abutment and on line with the north curb. (Note 16, p. 91.)

Elevation, 16.526 feet.

B. M. 641.—*Borden Avenue.* A ⅝-inch copper bolt set vertically in the granite copingstone on the north end of the west abutment of the highway bridge on Borden Avenue, over Dutch Kills Creek. The bolt is at the east corner of the iron railing, is 11 inches from the north end of the abutment and 3½ inches from the east end. (Note 5, p. 89.)

Elevation, 16.625 feet.

B. M. 642.—*Vernon Avenue.* A square (⊡) cut in the top of the southwest corner of a granite copingstone on top of the masonry bulkhead on the north side of Newtown Creek at Vernon Avenue. The bench mark is about 11 feet east of the west curb line of Vernon Avenue.

Elevation, 8.351 feet.

B. M. 643.—*Vernon Avenue.* A square (□) cut in the top of the southwest corner of the granite base block of the most southerly steel column under the westerly side of the highway bridge (Jackson Avenue Bridge) over Newtown Creek at Vernon Avenue. The bench mark is about 41 feet north of the north side of Newtown Creek and about 19 feet east of the west curb line of Vernon Avenue.

Elevation, 9.252 feet.

B. M. 644.—*Vernon Avenue.* A square (⊡) cut in the top of the south edge of the granite base block of steel column under the east side of the approach to the highway bridge over Newtown Creek (Jackson Avenue Bridge) at Vernon Avenue. The bench mark is about 5 feet north of the north line of Borden Avenue on the east side of Vernon Avenue.

Elevation, 13.216 feet.

B. M. 645.—*Borden Avenue.* A square (□) cut in the top of the east end of the bluestone door-sill at the main entrance to the Y. M. C. A. building on the north side of Borden Avenue about 27 feet west of West Avenue.

Elevation, 12.768 feet.

B. M. 646.—*Borden Avenue.* Coast and Geodetic Survey B. M. B., on Miller's Long Island City Hotel, on the northeast corner of Borden Avenue and Front Street. The entrance to the building is at the southwest corner, there being a triangular space or vestibule before the door. The bench mark is a cross (+) within a square (⊓), cut in the top of the bluestone sill at the northwest corner of this space, about 1 inch from the west edge of the sill and 3¼ inches from the south end of the brick wall. Established in May, 1887.

Elevation, 7.882 feet.

B. M. 647.—*Maspeth.* The top of a ¾-inch iron rod inside of a wrought-iron pipe casing in the Grand Street side of Mount Olivet Cemetery. The bench mark is 2.7 feet inside of the iron fence on the south side of Grand Street and about 10 feet west of the west line of Remsen Place. The top of the casing is 5 inches above the ground. (Note 21, p. 91.)

Elevation, 63.792 feet.

B. M. 648.—*Maspeth.* A square (□) lettered B. W. S., cut in the top of the east end of the concrete door-sill at the entrance to the Brooklyn Rapid Transit waiting room on the southeast corner of Grand Street and Juniper Avenue. The waiting room is in the northwest corner of the two-story red brick car barn and the bench mark is 6.6 feet from Juniper Avenue.

Elevation, 87.701 feet.

B. M. 649.—*Newtown.* The top of an iron ring bolt set vertically in the top of the west end of the south abutment of the highway bridge over the main line of the Long Island Railroad at Grand Street. The bolt is about 4.6 feet above the bridge seat.

Elevation, 56.428 feet.

B. M. 650.—*Newtown.* The top of a 1-inch square iron bolt set vertically in the east end of the bridge seat of the south abutment of highway bridge over the main line of the Long Island Railroad at Grand Street. The top of the bolt projects about ¾ inch and the letters B. M. are cut alongside. Established in 1909 by the Long Island Railroad.

Elevation, 51.950 feet.

B. M. 651.—*Newtown.* A ⅝-inch copper bolt set vertically in the top of the east end of the bridge seat of the north abutment of the highway bridge over the main line of the Long Island Railroad at Grand Street. The bolt is 2.6 feet from the east end of the bridge seat and 8½ inches from the south side of the abutment. (Note 5, p. 89.)

Elevation, 51.211 feet.

B. M. 652.—*Elmhurst.* A square (□) cut in the top of the bluestone door-sill at the main or corner entrance to the Elmhurst Branch of the Bank of Long Island (two-story yellow brick), on the westerly corner of Broadway and Fitch Avenue (Whitney Avenue). The bench mark is 9 inches from the south end of the sill, which is at the easterly corner of the building.

Elevation, 36.984 feet.

B. M. 653.—*Elmhurst.* A 2-inch brass tablet bolt set vertically in the top of the limestone base of the pilaster on the southerly side of the main entrance to Public School No. 89, on the northeasterly corner of Hampton Street (Fifth Street) and Britton Avenue (Orchard Avenue). The bolt is on the Britton Avenue side of the building. (Note 2, p. 89.)

Elevation, 69.318 feet.

B. M. 654.—*Queensboro Bridge.* A horizontal line in the end of a ¾-inch copper bolt set horizontally into the center of the second course of dressed granite on the west side of the Long Island City abutment of the Queensboro Bridge. The bench mark is on the east side of Ely Avenue and on the axis of the bridge. It is 1.42 feet above the sidewalk and sets back ⅜ inch. (Note 6, p. 90.)

Elevation, 20.427 feet.

B. M. 655.—*Queensboro Bridge.* The center of a level square cut in the top of the concave coping on the west side of the Long Island City abutment of the Queensboro Bridge. The bench mark is on the east side of Ely Avenue and on the axis of the bridge. It is 1.83 feet above the sidewalk. (Note 10, p. 90.)

Elevation, 20.833 feet.

B. M. 656.—*Queensboro Bridge.* A square (□) cut in the top of the northwesterly corner of the base block of the square granite booth on the northerly side of the Queensboro Bridge, between William Street and Crescent Street. The bench mark is on the most westerly of two small booths on the northerly side of the bridge, is about 100 feet east of William Street and 1.94 feet above the sidewalk.

Elevation, 29.117 feet.

B. M. 657.—*Jackson Avenue.* A ½-inch copper bolt set vertically in the center of the top of the granite window-sill of the building of Hook and Ladder Co. No. 66, on the northerly side of Jackson Avenue, between Ninth Avenue and Steinway Avenue (Tenth Avenue). The bolt is 3.5 feet from the west side of the building. (Note 8, p. 90.)

Elevation, 37.844 feet.

B. M. 658.—*Webster Avenue.* A square (□) cut in the top of the west end of bluestone door-sill at the house entrance to three-story red brick tenement at No. 507 Webster Avenue. The bench mark is on the northerly side of Webster Avenue between Jackson Avenue and Eighth Avenue.

Elevation, 33.552 feet.

B. M. 659.—*Prospect Street.* A square (□) cut in the top of the southerly corner of the fifth bluestone step or landing at the main entrance to Public School No. 4, on the northwesterly side of Prospect Street, between Payntor Avenue and Beebe Avenue.

Elevation, 46.458 feet.

B. M. 660.—*Jackson Avenue.* At the New York and Queens Railroad Company's car barn, on the southeast corner of Jackson Avenue and Woodside Avenue. The bench mark is a square (□) cut in the top of the horizontal surface at the east end of the bluestone sill of the small west window on the Jackson Avenue side of the brick clock tower. The point is 4.1 feet from the west side of the tower and 2.2 feet above the ground. Established by the Queens Topographical Bureau.

Elevation, 46.621 feet.

B. M. 661.—*Jackson Avenue.* A square (□) cut in the center of the top of a granite city monument set in concrete 3′ x 3′ x 4′, on the southeast corner of Jackson Avenue and Dickson Avenue. This monument was set by the Queens Topographical Bureau. It is 17.3 feet south of the south curb of Jackson Avenue and 11.3 feet east of the east curb of Dickson Street. The top of the monument is 2 inches below the ground.

Elevation, 47.706 feet.

B. M. 662.—*Astoria.* A square (◻) cut in the center of the top of the east edge of the granite basin head on the southwesterly corner of Broadway and Twelfth Avenue. Established by the Queens Topographical Bureau.

Elevation, 50.737 feet.

B. M. 663.—*Astoria.* The center of the top of the granite city monument on the northeasterly corner of Broadway and Ely Avenue. The top of the monument is 3 inches below the sidewalk.

Elevation, 16.668 feet.

B. M. 664.—*Astoria.* The center of the top of the granite city monument on the northeasterly corner of Broadway and Vernon Avenue. The top of the monument is 2 inches below the sidewalk.

Elevation, 9.779 feet.

B. M. 665.—*Astoria.* The top of a ¾-inch iron rod inside of a wrought-iron pipe casing in the public park on the westerly side of Vernon Avenue at the foot of Graham Avenue. The bench mark is on line with the southerly side of Graham Avenue and about 15 feet westerly from Vernon Avenue. The top of the casing is now 2 feet above the ground. (Note 21, p. 91.)

Elevation, 29.436 feet.

B. M. 666.—*Astoria.* A horizontal line on the face of a brass tablet bolt 2 inches in diameter set horizontally into the limestone at the northwesterly corner of Public School No. 83, on the easterly side of Vernon Avenue between Graham Avenue and Pierce Avenue. The bolt is on the Vernon Avenue side of the building, is 6 inches from the northerly side and 1.25 feet above the walk. (Note 1, p. 89.)

Elevation, 33.020 feet.

B. M. 667.—*Astoria.* A square (□) cut in the top of the easterly end of the lowest bluestone step at the northerly side of the main entrance to Public School No. 83, on the easterly side of Vernon Avenue between Graham Avenue and Pierce Avenue.

Elevation, 32.406 feet.

B. M. 668.—*Astoria.* A cross (+) cut in the westerly side of the granite base course at the northwesterly corner of Public School No. 83, on the easterly side of Vernon Avenue between Graham Avenue and Pierce Avenue. The cross is 6 inches from the northerly side of the building and 3¾ inches above the walk. (Note 15, p. 91.)

Elevation, 32.029 feet.

B. M. 669.—*Astoria.* A square (□) lettered B. W. S., cut in the top of the northeasterly corner of the second bluestone step at the main entrance to the Astoria Branch of the Corn Exchange Bank, on the northwesterly corner of Fulton Avenue and the Boulevard. The bench mark is at the southwesterly corner of the building, which is a four-story red brick.

Elevation, 13.263 feet.

B. M. 670.—*Astoria.* The top of a ¾-inch iron rod inside of a wrought-iron pipe casing in the north end of the Astoria Athletic Field (property of the Board of Education). The bench mark is just south of Hell Gate Light and 21 feet east of the concrete sea wall. (Note 21, p. 91.)

Elevation, 8.177 feet.

B. M. 671.—*Astoria.* A square (☐) cut in the top of the west end of the bluestone door-sill at the house entrance to the four-story yellow brick tenement at No. 385 Astoria Avenue (Flushing Avenue). The bench mark is on the north side of Astoria Avenue just west of Twelfth Avenue.

Elevation, 62.272 feet.

B. M. 672.—*Astoria.* A ⅝-inch copper bolt set vertically in the top of the granite sill of basement window at the southwest corner of Public School No. 84, on the easterly side of Eleventh Avenue between Potter Avenue and Ditmars Avenue. The bolt is 6 inches from the northerly end of the sill, is on the westerly side of the building, and 11.2 feet from the southerly side. (Note 5, p. 89.)

Elevation, 67.169 feet.

B. M. 673.—*Astoria.* A cross (+) cut in the granite foundation at the southwest corner of Public School No. 84, on the easterly side of Eleventh Avenue between Potter Avenue and Ditmars Avenue. The cross is on the westerly side of the building, is 6 inches north of the corner and 6 inches above the ground. (Note 15, p. 91.)

Elevation, 66.847 feet.

B. M. 674.—*Astoria.* A square (⊓) cut in the top of the center of the outer edge of the bluestone sill of the most easterly of three blind windows on the southerly side of the three-story brick tenement on the northeast corner of Wolcott Avenue and Steinway Avenue (Tenth Avenue). The bench mark is about 27 feet east of Tenth Avenue.

Elevation, 45.446 feet.

B. M. 675.—*Astoria.* Coast and Geodetic Survey B. M. No. 4a. A triangle cut in the top stone of the northerly corner of old "Polhemus Dock," on the Boulevard (River Road) about 255 feet northeasterly from the northeasterly line of Wolcott Avenue. This bench mark is now covered by an additional course of masonry. Established in 1886.

Elevation, 6.762 feet.

B. M. 676.—*Astoria.* Coast and Geodetic Survey B. M. No. 3. On the Boulevard (River Road), between Ditmars Avenue and Wolcott Avenue. The bench mark is a triangle cut in the lowest bluestone step or landing at the entrance to the Ditmars estate, formerly Barclay's. Two iron gate-posts rest on this step and the bench mark is 6 inches from the most easterly one and about 200 feet northeasterly from the northeasterly line of Ditmars Avenue. Established in 1886.

Elevation, 16.688 feet.

B. M. 677.—*Jackson Avenue.* A copper bolt in the top of a granite city monument set in concrete 3′ x 3′ x 4′, on the north side of Jackson Avenue, 66 feet west of the west side of Trains Meadow Road. This monument was set by the Queens Topographical Bureau. (Note 24, p. 91.)

Elevation, 40.400 feet.

B. M. 678.—*Trains Meadow Road.* A square (☐) cut in the top of the southwest corner of the large bluestone flag at the foot of the stone steps leading to Oak Hill Park, on the east side of Trains Meadow Road. The bench mark is about 175 feet south of the south side of Jackson Avenue and 5 inches from the south balustrade.

Elevation, 48.090 feet.

B. M. 679.—*Jackson Avenue.* The center of the top of a granite city monument on the northeast corner of Jackson Avenue and Junction Avenue. The top of the monument is flush with the sidewalk.

Elevation, 72.432 feet.

B. M. 680.—*Jackson Avenue.* A square (⊓) cut in the top of the center of the outer edge of the bluestone sill of west window of blacksmith shop (two-story red brick) on the south side of Jackson Avenue about 125 feet east of Junction Avenue. The bench mark is on the north side of the building, is 2.5 feet from the west side and 2.1 feet above the ground.

Elevation, 74.953 feet.

B. M. 681.—*Jackson Avenue.* A copper bolt in the top of a granite city monument set in concrete 3′ x 3′ x 4′, on the northeast corner of Jackson Avenue and Forty-seventh Street (Washington Avenue). The top of the monument is 1.5 feet below the ground. (Note 24, p. 91.)

Elevation, 65.879 feet.

B. M. 682.—*Jackson Avenue.* A square (□) cut in the top of the north end of the brownstone balustrade on the west side of the house entrance to No. 268 Jackson Avenue (two-story old frame dwelling) about 300 feet west of Pell Street (Shell Road). The bench mark is on the north side of Jackson Avenue and 2 feet above the sidewalk.

Elevation, 28.667 feet.

B. M. 683.—*Flushing Bridge.* A square (⊓) cut in the top of the west edge of the granite coping of the west abutment of the highway bridge over Flushing Creek at Jackson Avenue. The bench mark is on the south side of Jackson Avenue between two square granite posts on the west abutment and on the west side of the stone fence. It is 5.9 feet south of the northerly post and 7.15 feet north of the southerly post.

Elevation, 16.782 feet.

B. M. 684.—*Flushing Bridge.* A ⅝-inch copper bolt set vertically in the top of the north end of the east abutment of the highway bridge on Jackson Avenue, over Flushing Creek. The bolt is 9 inches from the west face of the abutment and 6 inches south of the granite post at the north end. (Note 5, p. 90.)

Elevation, 16.816 feet.

B. M. 685.—*Flushing.* A square (⊓) cut in the top of the outer edge of the concrete water-table in front of the Queens Borough garage (two-story cement block) at No. 66 Jackson Avenue (Broadway), between Prince Street and Main Street. The bench mark is about 14 feet from the west side of the building and 11 inches above the sidewalk. It is 3 inches from the east end of the water-table.

Elevation, 24.107 feet.

B. M. 686.—*Flushing.* A ½-inch copper bolt set vertically in the southwest corner of the granite base block of the limestone column at the south side of the main entrance to the Queens County Savings Bank, on the northeast corner of St. George's Place (Locust Street) and Main Street. The bolt is 3 feet above the sidewalk, is 6.5 feet from the south side of the building and almost on line with the west side. (Note 8, p. 90.)

Elevation, 38.665 feet.

B. M. 687.—*Flushing.* A square (⊓) cut in the top of the east end of the coping-stone in front of the Flushing Post Office (Marx Building) on the south side of Amity Street about 115 feet east of Main Street. The bench mark is 7 feet west of the east end of the building and 3½ inches above the sidewalk.

Elevation, 38.836 feet.

B. M. 688.—*Flushing.* A square (⊔) cut in the top of the northeasterly corner of the granite basin head on the northerly corner of Sanford Avenue and Kissena Road (Jamaica Road).

Elevation, 49.857 feet.

B. M. 689.—*Flushing.* A square (⊓) cut in the top of the west end of the bluestone door-sill at the southwesterly corner of Public School No. 22, on the northeasterly corner of Sanford Avenue and Murray Street. The bench mark is on the Sanford Avenue side of the building under the shed, is 2.25 feet from the west side and 4 inches above the walk. It is 8.5 feet north of the Sanford Avenue side of the main building.

Elevation, 72.713 feet.

B. M. 690.—*Flushing.* A horizontal line in the end of a ½-inch copper bolt set horizontally into the brick at the southwesterly corner of Public School No. 22, on the northeasterly corner of Sanford Avenue and Murray Street. The bolt is on the Murray Street side of the building, is 1.75 feet from the corner and 1.75 feet above the ground. (Note 18, p. 91.)

Elevation, 74.303 feet.

B. M. 691.—*Flushing.* A square (⊓) cut in the top of the west end of the bluestone sill of the eighth basement window from Murray Street on the south side of Public School No. 22, on the northeasterly corner of Murray Street and Sanford Avenue. The bench mark is 8 feet east of the boys' entrance and 8 inches above the walk.

Elevation, 72.810 feet.

B. M. 692.—*Kissena Road (Jamaica Road).* A square (□) cut in the top of the bluestone coping on the wing wall at the north end of the east parapet wall of culvert crossing Kissena Road about 200 feet north of Seventy-fifth Avenue (Quarrelsome Lane). The bench mark is 3.3 feet above the surface of the road.

Elevation, 73.909 feet.

B. M. 693.—*Kissena Road (Jamaica Road).* A square (⊔) cut in the top of the northeast corner of the bluestone coping of the east parapet wall of culvert crossing Kissena Road about 200 feet north of Seventy-fifth Avenue (Quarrelsome Lane). The bench mark is 4.5 feet above the surface of the road and the letters B. M. are cut in the stone on the west side of the point.

Elevation, 75.071 feet.

B. M. 694—*Kissena Road (Jamaica Road).* The center of a 2-inch brass bolt set vertically in the top of the northeast corner of the large bluestone door-sill at the main entrance to the New York City Parental School (Administration Building) on the west side of Kissena Road between the North Hempstead Turnpike and Seventy-fifth Avenue (Quarrelsome Lane). (Note 2, p. 89.)

Elevation, 117.612 feet.

B. M. 695.—*Kissena Road (Jamaica Road).* A square (□) cut in the top of a large partly buried boulder near the gutter on the east side of Kissena Road about 50 feet north of the north side of the north entrance to the grounds of the New York City Parental School. Established by the Queens Topographical Bureau.

Elevation, 85.397 feet.

B. M. 696.—*Kissena Road* (*Jamaica Road*). A square (⊔) cut in the top ot the northwest corner of the bluestone coping of the east parapet wall of culvert crossing Kissena Road about 400 feet north of the North Hempstead Turnpike.

Elevation, 13.996 feet.

B. M. 697.—*Kissena Road* (*Jamaica Road*). A square (□) cut in the top of the cement coping of balustrade on the south side of the stone steps leading to the basement of the two-story frame dwelling at No. 276 Kissena Road about 120 feet north of Holly Street. The bench mark is 6 inches west of the wooden column supporting the porch.

Elevation, 49.638 feet.

B. M. 698.—*Beechhurst Avenue* (*Whitestone Road*). A square (⊔) cut in the top of the southeast corner of the granite property monument at the corner of the fence on the northwest corner of Beechhurst Avenue and the North Side Division of the Long Island Railroad. The top of the monument is 4 inches above the ground.

Elevation, 65.737 feet.

B. M. 699.—*Bayside.* A ½-inch copper bolt set vertically in the top of the southeast corner of the cement door-sill at the main entrance to drug store on the northwest corner of Bell Avenue and the North Side Division of the Long Island Railroad. The bench mark is on the east side of the building (two-story cement block) and 3 inches from the east side of the sill. (Note 8, p. 90.)

Elevation, 81.325 feet.

B. M. 700.—*Bayside.* The center of a brass bolt marking the U. S. C. & G. S. triangulation station Bayside. The bench mark is on the estate of G. Howland Leavitt, on the south side of Poppenhusen Avenue (Little Bayside Road) and about 1,500 feet west of Rockingham Street (Bayside Road). It is in the lawn in the rear of the two-story frame house now occupied by Vincent Newman, and about 340 feet south of Poppenhusen Avenue. It is 16 feet west of the post and wire fence on the east side of the house. (Note 25, p. 91.)

Elevation, 109.979 feet.

B. M. 701.—*Bayside.* A square (⊓) cut in the top of the south edge of the bluestone coping of the wing at the south end of the east parapet wall of culvert crossing Bell Avenue and about 300 feet north of Poppenhusen Avenue (Little Bayside Road). Established by the Queens Topographical Bureau.

Elevation, 16.025 feet.

B. M. 702.—*Bayside.* The center of the top of a granite city monument set in concrete 3′ x 3′ x 4′, on the south side of Rocky Hill Road at Bell Avenue. This monument was set by the Queens Topographical Bureau. It is 4.5 feet north of the picket fence on the south side of Rocky Hill Road and about 54 feet west of the easterly line of Bell Avenue. The top of the monument is 1 foot below the ground.

Elevation, 73.644 feet.

B. M. 703.—*Bayside.* A square (⊓) cut in the top of the center of the east edge of the lowest marble step in front of the main entrance to Public School No. 31, on the west side of Bell Avenue between Jackson Avenue (Broadway) and Rocky Hill Road.

Elevation, 70.050 feet.

B. M. 704.—*Bayside.* A 2-inch cross (+) cut in the south face of foundation stone at the southeast corner of the one-story and attic dwelling on the north side of Rocky Hill Road, at the entrance to the Oakland Golf Club. The house containing the bench mark is about 1,500 feet north of Sixty-first Avenue (West Alley Road), at the right-angled turn in Rocky Hill Road. The cross is 1.5 feet above the ground and 5 inches from the east side of the house. (Note 15, p. 91.)

Elevation, 67.543 feet.

B. M. 705.—*Springfield Boulevard (Rocky Hill Road).* A square (□) cut in the top of the large brownstone boulder on the northeast corner of Springfield Boulevard (Rocky Hill Road) and Seventy-third Avenue (Cedar Lane).

Elevation, 121.957 feet.

B. M. 706.—*Douglaston.* A square (□) cut in the top of a granite city monument set in concrete 3' x 3' x 4', on the west side of Main Avenue, about 300 feet south of the North Side Division of the Long Island Railroad. This monument was set by the Queens Topographical Bureau. It is in front of the Douglaston Hose Company building, is 10.5 feet easterly from the southeast corner of the building and 23.0 feet southeasterly from the northeast corner. The top of the monument is 1 foot below the ground.

Elevation, 40.491 feet.

B. M. 707.—*Douglaston.* A square (⊓) cut in the top of the retaining wall on the east side of Main Avenue, about 16 feet south of the south side of the Douglaston Hose Company building. The bench mark is 2.3 feet above the sidewalk and an arrowhead cut in the stone below points to the bench. Established by the Department of Docks and Ferries.

Elevation, 44.557 feet.

B. M. 708.—*Little Neck.* The top of the fire hydrant on the westerly side of Two Hundred and Fiftieth Street (Old House Landing Road), about 50 feet north of the North Side Division of the Long Island Railroad.

Elevation, 33.787 feet.

B. M. 709.—*Little Neck.* A square (□) cut in the top of the west end of the bluestone door-sill on the south side of the Long Island Railroad station, just east of Two Hundred and Fiftieth Street (Old House Landing Road).

Elevation, 36.739 feet.

B. M. 710.—*Little Neck.* The top of a railroad spike driven vertically in the root of a large oak tree on the north side of Jackson Avenue (Broadway), about 150 feet west of Little Neck Road. The bench mark is on the north side of the tree.

Elevation, 83.122 feet.

B. M. 711.—*Little Neck.* A square (□) cut in the top of the northeast corner of the bluestone coping on the west parapet wall of culvert crossing Little Neck Road, about 2,000 feet north of Sixty-first Avenue (Hyde Park Road). The bench mark is just north of Van Nostrand's farm and about 750 feet north of two masonry columns surmounted by lions at the entrance to the Vanderbilt estate. The bench mark is 3 feet above the roadway.

Elevation, 185.143 feet.

B. M. 712.—*Little Neck.* A quadrant (∟) cut in the top of a projecting foundation stone at the northwest corner of Public School No. 32 (one-story and attic frame), on the east side of Little Neck Road about 50 feet south of Lakeville Road (Floral Park Road). The bench mark is on the west side of the building and 6 inches from the north side.

Elevation, 157.876 feet.

B. M. 713.—*Little Neck.* A quadrant (⌊) cut in the top of the bluestone coping of the second step of wing wall at the north end of the west parapet wall of culvert crossing Little Neck Road, about 2,000 feet north of Sixty-first Avenue (Hyde Park Road). The bench mark is on the first culvert north of Hyde Park Road and is in bad condition. Established by the Queens Topographical Bureau.

Elevation, 183.809 feet.

B. M. 714.—*Little Neck.* The top of a 1-inch brass bolt marking the U. S. C. & G. S. triangulation station Payne. The bench mark is on the Schenck farm, is about 810 feet east of Little Neck Road and about 435 feet south of Sixty-first Avenue (Hyde Park Road). The regular U. S. C. & G. S. brass marking bolt has been broken off leaving the 1-inch shank in the monument, which is set in concrete 3′ x 3′ x 4′. The bench mark is flush with the ground and 16 feet west of the post and wire fence on the west side of the Vanderbilt estate.

Elevation, 266.488 feet.

B. M. 715.—*Little Neck.* The center of a granite city monument on the Schenck farm, 12.92 feet southeast of U. S. C. & G. S. triangulation station Payne. The monument was set by the Queens Topographical Bureau and is 7 feet west of the post and wire fence on the west side of the Vanderbilt estate.

Elevation, 266.460 feet.

B. M. 716.—*Little Neck.* The center of the granite city monument at the intersection of Little Neck Road and Sixty-first Avenue (Hyde Park Road). The monument is about 14 feet west of the fence line on the east side of Little Neck Road and on the center line of Sixty-first Avenue. The top of the monument is 1 inch below the ground.

Elevation, 237.046 feet.

B. M. 717.—*Flushing.* The center of the top of the granite city monument on the southeasterly corner of Connorton Avenue (Myrtle Avenue) and Tallman Avenue (Lawrence Avenue). The top of the monument is flush with the ground.

Elevation, 8.770 feet.

B. M. 718.—*College Point.* A horizontal line in the end of a ½-inch copper bolt set horizontally into the second course of masonry at the northeast corner of the four-story brick factory building on the southwest corner of Schlesinger Street (Nineteenth Street) and Kinney Avenue (Fifth Avenue). The bolt is on the east side of the building, is 1.2 feet above the ground and 6 inches from the corner. (Note 18, p. 91.)

Elevation, 43.631 feet.

B. M. 719.—*College Point.* A square (⊓) cut in the top of the northeast corner of the lowest bluestone step at the main entrance to the five-story brick factory building on the southwest corner of Schlesinger Street (Nineteenth Street) and Kinney Avenue (Fifth Avenue). The bench mark is on the north side of the building and 19.9 feet from the east side.

Elevation, 44.632 feet.

B. M. 720.—*College Point.* Coast and Geodetic Survey tidal station No. 68. A cross (+) cut in the foundation stone on the west side of the two-story and attic frame house on the south side of Gower Avenue (Third Avenue) and 25 feet west of the west line of Ansonia Street (First Street). The bench mark is 10 inches north of the rear of the house and 6 inches above the ground. Established in 1886. (Note 15, p. 91.)

Elevation, 9.588 feet.

B. M. 721.—*College Point.* The bottom of a square cut in the projecting foundation stone at the southwest corner of the three-story brick factory building on the northwest corner of Ansonia Street (First Street) and Gower Avenue (Third Avenue). The bench mark is on the south side of the building, is 3 feet east of the west side and 1.64 feet above the sidewalk. An arrowhead is cut in the first course of brick above the foundation and points to the bench mark. (Note 10, p. 90.)

Elevation, 10.226 feet.

B. M. 722.—*College Point.* A square (□) cut in the top of the north end of the granite basin head on the northeast corner of Ansonia Street (First Street) and Gower Avenue (Third Avenue).

Elevation, 9.686 feet.

B. M. 723.—*College Point.* A square (□) cut in the top of the southwest corner of the granite basin head on the southeast corner of Chelsea Street (Third Street) and Villaview Road (First Avenue). Established by the Department of Docks and Ferries.

Elevation, 21.664 feet.

B. M. 724.—*College Point.* The center of the top of the granite city monument on the northwest corner of Mamaroneck Street (Thirteenth Street) and Gower Avenue (Third Avenue). The top of the monument is flush with the sidewalk.

Elevation, 42.021 feet.

B. M. 725.—*College Point.* A square (⊓) cut in the top of the east end of the bluestone sill of the most easterly basement window on the north side of Public School No. 27, on the west side of Mamaroneck Street (Thirteenth Street) between Elf Road (High Street) and Villaview Avenue (First Avenue). The bench mark is 4.3 feet from the east side of the building.

Elevation, 56.598 feet.

B. M. 726.—*College Point.* A ½-inch copper bolt set vertically in the top of the northeast corner of the bluestone sill of basement window on the east side of Public School No. 27, on the west side of Mamaroneck Street (Thirteenth Street) between Elf Road (High Street) and Villaview Avenue (First Avenue). The bolt is in the sill of the first basement window south of the main entrance and 1 foot above the ground. (Note 8, p. 90.)

Elevation, 56.541 feet.

B. M. 727.—*College Point.* A square (□) cut in the top of the northeast corner of the granite basin head on the northwest corner of Kinney Avenue (Fifth Avenue) and Schlesinger Street (Nineteenth Street). This point was formerly Coast and Geodetic Survey B. M. No. 11. During the past few years, however, the position of the basin head has been changed, thus destroying the original elevation.

Elevation, 41.972 feet.

B. M. 728.—*College Point.* A square (⊓) cut in the top of the northwest corner of the granite basin head on the southwest corner of Elf Road (High Street) and Rockville Street (North Eighteenth Street), at the gore-shaped corner formed by the intersection of Elf Road and Gower Avenue (Third Avenue).

Elevation, 57.450 feet.

B. M. 729.—*Whitestone.* A horizontal line in the end of a ⅝-inch copper bolt set horizontally into the granite base course at the northeast corner of Public School No. 79, on the west side of Ziegler Avenue (Seventh Avenue) between Arndt Avenue

(Fourteenth Street) and Iredell Street (Fifteenth Street). The bolt is on the east side of the building, is 1 foot from the north side and 9 inches above the ground. (Note 18, p. 91.)

Elevation, 51.388 feet.

B. M. 730.—*Whitestone.* A square (ㄣ) cut in the top of the northwest corner of the granite balustrade on the east side of the north basement entrance to Public School No. 79, on the west side of Ziegler Avenue (Seventh Avenue) between Arndt Avenue (Fourteenth Street) and Iredell Street (Fifteenth Street). The bench mark is on the north side of the building near the base of the iron railing and 1.3 feet above the ground.

Elevation, 51.778 feet.

B. M. 731.—*Whitestone.* A quadrant (ㄣ) cut in the top of the northeast corner of a projecting stone below the top course of masonry in the stone pier north of the White Rock Lake Hotel, on the south side of Little Bay. The pier containing the bench mark is said to be the base of the Harvey W. Peace monument. It is on the shore between the north side of Little Bay Avenue (Willets Point Road) and the edge of Little Bay, is about 16 feet west of the west side of the White Rock Lake Hotel and 3.7 feet above the ground. The letter B is cut in the east side of the stone containing the bench mark.

Elevation, 11.827 feet.

B. M. 732.—*Willets Point.* A square (ㄇ) cut in the top of the northwest corner of the lowest bluestone step of old unused entrance to the grounds of the Sacred Heart Convent, on the south side of Little Bay Avenue (Willets Point Road) about 1,500 feet west of Bell Avenue. The bench mark is about 28 feet east of the west end of the stone wall on the south side of Little Bay Avenue and 1.6 feet above the ground.

Elevation, 11.689 feet.

B. M. 733.—*Willets Point.* A cross (+) cut in the top of the west nut on the top of the fixed rim of the fire hydrant at the intersection of the center lines of Bell Avenue and Little Bay Avenue (Willets Point Road). The hydrant is on the south side of a small grass plot and 2 feet above the ground.

Elevation, 13.440 feet.

B. M. 734.—*Willets Point.* A small cross (+), ¾-inch, cut in the top of the horizontal surface at the east end of the bluestone sill of the east window on the north side of the two-story red brick casemate storehouse building, just south of the long dock, in the military reservation at Fort Totten. The cross is 3½ inches from the east end of the sill and the figure 19.5205 is scratched below, on the vertical face of the sill. This point was established in 1907 by Mr. Mayhew, post engineer, by running from the United States Engineer's bench mark on the face of the old retaining wall at the long dock. The Engineer's bench mark was a cross (+), lettered 18.0555, cut in a stone of the retaining wall, on the left when coming from the dock into the post grounds. This point was also a bench mark of the Coast and Geodetic Survey, and has since been destroyed. Using the above figure, Mr. Mayhew obtained the elevation 19.5205 for the cross on the window-sill of the casemate storehouse building.

Elevation, 15.596 feet.

B. M. 735.—*Willets Point.* A small cross (+), 1-inch, cut in the top of the horizontal surface at the east end of the bluestone sill of the west window on the north side of the two-story red brick casemate storehouse building, just south of the long dock, in the military reservation at Fort Totten. The cross is 2¾ inches from the east end of the sill and the figure 19.454 is scratched below, on the vertical face.

Elevation, 15.599 feet.

B. M. 736.—*Willets Point.* The center of the top of a ¾-inch copper bolt set vertically in the inclined top of the granite base course at the northeast corner of the east wing of the Quartermaster's building (two-story red brick) in the military reservation at Fort Totten. The building containing the bench mark was formerly known as the "Artillery School" or "Officers' School Building," and is just north of the Headquarters or Administration Building. The bench mark is about 75 feet north of the north side of the main road, is 11 inches above the ground and 10½ inches west of the corner of the building. This point was established by Mr. Mayhew, post engineer, who ran a line of levels from a granite monument in front of the Officers' School Building. The monument had an elevation marked upon it and, by using this figure, Mr. Mayhew obtained for the bolt an elevation of 52.827 feet, which figure is cut in the limestone block above the bench mark.

Elevation, 48.992 feet.

B. M. 737.—*Willets Point.* A ¾-inch copper bolt set vertically in the top of the southwest corner of a massive granite block in the top of the sea wall in the military reservation at Fort Totten. The bolt is about 100 feet south of the south side of the long dock and on the west side of the casemate storehouse building (two-story red brick), about 38 feet south of the north side of the building. It is opposite an arrowhead cut in the first course of brick above the concrete foundation and 5 inches from the face of the wall. (Note 5, p. 90.)

Elevation, 11.005 feet.

# STANDARD ELEVATIONS OF BENCH MARKS

IN THE

# BOROUGH OF MANHATTAN.

| No. of Bench Mark. | LOCALITY. | Elevation Above Mean Sea Level. | |
|---|---|---|---|
| | | Meters. | Feet. |
| 738 | Governors Island | 2.1741 | 7.133 |
| 739 | " " | 5.0413 | 16.540 |
| 740 | " " | 2.6028 | 8.539 |
| 741 | " " | 2.4382 | 7.999 |
| 742 | Battery | 2.1460 | 7.041 |
| 743 | " | 3.1943 | 10.480 |
| 744 | " | 2.7669 | 9.078 |
| 745 | " | 4.0021 | 13.130 |
| 746 | " | 3.7692 | 12.366 |
| 747 | " | 2.0307 | 6.662 |
| 748 | " | 5.1996 | 17.059 |
| 749 | " | 5.1660 | 16.949 |
| 750 | Bowling Green | 6.6291 | 21.749 |
| 751 | Exchange Place | 11.6472 | 38.212 |
| 752 | " " | 6.8308 | 22.411 |
| 753 | Pine Street | 12.4310 | 40.784 |
| 754 | Cedar Street | 11.0954 | 36.402 |
| 755 | " " | 11.0897 | 36.384 |
| 756 | " " | 11.0911 | 36.388 |
| 757 | Liberty Street | 11.5826 | 38.001 |
| 758 | Broadway | 11.3076 | 37.098 |
| 759 | Mail Street | 11.9906 | 39.339 |
| 760 | City Hall | 13.6369 | 44.740 |
| 761 | Chambers Street | 12.4100 | 40.715 |
| 762 | " " | 12.2800 | 40.289 |
| 763 | Municipal Building | 12.6835 | 41.612 |
| 764 | Franklin Square | 7.8789 | 25.849 |
| 765 | Cherry Street | 5.4013 | 17.721 |
| 766 | Pike Street | 9.8693 | 32.380 |
| 767 | Madison Street | 11.3370 | 37.195 |
| 768 | " " | 10.3207 | 33.860 |
| 769 | " " | 9.7855 | 32.105 |
| 770 | Corlears Hook Park | 7.5537 | 24.782 |
| 771 | Jackson Street | 2.9926 | 9.818 |
| 772 | " " | 3.3134 | 10.871 |
| 773 | Gouverneur Slip | 2.7511 | 9.026 |
| 774 | Water Street | 2.0175 | 6.619 |
| 775 | " " | 3.5445 | 11.629 |
| 776 | Cherry Street | 2.3113 | 7.583 |
| 777 | Mangin Street | 2.6997 | 8.857 |
| 778 | " " | 3.0369 | 9.964 |
| 779 | Leonard Street | 11.6416 | 38.194 |
| 780 | Catharine Lane | 11.3727 | 37.312 |
| 781 | " " | 11.6742 | 38.302 |
| 782 | White Street | 8.0361 | 26.365 |
| 783 | Prince Street | 12.1383 | 39.824 |
| 784 | Broadway | 13.7197 | 45.012 |
| 785 | " | 13.1694 | 43.207 |
| 786 | Jersey Street | 14.0094 | 45.963 |

| No. of Bench Mark. | Locality. | Elevation Above Mean Sea Level. | |
|---|---|---|---|
| | | Meters. | Feet. |
| 787 | Jersey Street | 13.2134 | 43.351 |
| 788 | Great Jones Street | 14.1222 | 46.333 |
| 789 | " " " | 14.1027 | 46.269 |
| 790 | Astor Place | 12.1354 | 39.814 |
| 791 | " " | 13.1099 | 43.011 |
| 792 | Fourth Avenue | 13.6474 | 44.775 |
| 793 | Seventh Street | 13.8750 | 45.522 |
| 794 | Union Square, East | 13.9403 | 45.736 |
| 795 | " " West | 13.4822 | 44.233 |
| 796 | West Eighteenth Street | 13.3482 | 43.793 |
| 797 | Fifth Avenue | 13.3326 | 43.742 |
| 798 | Broadway | 13.3515 | 43.804 |
| 799 | Fifth Avenue | 13.0759 | 42.900 |
| 800 | " " | 14.2086 | 46.616 |
| 801 | West Twenty-sixth Street | 12.3573 | 40.542 |
| 802 | Fifth Avenue | 14.1110 | 46.296 |
| 803 | East Twenty-sixth Street | 11.5226 | 37.803 |
| 804 | " " " " | 10.2419 | 33.602 |
| 805 | East Twenty-seventh Street | 6.2628 | 20.547 |
| 806 | " " " " | 6.7390 | 22.110 |
| 807 | East Twenty-fourth Street | 5.2781 | 17.317 |
| 808 | West Twenty-ninth Street | 13.8966 | 45.592 |
| 809 | Fifth Avenue | 15.5918 | 51.154 |
| 810 | West Thirty-third Street | 13.3222 | 43.708 |
| 811 | East Thirty-seventh Street | 24.4907 | 80.350 |
| 812 | West Forty-second Street | 22.7879 | 74.763 |
| 813 | " " " " | 25.3976 | 83.325 |
| 814 | West Forty-first Street | 24.8731 | 81.604 |
| 815 | " " " " | 13.8478 | 45.432 |
| 816 | West Forty-second Street | 17.9632 | 58.934 |
| 817 | " " " " | 14.9291 | 48.980 |
| 818 | " " " " | 15.0637 | 49.422 |
| 819 | " " " " | 19.7781 | 64.889 |
| 820 | " " " " | 4.4334 | 14.545 |
| 821 | " " " " | 4.6424 | 15.231 |
| 822 | " " " " | 4.4617 | 14.638 |
| 823 | Eleventh Avenue | 5.1352 | 16.848 |
| 824 | Fifth Avenue | 24.5066 | 80.402 |
| 825 | " " | 24.6477 | 80.865 |
| 826 | " " | 16.7894 | 55.083 |
| 827 | " " | 16.2036 | 53.161 |
| 828 | Columbus Circle | 26.4071 | 86.637 |
| 829 | Eleventh Avenue | 8.6487 | 28.375 |
| 830 | " " | 8.3290 | 27.326 |
| 831 | " " | 8.4924 | 27.862 |
| 832 | Central Park West | 24.1075 | 79.093 |
| 833 | " " " | 24.0256 | 78.824 |
| 834 | " " " | 27.4061 | 89.915 |
| 835 | West Seventy-second Street | 27.9551 | 91.716 |
| 836 | Central Park | 20.3094 | 66.631 |
| 837 | East Sixty-eighth Street | 21.9759 | 72.099 |
| 838 | East Sixty-seventh Street | 24.5315 | 80.484 |
| 839 | East Sixty-eighth Street | 24.7150 | 81.083 |
| 840 | " " " " | 26.3320 | 86.391 |
| 841 | " " " " | 14.2170 | 46.644 |
| 842 | East Seventy-first Street | 11.8455 | 38.863 |
| 843 | East Seventy-fourth Street | 3.4210 | 11.224 |
| 844 | East Seventy-sixth Street | 3.2450 | 10.646 |
| 845 | " " " " | 8.7238 | 28.622 |

| No. of Bench Mark. | Locality. | Elevation Above Mean Sea Level. | |
|---|---|---|---|
| | | Meters. | Feet. |
| 846 | Fifth Avenue | 25.9221 | 85.046 |
| 847 | East Seventy-eighth Street | 10.6333 | 34.886 |
| 848 | " " " " | 10.7568 | 35.291 |
| 849 | Central Park West | 27.6823 | 90.821 |
| 850 | " " " | 28.7210 | 94.229 |
| 851 | Manhattan Square | 25.5449 | 83.809 |
| 852 | " " | 25.6171 | 84.045 |
| 853 | " " | 25.6434 | 84.132 |
| 854 | " " | 26.1733 | 85.870 |
| 855 | " " | 26.4599 | 86.811 |
| 856 | Central Park | 24.0339 | 78.851 |
| 857 | " " | 27.7462 | 91.031 |
| 858 | " " | 27.2638 | 89.448 |
| 859 | " " | 27.2812 | 89.505 |
| 860 | East Eighty-sixth Street | 28.9235 | 94.893 |
| 861 | " " " " | 23.7385 | 77.882 |
| 862 | Third Avenue | 25.2597 | 82.873 |
| 863 | East Eighty-sixth Street | 15.6452 | 51.329 |
| 864 | " " " " | 15.7752 | 51.756 |
| 865 | " " " " | 12.2894 | 40.319 |
| 866 | " " " " | 11.1000 | 36.417 |
| 867 | " " " " | 8.9113 | 29.236 |
| 868 | " " " " | 2.6960 | 8.845 |
| 869 | East Eighty-fourth Street | 1.7863 | 5.861 |
| 870 | " " " " | 1.6380 | 5.374 |
| 871 | Carl Schurz Park | 2.2991 | 7.543 |
| 872 | " " " | 2.6146 | 8.578 |
| 873 | " " " | 2.6431 | 8.672 |
| 874 | Central Park West | 33.7533 | 110.739 |
| 875 | " " " | 34.7245 | 113.925 |
| 876 | " " " | 31.7920 | 104.304 |
| 877 | " " " | 31.8070 | 104.353 |
| 878 | " " " | 31.1668 | 102.253 |
| 879 | Central Park | 34.7915 | 114.145 |
| 880 | " " | 33.8853 | 111.172 |
| 881 | West One-hundred and First Street | 26.5038 | 86.955 |
| 882 | " " " " " " | 26.1706 | 85.861 |
| 883 | Fifth Avenue | 17.3477 | 56.915 |
| 884 | Central Park | 15.1217 | 49.612 |
| 885 | West One Hundred and Sixth Street | 24.5598 | 80.577 |
| 886 | Seventh Avenue | 10.8212 | 35.503 |
| 887 | " " | 10.7463 | 35.257 |
| 888 | West One Hundred and Tenth Street | 6.1283 | 20.106 |
| 889 | Fifth Avenue | 5.7396 | 18.831 |
| 890 | Park Avenue | 4.5694 | 14.991 |
| 891 | West One Hundred and Fifteenth Street | 10.5836 | 34.723 |
| 892 | " " " " " " | 10.6686 | 35.002 |
| 893 | " " " " " " | 11.0149 | 36.138 |
| 894 | Fifth Avenue | 6.5049 | 21.342 |
| 895 | Mount Morris Park | 8.8019 | 28.878 |
| 896 | " " " | 9.1202 | 29.922 |
| 897 | Madison Avenue | 9.7954 | 32.137 |
| 898 | Hancock Place | 9.7137 | 31.869 |
| 899 | Manhattan Street | 11.5787 | 37.988 |
| 900 | " " | 4.2335 | 13.889 |
| 901 | " " | 3.9879 | 13.084 |
| 902 | St. Nicholas Avenue | 11.0739 | 36.332 |
| 903 | " " | 10.8559 | 35.616 |
| 904 | Madison Avenue | 7.0639 | 23.176 |

| No. of Bench Mark. | LOCALITY. | Elevation Above Mean Sea Level. | |
|---|---|---|---|
| | | Meters. | Feet. |
| 905 | East One Hundred and Thirtieth Street | 4.5473 | 14.919 |
| 906 | " " " " " " | 3.2518 | 10.669 |
| 907 | " " " " " " | 3.1014 | 10.175 |
| 908 | " " " " " " | 3.1067 | 10.193 |
| 909 | " " " " " " | 4.5713 | 14.998 |
| 910 | St. Nicholas Avenue | 12.2102 | 40.060 |
| 911 | St. Nicholas Park | 16.7694 | 55.018 |
| 912 | St. Nicholas Avenue | 15.8669 | 52.057 |
| 913 | " " | 23.2654 | 76.330 |
| 914 | " " | 24.3937 | 80.032 |
| 915 | " " | 30.5933 | 100.372 |
| 916 | " " | 32.1054 | 105.333 |
| 917 | " " | 41.3953 | 135.811 |
| 918 | West One Hundred and Fifty-fifth Street | 45.3326 | 148.729 |
| 919 | St. Nicholas Avenue | 45.0950 | 147.949 |
| 920 | " " | 44.6811 | 146.591 |
| 921 | " " | 50.4428 | 165.494 |
| 922 | West One Hundred and Sixty-ninth Street | 53.7327 | 176.288 |
| 923 | " " " " " " " | 53.8135 | 176.553 |
| 924 | West One Hundred and Seventieth Street | 54.3203 | 178.216 |
| 925 | West One Hundred and Seventy-fourth Street | 56.6672 | 185.916 |
| 926 | Highbridge Water Tower | 61.6142 | 202.146 |
| 927 | " " " | 61.9494 | 203.246 |
| 928 | " " " | 61.5059 | 201.791 |
| 929 | West One Hundred and Eighty-first Street | 50.9092 | 167.025 |
| 930 | " " " " " " " | 51.6731 | 169.531 |
| 931 | Wadsworth Avenue | 57.4042 | 188.334 |
| 932 | " " | 56.4862 | 185.322 |
| 933 | " " | 56.3184 | 184.771 |
| 934 | Broadway | 22.8120 | 74.842 |
| 935 | " | 22.8077 | 74.828 |
| 936 | " | 12.7003 | 41.668 |
| 937 | Dyckman Street | 7.3754 | 24.198 |
| 938 | " " | 6.8774 | 22.564 |
| 939 | " " | 3.4408 | 11.289 |
| 940 | West Two Hundred and Seventh Street | 12.4881 | 40.971 |
| 941 | " " " " " " | 4.3731 | 14.347 |
| 942 | " " " " " " | 2.8325 | 9.293 |
| 943 | " " " " " " | 9.1332 | 29.965 |
| 944 | Broadway | 9.9480 | 32.638 |
| 945 | Amsterdam Avenue | 9.4321 | 30.945 |
| 946 | West Two Hundred and Sixteenth Street | 5.8574 | 19.217 |
| 947 | " " " " " " | 4.2554 | 13.961 |
| 948 | United States Ship Canal Bridge | 10.3335 | 33.903 |
| 949 | " " " " " | 9.9376 | 32.604 |
| 950 | " " " " " | 2.1955 | 7.203 |
| 951 | " " " " " | 6.1941 | 20.322 |
| 952 | Broadway | 5.9687 | 19.582 |

# DESCRIPTIONS OF BENCH MARKS

IN THE

## BOROUGH OF MANHATTAN.

B. M. 738.—*Governors Island.* The bottom of a 2½-inch square cut in the coping of the granite sea wall on the east side of Governors Island, opposite the foot of Hamilton Avenue, Brooklyn. The bench mark is near the southeast corner of the coping stone, is south of the transformer house and just east of the Quartermaster's building (two-story and attic red brick), opposite a point about 11 feet south of the north end. The letter B is cut in the stone on the west side of the point. (Note 4, p. 89.)

Elevation, 7.133 feet.

B. M. 739.—*Governors Island.* The center of a 3/16-inch drill hole in the center of the cross in the cornerstone of the Chapel of Saint Cornelius the Centurian, on the easterly side of Governors Island. The Chapel is a white limestone structure, is about 150 feet north of Corbin Hall, and the cornerstone contains the inscription A. D. 1905 below the cross, which is on the north side of the Chapel about 3 feet above the ground.

Elevation, 16.540 feet.

B. M. 740.—*Governors Island.* Coast and Geodetic Survey B. M. I. A cross (+) cut in the south end of the bluestone door-sill at the main entrance into Castle William. The bench mark is on the east side of the old fort, is 3½ inches from the south end of the sill and 3 inches from the east edge. Established in 1886.

Elevation, 8.539 feet.

B. M. 741.—*Governors Island.* Coast and Geodetic Survey B. M. H. A cross (+) cut in the extreme northeast corner of the large stone having a ring bolt in the top and forming part of the coping of the sea wall on the north side of Governors Island in front of Castle William. It is about 100 feet west of the long dock. Established in 1886.

Elevation, 7.999 feet.

B. M. 742.—*Battery.* A ⅞-inch copper bolt leaded vertically in the granite coping of the sea wall on the north side of the Battery Basin, on the west side of the new United States Barge Office. The bolt is about 50 feet west of the northeast corner of the basin and 1.6 feet east of a stone post. The letter B is cut in the coping just south of the bolt.

Elevation, 7.041 feet.

B. M. 743.—*Battery.* A 2-inch square (□) cut in the top of the third or top granite step at the west side of the passenger entrance to the Thirty-ninth Street Ferry, just east of the foot of Whitehall Street. The bench mark is at the east side of the wagon entrance to the Stapleton Ferry, and 2.1 feet from the south end of the step.

Elevation, 10.480 feet.

B. M. 744.—*Battery.* The center of the horizontal bar of a T cut in the south side of the 16-inch square stone column on the northeast corner of Whitehall Street and South Street. The column containing the bench mark is at the southwest corner of the Eastern Hotel (five-story brick) and the point is about 2 feet above the sidewalk.

Elevation, 9.078 feet.

B. M. 745.—*Battery.* The center of a square (ㅁ) cut in the top of the most northerly corner of the second granite step at the base of the Ericsson monument, in the west side of Battery Park between the Aquarium and the new United States Barge Office.

Elevation, 13.130 feet.

B. M. 746.—*Battery.* The center of the horizontal bar of a T cut in the brownstone base supporting two pilasters on the south side of the easterly entrance to the Aquarium (Castle Garden). The bench mark is about 5 feet south of the entrance and 1.75 feet above the sidewalk.*

Elevation, 12.366 feet.

B. M. 747.—*Battery.* A square (ㅁ) cut in the top of the northwest corner of a granite block in the coping of the stone bulkhead between Pier A and Pier 1, Hudson River. The bench mark is about 32 feet north of the north side of Pier A.†

Elevation, 6.662 feet.

B. M. 748.—*Battery.* A horizontal line in the end of a ⅞-inch copper bolt leaded horizontally into the north side of the south buttress on the west side of the United States Custom House. The bolt is on the State Street side of the building, about 50 feet north of Bridge Street. It is 2.75 feet above the sidewalk and 6 inches from the west side of the buttress. (Note 6, p. 90.)

Elevation, 17.059 feet.

B. M. 749.—*Battery.* A 2-inch square (□) cut in the top of the southeast corner of the fifth granite step of the south entrance on the State Street side of the United States Custom House. The bench mark is about 52 feet north of Bridge Street.

Elevation, 16.949 feet.

B. M. 750.—*Bowling Green.* A square (ㅁ) cut in the top of the northwest corner of the granite plinth of the most northerly granite column on the Whitehall Street side of the New York Produce Exchange, between Stone Street and Beaver Street. The bench mark is about 30 feet north of the prolongation of the north line of Battery Place and 3.7 feet above the sidewalk.‡

Elevation, 21.749 feet.

B. M. 751.—*Exchange Place.* A cross (+) cut in the top of the southwest corner of the square limestone base of large limestone pilaster on the west side of the Downtown building (twenty-story stone) on the northeast corner of Exchange Place and Broadway. The bench mark is about 20 feet north of Exchange Place, is 4.15 feet above the sidewalk and 5 inches northeast of the corner of the stone. (Note 17, p. 91.)

Elevation, 38.212 feet.

* This bench mark was established by the Department of Docks and Ferries and was formerly the base of the departmental system of levels in the Borough of Manhattan. See p. 76 for history of Dock Department Datum.

† Established by the Department of Docks and Ferries. See p. 76 for the history of the Dock Department Datum and this initial bench mark.

‡ See p. 76 for the history of this bench mark, which was established by the Rapid Transit Railroad Commission.

B. M. 752.—*Exchange Place.* A square (ㄣ) cut in the top of the northeast corner of the heavy granite water-table of the Exchange Court building (twelve-story stone) on the southwest corner of Exchange Place and New Street. The bench mark is 2.1 feet above the sidewalk.

Elevation, 22.411 feet.

B. M. 753.—*Pine Street.* The center of a square (ㄣ) cut in the top of the granite water-table at the northwest corner of the American Surety Company building, on the southeast corner of Pine Street and Broadway. The point is 3.1 feet above the sidewalk or 6¾ inches above the large granite base block.

Elevation, 40.784 feet.

B. M. 754.—*Cedar Street.* This is the top of the southwest corner of the wide granite water-table of the Title Insurance Company building, on the northeast corner of Cedar Street and Temple Street. The bench mark is 7.2 feet above the sidewalk, and an arrowhead cut in the Cedar Street side of the coping points to the exact spot.

Elevation, 36.402 feet.

B. M. 755.—*Cedar Street.* The center of a square (⊓) cut in the top of the south edge of the wide granite water-table of the Title Insurance Company building, on the north side of Cedar Street between Broadway and Temple Street. The bench mark is about 50 feet west of Broadway and about 3 feet above the sidewalk.

Elevation, 36.384 feet.

B. M. 756.—*Cedar Street.* The center of a quadrant (ㄣ) cut in the top of the extreme southeast corner of the granite water-table of the Title Insurance Company building, on the northwest corner of Cedar Street and Broadway. The bench mark is 2 inches above the sidewalk.

Elevation, 36.388 feet.

B. M. 757.—*Liberty Street.* The center of the horizontal bar of a **T** cut in the granite foundation of the Washington Life building, on the southwest corner of Liberty Street and Broadway. The bench mark is on the Liberty Street side of the building, is ¾ inch from the Broadway side and 2.2 feet above the sidewalk.

Elevation, 38.001 feet.

B. M. 758.—*Broadway.* A ⅞-inch copper bolt leaded vertically in the top of the wide granite step at the entrance to No. 141 Broadway (Washington Life building). The bolt is 11½ inches from the south end of the step and about 50 feet south of Liberty Street.

Elevation, 37.098 feet.

B. M. 759.—*Mail Street.* A square (⊓) cut in the top of the northeast corner of the wide granite step at the unused north entrance to the General Post Office, on the southwesterly side of Mail Street. The bench mark is at the extreme northwesterly corner of the building and on line with the southerly curb line of Park Place.

Elevation, 39.339 feet.

B. M. 760.—*City Hall.* The center of the horizontal bar of a large **T** cut in the brownstone water-table on the south side of the City Hall, 6 inches from the southwest corner. The bench mark is 4.2 feet above the area.*

Elevation, 44.740 feet.

* See p. 72 for history of this bench mark, which was established prior to 1840.

B. M. 761.—*Chambers Street.* A 2-inch square (□) cut in the top of the granite block at the southwest corner of the sixth or top step at the west entrance to the Hall of Records, on the northwest corner of Chambers Street and Centre Street. The bench mark is on the Chambers Street side, about 120 feet west of Centre Street.

Elevation, 40.715 feet.

B. M. 762.—*Chambers Street.* The bottom of a square cut in the extreme southwest corner of the Hall of Records. The bench mark is on the north side of Chambers Street and on the east side of the alley connecting Chambers Street and Reade Street. The point is on the top of the vertical foundation, just where it curves in toward the building. It is about 2 feet above the sidewalk. (Note 10, p. 90.)

Elevation, 40.289 feet.

B. M. 763.—*Municipal Building.* A 1-inch copper bolt in the top of the southwest corner of the granite pedestal of granite column at the southwest corner of the Municipal Building, at the intersection of Tryon Row and Centre Street. The bolt is on the southerly side of the building, at the base of the first column east of the cornerstone.

Elevation, 41.612 feet.

B. M. 764.—*Franklin Square.* The bottom of a square cut in the top of the granite water-table on the south side of the abutment of the Brooklyn Bridge, on the northwest corner of Pearl Street and Frankfort Street. The bench mark is on the south corner of the water-table, is 2.5 feet above the sidewalk and lettered B. M. (Note 4, p. 89.)

Elevation, 25.849 feet.

B. M. 765.—*Cherry Street.* The bottom of a square cut in the top of a large granite block in the anchorage of the Manhattan Bridge. The bench mark is on the south side of Cherry Street and on the west side of the arch. It is 3.1 feet above the sidewalk. (Note 4, p. 89.)

Elevation, 17.721 feet.

B. M. 766.—*Pike Street.* A square (ㅁ) cut in the top of the southeast corner of the bluestone coping under the iron railing on the south lot line of No. 36 Pike Street, about 46 feet north of Madison Street.

Elevation, 32.380 feet.

B. M. 767.—*Madison Street.* A horizontal line on the face of a brass tablet bolt 2 inches in diameter, set horizontally into the east face of the buttress on the west side of the girls' entrance to Public School No. 12, on the northwest corner of Madison Street and Jackson Street. The bolt is on the Madison Street side of the building in the limestone base of the buttress, is about 57 feet west of Jackson Street and 5.8 feet above the area. (Note 1, p. 89.)

Elevation, 37.195 feet.

B. M. 768.—*Madison Street.* A ¾-inch copper bolt in the southwest corner of the fifth bluestone step or landing at the girls' entrance to Public School No. 12, on the northwest corner of Madison Street and Jackson Street. The bolt is on the Madison Street side, about 53 feet west of Jackson Street. (Note 5, p. 89.)

Elevation, 33.860 feet.

B. M. 769.—*Madison Street.* A cross (+) cut in the east face of the buttress on the west side of the girls' entrance to Public School No. 12, on the northwest corner of Madison Street and Jackson Street. The cross is on the Madison Street side of the building in the first course of granite, is about 57 feet west of Jackson Street and about 6½ inches above the area. (Note 15, p. 91.)

Elevation, 32.105 feet.

B. M. 770.—*Corlears Hook Park.* A square (⊔) cut in the top of the southwest corner of the limestone balustrade on the west side of the stone steps at the south entrance to the summer house in Corlears Hook Park. The bench mark is just south of Cherry Street, is between Jackson Street and Corlears Street and 2.5 feet above the sidewalk.

Elevation, 24.782 feet.

B. M. 771.—*Jackson Street.* The highest point within a square (⊔) cut in the top of the southwest corner of the bluestone coping on the north side of the south cellar entrance to the six-story tenement at No. 91 Jackson Street. The bench mark is about 11 feet north of South Street.

Elevation, 9.818 feet.

B. M. 772.—*Jackson Street.* A square (⊔) cut in the top of the southwest corner of the granite watering trough on the east side of Jackson Street, about 60 feet north of South Street.

Elevation, 10.871 feet.

B. M. 773.—*Gouverneur Slip.* The center of the horizontal bar of a T cut in the south side of a stone block supporting the iron column at the southwest corner of the five-story loft building on the northeast corner of Gouverneur Slip and South Street. The bench mark is 11½ inches from the corner and about 1.7 feet above the sidewalk. Established by the Department of Docks and Ferries.

Elevation, 9.026 feet.

B. M. 774.—*Water Street.* A square (⊔) cut in the top of the southwest corner of the granite sill of the wide door on the south side of Coe's warehouse (five-story brick) on the north side of Water Street between Corlears Street and East Street. The bench mark is at the most westerly door, about 287 feet east of Corlears Street, and about 2 feet from the west side of the building.

Elevation, 6.619 feet.

B. M. 775.—*Water Street.* Coast and Geodetic Survey B. M. E. At Coe's bonded warehouse (five-story brick), on the north side of Water Street, between Corlears Street and East Street. The bench mark is the intersection of a cross (+) cut in the south side of the granite door-jamb of the first door from the southwest corner of the building. The cross is about 285 feet east of Corlears Street, is 4.8 feet above the sidewalk and 1½ inches from the west side of the building. Established in May, 1887.

Elevation, 11.629 feet.

B. M. 776.—*Cherry Street.* The bottom of a square cut in the top of the southeast corner of the water-table of the four-story yellow brick loft building on the northwest corner of Cherry Street and East Street. The bench mark is 4 inches above the sidewalk. (Note 10, p. 90.)

Elevation, 7.583 feet.

B. M. 777.—*Mangin Street.* A square (□) cut in the top of the south side of the granite base block of the most northerly column of the Williamsburg Bridge, on the west side of Mangin Street, corner of Delancey Street.

Elevation, 8.857 feet.

B. M. 778.—*Mangin Street.* A square, lettered □ U. S. E., cut in the top of the southeast corner of the granite base block of the most southerly column of the Williamsburg Bridge on the west side of Mangin Street, about 95 feet south of Delancey Street.

Elevation, 9.964 feet.

B. M. 779.—*Leonard Street.* A ¾-inch copper bolt in the top of the southeast corner of the square granite base of the most southerly granite column on the west side of the New York Life building at No. 346 Broadway, corner of Leonard Street. The bolt is 1.3 feet above the sidewalk and 4 feet north of Catharine Lane. (Note 5, p. 89.)

Elevation, 38.194 feet.

B. M. 780.—*Catharine Lane.* A square (⧠) cut in the top of the southeast corner of the granite coping on the Catharine Lane side of the New York Life building at No. 346 Broadway. The bench mark is 9.1 feet east of the Broadway side and 1 foot above the sidewalk.

Elevation, 37.312 feet.

B. M. 781.—*Catharine Lane.* The bottom of a square cut in the top of the southeast corner of the flat water-table above the 10-inch curved coping on the Catharine Lane side of the New York Life building at No. 346 Broadway. The bench mark is about 60.5 feet east of the Broadway side of the building and 4.1 feet above the sidewalk. An arrowhead cut in the east side of the curved coping points to the bench mark. (Note 10, p. 90.)

Elevation, 38.302 feet.

B. M. 782.—*White Street.* A square (⧠) cut in the iron door-sill at the entrance to the five-story loft building at No. 66 White Street. The bench mark is about 90 feet west of Broadway and at the west side of the entrance.

Elevation, 26.365 feet.

B. M. 783.—*Prince Street.* A square (⧠) cut in the top of the northeast corner of a large copingstone on the north side of the nine-story loft building on the southwest corner of Prince Street and Broadway. The bench mark is about 25 feet west of Broadway and about 1.1 feet above the sidewalk.

Elevation, 39.824 feet.

B. M. 784.—*Broadway.* A ⅝-inch copper bolt in the center of the top of the granite base of the granite column on the south side of the main entrance to Nos. 594-596 Broadway, between East Houston Street and Prince Street. The bolt is behind the column and 1.3 feet above the sidewalk. (Note 8, p. 90.)

Elevation, 45.012 feet.

B. M. 785.—*Broadway.* A 3-inch square (⧠) cut in the top of the southeast corner of the red granite step at the main entrance to No. 611 Broadway, corner of West Houston Street. The bench mark is about 41 feet north of West Houston Street and 4 inches above the sidewalk.

Elevation, 43.207 feet.

B. M. 786.—*Jersey Street.* The center of the horizontal bar of a **T** cut in the northerly side of a granite block at the northeasterly corner of George W. Miller & Co.'s paper warehouse, on the southwesterly corner of Jersey Street and Lafayette Street. The bench mark is on the Jersey Street side, is 3.8 feet above the sidewalk and 1 foot from the corner.

Elevation, 45.963 feet.

B. M. 787.—*Jersey Street.* A level rectangular surface 2½" x ½", cut in the top of the granite water-table on the Jersey Street side of George W. Miller & Co.'s paper warehouse, on the southwesterly corner of Lafayette Street. The bench mark is 1.15 feet above the sidewalk.

Elevation, 43.351 feet.

B. M. 788.—*Great Jones Street.* A square (⊡) cut in the top of the southeast corner of the large granite block at the southwest corner of the twelve-story loft building on the northeast corner of Great Jones Street and Broadway. The bench mark is 5.3 feet above the sidewalk and 5.54 feet east of Broadway.

Elevation, 46.333 feet.

B. M. 789.—*Great Jones Street.* A square (⊡) cut in the top of the southwest corner of the large granite block at the southeast corner of the twelve-story loft building on the northeast corner of Great Jones Street and Broadway. The bench mark is about 132 feet east of Broadway, on the east side of the store entrance. It is 4.6 feet above the sidewalk.

Elevation, 46.269 feet.

B. M. 790.—*Astor Place.* A square (□) cut in the top of the cement curb on the south side of Astor Place, about 3 feet west of the east line of Broadway.

Elevation, 39.814 feet.

B. M. 791.—*Astor Place.* A square (□) cut in the top of the granite block at the base of the iron column at the northwest corner of Cooper Union, on the southeast corner of Astor Place and Fourth Avenue. The bench mark is on the Fourth Avenue side of the building, is about 2 feet from the corner and 6 inches above the sidewalk.

Elevation, 43.011 feet.

B. M. 792.—*Fourth Avenue.* The center of a square (⊡) cut in the top of the brownstone water-table at the southwest corner of Cooper Union, on the northeast corner of Fourth Avenue and East Seventh Street.

Elevation, 44.775 feet.

B. M. 793.—*East Seventh Street.* A square (⊡) cut in the top of the northwest corner of the third granite step at the base of the Peter Cooper monument, on the south side of East Seventh Street, between Third Avenue and Fourth Avenue.

Elevation, 45.522 feet.

B. M. 794.—*Union Square East.* A ⅝-inch copper bolt in the top of a large granite block at the southwest corner of the Union Square Savings Bank, on the northeast corner of East Fifteenth Street and Fourth Avenue. The bolt is on the west side of the building, is 1 foot from the south and west edges of the block and 2.83 feet from the south side of the building. (Note 8, p. 90.)

Elevation, 45.736 feet.

B. M. 795.—*Union Square West.* A square (⊡) cut in the top of the southeast corner of the granite plinth of the marble column on the south side of the main entrance to the Spingler Building, at Nos. 5 to 9 Union Square West, between East Fourteenth Street and East Fifteenth Street. The bench mark is 2.5 feet above the sidewalk.

Elevation, 44.233 feet.

B. M. 796.—*East Eighteenth Street.* A ¾-inch copper bolt in the top of the granite wagon guard on the east side of the wagon entrance to Engine House No. 14, on the south side of East Eighteenth Street between Broadway and Fifth Avenue. The top of the bolt is 1.3 feet above the sidewalk. (Note 8, p. 90.)

Elevation, 43.793 feet.

B. M. 797.—*Fifth Avenue.* A ¾-inch copper bolt in the top of the southeast corner of the granite plinth of the limestone column on the south side of the Fifth Avenue entrance to the Fuller Building, at No. 175 Fifth Avenue, between East Twenty-second Street and East Twenty-third Street. The bolt is 10 inches above the sidewalk. (Note 5, p. 89.)

Elevation, 43.742 feet.

B. M. 798.—*Broadway.* A square (ㅂ) cut in the top of the northeast corner of the granite plinth of limestone column on the north side of the Broadway entrance to the Fuller building, at No. 949 Broadway between East Twenty-second Street and East Twenty-third Street. The bench mark is 11 inches above the sidewalk.

Elevation, 43.804 feet.

B. M. 799.—*Fifth Avenue.* A square (ㅂ) cut in the top of the southwest corner of the wide granite door-sill at the Fifth Avenue entrance to the Fuller building, at No. 175 Fifth Avenue between East Twenty-second Street and East Twenty-third Street. The sill is level with the sidewalk.

Elevation, 42.900 feet.

**B. M. 800.**—***Fifth Avenue.*** **The** top of the southeast corner of the granite plinth of the granite column on the south side of the main entrance to the Fifth Avenue building, at No. 200 Fifth Avenue, between West Twenty-third Street and West Twenty-fourth Street. The bench mark is on the flat polished surface, is about 87 feet north of West Twenty-third Street and 4.7 feet above the sidewalk. (Note 26, p. 91.)

Elevation, 46.616 feet.

B. M. 801.—*West Twenty-sixth Street.* A ⅞-inch copper bolt in the top of the south side of the bluestone basin head on the northwest corner of West Twenty-sixth Street and Fifth Avenue.

Elevation, 40.542 feet.

B. M. 802.—*Fifth Avenue.* The top of the southeast corner of the polished granite pedestal of pilaster on the south side of the north entrance to the Croisic building at No. 220 Fifth Avenue. The bench mark is about 43 feet north of West Twenty-sixth Street and 4.73 feet above the sidewalk. The exact point is the southeast corner of the triangular space. (Note 26, p. 91.)

Elevation, 46.296 feet.

B. M. 803.—*East Twenty-sixth Street.* A ⅝-inch iron bolt set horizontally into the water-table of the four-story brick tenement on the southwest corner of East Twenty-sixth Street and Third Avenue. The bolt is on the north side of the building, is 23.8 feet west of Third Avenue and 8 inches above the sidewalk. (Note 11, p. 90.)

Elevation, 37.803 feet.

B. M. 804.—*East Twenty-sixth Street.* A ⅝-inch iron bolt set horizontally into the brownstone water-table of the four-story brick tenement on the southeast corner of East Twenty-sixth Street and Second Avenue. The bolt is on the north side of the building about 33 feet east of Second Avenue and 1.1 feet above the sidewalk. (Note 11, p. 90.)

Elevation, 33.602 feet.

B. M. 805.—*East Twenty-seventh Street.* A 1-inch copper bolt set in the center of the top of a granite monument in the grounds of Bellevue Hospital. The bolt is about 6 feet north of the center line of East Twenty-seventh Street and 2.5 feet east of the iron railing on the east side of First Avenue. The top of the monument is 3 inches above the ground.

Elevation, 20.547 feet.

B. M. 806.—*East Twenty-seventh Street.* A square (ᒪ) cut in the top of the corner of the granite base block or plinth at the extreme southeast corner of the Cornell Medical College, on the northwest corner of East Twenty-seventh Street and First Avenue. The bench mark is 1.25 feet above the area.

Elevation, 22.110 feet.

B. M. 807.—*East Twenty-fourth Street.* The center of the horizontal bar of a T cut in the east side of the brownstone water-table of the four-story tenement on the northwest corner of East Twenty-fourth Street and First Avenue. The bench mark is 5½ inches from the south side of the building. Established by the Department of Docks and Ferries.

Elevation, 17.317 feet.

B.M. 808.—*West Twenty-ninth Street.* A square (ᒪ) cut in the top of the southwest corner of the granite gate-sill at the south entrance to The Marble Collegiate Church, on the northwest corner of West Twenty-ninth Street and Fifth Avenue. The bench mark is about 25 feet west of Fifth Avenue and 4½ inches above the sidewalk.

Elevation, 45.592 feet.

B. M. 809.—*Fifth Avenue.* A ⅝-inch copper bolt in the top of the north end of the lowest bluestone step at the store entrance to No. 345 Fifth Avenue. The bench mark is at the base of the granite block supporting the pilaster, is about 80 feet north of East Thirty-third Street and 4 inches above the sidewalk. (Note 8, p. 90.)

Elevation, 51.154 feet.

B. M. 810.—*West Thirty-third Street.* The center of a 2-inch brass bolt set vertically in the top of the northwest corner of the square granite base of the most northerly granite column on the Seventh Avenue side of the Pennsylvania Railroad Station at West Thirty-third Street. The bolt is 3.1 feet above the sidewalk. (Note 2, p. 89.)

Elevation, 43.708 feet.

B. M. 811.—*East Thirty-seventh Street.* This point is the center of the 5-inch square space on the top of the northwest corner of the large marble block at the extreme northwest corner of the Tiffany building, on the southeast corner of East Thirty-seventh Street and Fifth Avenue. The bench mark is 2.2 feet above the sidewalk. (Note 26, p. 91.)

Elevation, 80.350 feet.

B. M. 812.—*West Forty-second Street.* A ⅞-inch copper bolt in the top of the north end of the lowest granite step at the Fifth Avenue and West Forty-second Street entrance to the New York Public Library. The bolt is about 25 feet from the south curb of West Forty-second Street and about 25 feet from the west curb of Fifth Avenue. (Note 5, p. 89.)

Elevation, 74.763 feet.

B. M. 813.—*West Forty-second Street.* The center of the horizontal bar of an ornamental cross with bars 1¾ inches long, near the east end of the cornerstone of the New York Public Library at West Forty-second Street and Fifth Avenue. The bench mark is on the north side of the building, is 1.33 feet west of the east side and 6.67 feet above the area. The cornerstone contains the inscription + MDCCCCII + and the cross adjacent to the letter M is the bench mark.

Elevation, 83.325 feet.

B. M. 814.—*West Forty-first Street.* The center of the horizontal bar of a **T** cut in the north side of a massive granite block surmounted by the figure of a lion, on the north side of the Fifth Avenue entrance to the New York Public Library opposite East Forty-first Street. The bench mark is 2.5 feet above the walk and 3½ inches from the east side of the block.

Elevation, 81.604 feet.

B. M. 815.—*West Forty-first Street.* A ⅝-inch iron bolt set horizontally into the brownstone water-table on the north side of the four-story brick tenement on the southeast corner of West Forty-first Street and Eighth Avenue. The bolt is about 8.5 feet east of the corner. (Note 11, p. 90.)

Elevation, 45.432 feet.

B. M. 816.—*West Forty-second Street.* A 1-inch square (□) cut in the top of the projecting granite block at the extreme northwest corner of the Knickerbocker Hotel, on the southeast corner of West Forty-second Street and Broadway. The bench mark is 8½ inches above the sidewalk.

Elevation, 58.934 feet.

B. M. 817.—*West Forty-second Street.* A ¾-inch copper bolt in the top of the northeast corner of the fifth granite step or landing at the main entrance to the Franklin Savings Bank, on the southeast corner of West Forty-second Street and Eighth Avenue. The bolt is on the west side of the building, about 18 feet south of West Forty-second Street. (Note 5, p. 89.)

Elevation, 48.980 feet.

B. M. 818.—*West Forty-second Street.* The center of the horizontal bar of a **T** cut in the granite foundation of the Franklin Savings Bank, on the southeast corner of West Forty-second Street and Eighth Avenue. The bench mark is on the west side of the building, is 8½ inches from the north side and 2.5 feet above the sidewalk.

Elevation, 49.422 feet.

B. M. 819.—*West Forty-second Street.* A square (□) cut 6 inches from the southwest corner of the subway light grating in Bryant Park, about 15 feet south of the iron railing on the south side of West Forty-second Street and 100 feet east of the east line of Sixth Avenue.

Elevation, 64.889 feet.

B. M. 820.—*West Forty-second Street.* The center of the enclosed surface within the figure 4, cut in the cement sidewalk on the southeast corner of West Forty-second Street and Eleventh Avenue. The point is about 5 feet south of the south curb line of West Forty-second Street and about 7.5 feet east of the east curb line of Eleventh Avenue.

Elevation, 14.545 feet.

B. M. 821.—*West Forty-second Street.* A square (⊡) cut in the top of the curved coping under the iron railing on the southwest corner of West Forty-second Street and Eleventh Avenue. The bench mark is on the projecting copingstone at the middle of the arc.

Elevation, 15.231 feet.

B. M. 822.—*West Forty-second Street.* Coast and Geodetic Survey tidal bench mark of 1886. The bench mark is the intersection of a cross (+) cut in the brick pilaster supporting the north end of the iron arch over the passageway at the southwest corner of the Consolidated Gas Company's building on West Forty-second Street, near the Hudson River. The bench mark is about 200 feet east of the entrance to the West Shore Ferry, is about 85 feet south of West Forty-second Street and on the thirty-second course of brick above the ground. Established in 1886.

Elevation, 14.638 feet.

B. M. 823.—*Eleventh Avenue.* A square (□) cut in the top of the northwest corner of the granite door-sill at the house entrance to the five-story brick tenement at No. 592 Eleventh Avenue. The bench mark is about 25 feet north of West Forty-fourth Street.

Elevation, 16.848 feet.

B. M. 824.—*Fifth Avenue.* A ¾-inch copper bolt in the top of the northwesterly corner of the third granite step or platform at the main entrance to St. Patrick's Cathedral on Fifth Avenue, between East Fiftieth Street and East Fifty-first Street. (Note 5, p. 89.)

Elevation, 80.402 feet.

B. M. 825.—*Fifth Avenue.* The center of the horizontal bar of a **T** cut in the south face of a granite pilaster on the south side of the main entrance to St. Patrick's Cathedral on Fifth Avenue, between East Fiftieth Street and East Fifty-first Street. The bench mark is about 83 feet north of the north line of East Fiftieth Street, is 2.3 feet above the walk and 1.8 feet from the west side of the pilaster.

Elevation, 80.865 feet.

B. M. 826.—*Fifth Avenue.* A ⅝-inch copper bolt in the top of the granite pedestal supporting two limestone columns on the north side of the main entrance to the ten-story building at No. 786 Fifth Avenue. The bolt is about 18 feet south of East Sixtieth Street, is between the plinths of both columns and 3.35 feet above the sidewalk. (Note 5, p. 89.)

Elevation, 55.083 feet.

B. M. 827.—*Fifth Avenue.* The top of the southeast corner of the lowest granite step at the base of the Sherman monument at the Fifth Avenue and Fifty-ninth Street entrance to Central Park. The bench mark is 9 inches above the sidewalk. (Note 26, p. 91.)

Elevation, 53.161 feet.

B. M. 828.—*Columbus Circle.* A ⅞-inch copper bolt in the center of the top of the highest granite step on the north side of the Columbus monument at the intersection of West Fifty-ninth Street and Eighth Avenue. (Note 5, p. 89.)

Elevation, 86.637 feet.

B. M. 829.—*Eleventh Avenue.* The bottom of a square cut in the center of the 6-inch wide granite water-table on the east side of the Interborough Rapid Transit Company's power house, on the northwest corner of Eleventh Avenue and West Fifth-eighth Street. The bench mark is 1 foot from the extreme southeast corner of the stone, and 2.5 feet above the sidewalk. (Note 10, p. 90.)

Elevation, 28.375 feet.

B. M. 830.—*Eleventh Avenue.* A square (⊓) cut in the edge of the brownstone door-sill of the two-story brick building on the southwest corner of Eleventh Avenue and West Fifty-seventh Street. The bench mark is 2.5 feet west of the corner.

Elevation, 27.326 feet.

B. M. 831.—*Eleventh Avenue.* A square (⊔) cut in the top of the northwest corner of the granite basin head on the southeast corner of Eleventh Avenue and West Sixtieth Street. The bench mark is on line with the east line of Eleventh Avenue.

Elevation, 27.862 feet.

B. M. 832.—*Central Park West.* The center of a circle cut in the top of the southwest corner of the bluestone basin head on the northeast corner of Central Park West and Columbus Circle.

Elevation, 79.093 feet.

B. M. 833.—*Central Park West.* A ⅝-inch iron bolt in the top of the granite basin head on the northwest corner of Central Park West and West Sixty-fifth Street. The bolt is near the south side of the basin head and about 2 feet east of Central Park West.

Elevation, 78.824 feet.

B. M. 834.—*Central Park West.* A square (□) cut in the top of the north end of the seventh granite step at the main entrance to the Second Church of Christ Scientist, on the southwest corner of Central Park West and West Sixty-eighth Street. The bench mark is on the east side of the church, about 26 feet south of West Sixty-eighth Street.

Elevation, 89.915 feet.

B. M. 835.—*West Seventy-second Street.* A square (□) cut in the top of the bluestone coping at the base of iron fence on line with the west line of the Dakota Apartments, at No. 1 West Seventy-second Street. The bench mark is about 205 feet west of Central Park West.

Elevation, 91.716 feet.

B. M. 836.—*Central Park.* The bottom of a square cut in the top of the natural rock in Central Park, about 12 feet west of the park wall on Fifth Avenue and about 8 feet north of the north house line of East Sixty-sixth Street. The letter B is cut in the rock on the north side of the point, which is opposite a straight heavy cut on the coping of the park wall. (Note 4, p. 89.)

Elevation, 66.631 feet.

B. M. 837.—*East Sixty-eighth Street.* The center of the enclosed surface within the figure 4, cut in the bluestone sidewalk on the northeast corner of East Sixty-eighth Street and Fifth Avenue. The bench mark is 4.8 feet east of the Fifth Avenue curb line and 4 feet north of the East Sixty-eighth Street curb line.

Elevation, 72.099 feet.

B. M. 838.—*East Sixty-seventh Street.* A 3-inch square (⊔) cut in the top of the southwest corner of the granite water-table at the southwest corner of the Fire Headquarters building, at No. 157 East Sixty-seventh Street. The bench mark is about 215 feet west of Third Avenue, is at the east side of the iron gate at the garage entrance and 2.75 feet above the sidewalk.

Elevation, 80.484 feet.

B. M. 839.—*East Sixty-eighth Street.* The center of the top of a marble city monument in the sidewalk on the northwest corner of East Sixty-eighth Street and Third Avenue. The top of the monument is ½ inch above the sidewalk.

Elevation, 81.083 feet.

B. M. 840.—*East Sixty-eighth Street.* A 3-inch square (⊔) cut in the top of the northwest corner of the wide granite water-table in front of the Fire Headquarters building on the south side of East Sixty-eighth Street, between Third Avenue and Lexington Avenue. The bench mark is about 270 feet west of Third Avenue, is 2.1 feet above the sidewalk and almost on line with the west side of the building.

Elevation, 86.391 feet.

B. M. 841.—*East Sixty-eighth Street.* A square (⊔) cut at the southeast corner of the bluestone coping under the iron fence in front of the three-story brick house of the Dominican Fathers, at No. 411 East Sixty-eighth Street. The bench mark is about 25 feet west of the east side of the building.

Elevation, 46.644 feet.

B. M. 842.—*East Seventy-first Street.* A square (⊔) cut in the southwest corner of the stone base of the iron column on the west side of the six-story red brick factory building on the southeast corner of East Seventy-first Street and Avenue A. The bench mark is about 38 feet south of East Seventy-first Street and on the north side of the entrance.

Elevation, 38.863 feet.

B. M. 843.—*East Seventy-fourth Street.* A square (□) cut in the top of the granite fender post on the north side of the south wagon entrance to the power house of the Manhattan Railway Company, on the northwest corner of East Seventy-fourth Street and Avenue B (Exterior Street). The bench mark is on the east side of the building, about 57 feet north of East Seventy-fourth Street.

Elevation, 11.224 feet.

B. M. 844.—*East Seventy-sixth Street.* A square (⊓) cut in the top of the northeast corner of the lowest cement step at the basement entrance to the "East Side House Settlement," on the southwest corner of East Seventy-sixth Street and Avenue B (Exterior Street). The bench mark is under the brick porch, is about 68 feet south of East Seventy-sixth Street and 7½ inches above the sidewalk.

Elevation, 10.646 feet.

B. M. 845.—*East Seventy-sixth Street.* The top of a ⅝-inch square iron bolt in the granite basin head on the northwest corner of East Seventy-sixth Street and Avenue A.

Elevation, 28.622 feet.

B. M. 846.—*Fifth Avenue.* A cross (+) cut in the top of the northeast corner of the third or highest granite step at the base of the Richard Morris Hunt monument on the west side of Fifth Avenue between East Seventieth Street and East Seventy-first Street. (Note 17, p. 91.)

Elevation, 85.046 feet.

B. M. 847.—*East Seventy-eighth Street.* The top of a galvanized iron bolt in the granite water-table on the west side of Public School No. 158, on the southeast corner of East Seventy-eighth Street and Avenue A. The bench mark is 6.7 feet south of East Seventy-eighth Street and 6½ inches above the area. (Note 13, p. 90.)

Elevation, 34.886 feet.

B. M. 848.—*East Seventy-eighth Street.* The center of the horizontal bar of a **T** cut in the granite water-table on the west side of Public School No. 158, on the southeast corner of East Seventy-eighth Street and Avenue A. The bench mark is 6.7 feet south of East Seventy-eighth Street and 11½ inches above the area.

Elevation, 35.291 feet.

B. M. 849.—*Central Park West.* A ¾-inch copper bolt in the top of the south end of the lowest granite step at the main entrance to the New York Historical Society building, on the west side of Central Park West between West Seventy-sixth Street and West Seventy-seventh Street. (Note 5, p. 89.)

Elevation, 90.821 feet.

B. M. 850.—*Central Park West.* The center of the horizontal bar of the figure 4 in the inscription "1804 · 1903," on the cornerstone of the New York Historical Society building, on the west side of Central Park West between West Seventy-sixth Street and West Seventy-seventh Street. The bench mark is about 45 feet south of West Seventy-seventh Street, is 11.33 feet from the north side of the building and 3.6 feet above the sidewalk.

Elevation, 94.229 feet.

B. M. 851.—*Manhattan Square.* A ⅞-inch copper bolt in the west end of the granite door-sill at the extreme southeasterly entrance to the American Museum of Natural History in Manhattan Square at West Seventy-seventh Street. The bolt is about 55 feet west of the west line of Central Park West. (Note 5, p. 89.)

Elevation, 83.809 feet.

B. M. 852.—*Manhattan Square.* A cross (+) cut in the south end of the granite door-jamb on the west side of the extreme southeasterly entrance to the American Museum of Natural History in Manhattan Square at West Seventy-seventh Street. The cross is 0.236 foot above the door-sill and about 55 feet west of the west line of Central Park West. (Note 15, p. 91.)

Elevation, 84.045 feet.

B. M. 853.—*Manhattan Square.* A ¾-inch copper bolt in the west end of the second granite step at the center door of the West Seventy-seventh Street or main entrance to the American Museum of Natural History. The bolt is under the arch and is 4½ inches from the west end of the step. (Note 5, p. 89.)

Elevation, 84.132 feet.

B. M. 854.—*Manhattan Square.* The center of the top of a bronze plate 8½" x 8½", set in the top of a 16-inch square granite monument under the center of the brick arch at the West Seventy-seventh Street or main entrance to the American Museum of Natural History. The bench mark is 2.4 feet above the ground.

Elevation, 85.870 feet.

B. M. 855.—*Manhattan Square.* The center of the top of the 3-inch brass tablet bolt marking the U. S. C. & G. S. astronomic station in Manhattan Square. The bolt is set in the top of a concrete pier 14" x 26" and 3 feet high, is about 65 feet north of the north line of West Seventy-ninth Street and about 250 feet west of the west line of Central Park West.

Elevation, 86.811 feet.

B. M. 856.—*Central Park.* The top of a ¾-inch iron rod inside of a wrought-iron pipe casing in the east side of Central Park, about 78 feet north of the north house line of East Seventy-ninth Street. The bench mark is 4.6 feet west of the west side of the park wall on Fifth Avenue and the top of the casing is 1 foot above the ground. (Note 21, p. 91.)

Elevation, 78.851 feet.

B. M. 857.—*Central Park.* A ⅞-inch copper bolt 6 inches from the north end of the eighth granite step at the Fifth Avenue entrance to the Metropolitan Museum of Art, opposite East Eighty-second Street. (Note 5, p. 89.)

Elevation, 91.031 feet.

B. M. 858.—*Central Park.* A cross (+) cut in the granite base course of the pedestal supporting two limestone columns about 80 feet north of the Fifth Avenue entrance to the Metropolitan Museum of Art opposite East Eighty-second Street. The cross is on the east face of the stone, is 1.3 feet above the ground and 6 inches from the north end of the stone. (Note 15, p. 91.)

Elevation, 89.448 feet.

B. M. 859.—*Central Park.* A ¾-inch copper bolt in the top of the granite base course at the northeast corner of the Metropolitan Museum of Art in the east side of Central Park, opposite East Eighty-fourth Street. The bolt is 9 inches from the corner of the building and 1.5 feet above the ground. It is about 1 foot north of the south house line of East Eighty-fourth Street and about 30 feet west of the park wall. (Note 5, p. 89.)

Elevation, 89.505 feet.

B. M. 860.—*East Eighty-sixth Street.* A ⅝-inch square iron bolt in the granite basin head on the northeast corner of East Eighty-sixth Street and Fifth Avenue. The bolt is 8 inches from the west edge of the basin head.

Elevation, 94.893 feet.

B. M. 861.—*East Eighty-sixth Street.* The center of the enclosed surface within the figure 4, cut in the cement sidewalk on the southeast corner of East Eighty-sixth Street and Third Avenue. The bench mark is 5 feet south of the south curb line.

Elevation, 77.882 feet.

B. M. 862.—*Third Avenue.* The top of the southwest corner of the 3-foot square granite plinth of the granite column on the south side of the west entrance to the Yorkville Bank (four-story stone and brick) on the northeast corner of Third Avenue and East Eighty-fifth Street. The bench mark is about 16 feet north of East Eighty-fifth Street and 3.8 feet above the sidewalk. (Note 26, p. 91.)

Elevation, 82.873 feet.

B. M. 863.—*East Eighty-sixth Street.* The top of a galvanized iron bolt in the granite water-table on the north side of Public School No. 77, on the southeast corner of East Eighty-sixth Street and First Avenue. The bolt is 2.5 feet from the west side of the building and 1.4 feet above the area. (Note 13, p. 90.)

Elevation, 51.329 feet.

B. M. 864.—*East Eighty-sixth Street.* The center of the horizontal bar of a T cut in the granite water-table on the north side of Public School No. 77, on the southeast corner of East Eighty-sixth Street and First Avenue. The bench mark is 2.5 feet from the west side of the building and 1.8 feet above the area.

Elevation, 51.756 feet.

B. M. 865.—*East Eighty-sixth Street.* A square (□) cut in the bluestone sidewalk on the southeast corner of East Eighty-sixth Street and Avenue A.

Elevation, 40.319 feet.

B. M. 866.—*East Eighty-sixth Street.* A square (□) cut in the center of the sidewalk on the south side of East Eighty-sixth Street, about 277 feet east of Avenue A.

Elevation, 36.417 feet.

B. M. 867.—*East Eighty-sixth Street.* A square (□) cut in the center of the sidewalk on the south side of East Eighty-sixth Street, about 540 feet east of Avenue A.

Elevation, 29.236 feet.

B. M. 868.—*East Eighty-sixth Street.* A square (□) cut in the top of a large boulder on the south side of East Eighty-sixth Street and on the west side of the river promenade on the east side of Carl Schurz Park.

Elevation, 8.845 feet.

B. M. 869.—*East Eighty-fourth Street.* The top of a heavy iron ring-bolt leaded vertically in the natural rock at the foot of East Eighty-fourth Street. The ring-bolt is 8.5 feet south of the south end of the sea wall on the east side of Carl Schurz Park and about 35.7 feet east of the high retaining wall at the foot of the street.

Elevation, 5.861 feet.

B. M. 870.—*East Eighty-fourth Street.* A 1-inch copper bolt leaded vertically in the natural rock at the foot of East Eighty-fourth Street. The bolt is 8 feet south of the south end of the sea wall on the east side of Carl Schurz Park, and about 36 feet east of the high retaining wall at the foot of the street. The bench mark is 7 inches northeast of the iron ring-bolt. (Note 5, p. 89.)

Elevation, 5.374 feet.

B. M. 871.—*Carl Schurz Park.* Coast and Geodetic Survey B. M. 5. This point is on the extreme southeast corner of the sea wall on the east side of Carl Schurz Park, and is on the top course of masonry under the 1-foot granite coping. Established in 1886.

Elevation, 7.543 feet.

B. M. 872.—*Carl Schurz Park.* The bottom of a square cut near the southeast corner of the granite copingstone on the southwest corner of the sea wall on the east side of Carl Schurz Park. The bench mark is at the south end of the promenade and about on line with the north line of East Eighty-fourth Street. An arrowhead cut in the south side of the coping points to the bench mark. (Note 4, p. 89.)

Elevation, 8.578 feet.

B. M. 873.—*Carl Schurz Park.* A square (□) cut in the granite coping of the sea wall on the east side of Carl Schurz Park. The bench mark is about on line with the center line of East Eighty-fifth Street or about 230 feet north of the south end of the sea wall. It is 2 inches from the west edge of the coping.

Elevation, 8.672 feet.

B. M. 874.—*Central Park West.* A ¾-inch iron bolt in the top of the granite basin head on the northeast corner of Central Park West and West Eighty-sixth Street. The bolt is 7 inches from the west side of the basin head.

Elevation, 110.739 feet.

B. M. 875.—*Central Park West.* A level triangular surface 2" x 2" x 2¾", on top of the extreme northeast corner of the concave limestone water-table of the eleven-story apartment house on the southeast corner of Central Park West and West Eighty-sixth Street. The bench mark is 4.1 feet above the area.

Elevation, 113.925 feet.

B. M. 876.—*Central Park West.* A cross (+) cut in the top of the northeast corner of the granite pedestal of granite column on the north side of the main entrance to the First Church of Christ Scientist on the northwest corner of Central Park West and West Ninety-sixth Street. The cross is about 58 feet north of the corner. (Note 17, p. 91.)

Elevation, 104.304 feet.

B. M. 877.—*Central Park West.* The center of the horizontal bar of the final letter H in CHURCH, in the inscription on the cornerstone of the "FIRST CHURCH OF CHRIST SCIENTIST," on the northwest corner of Central Park West and West Ninety-sixth Street. The bench mark is on the south side of the building, is 2.4 feet from the east side and 2 feet above the sidewalk.

Elevation, 104.353 feet.

B. M. 878.—*Central Park West.* A ⅝-inch iron bolt in the granite basin head on the southeast corner of Central Park West and West Ninety-sixth Street. The bolt is 7 inches from the north side of the basin head.

Elevation, 102.253 feet.

B. M. 879.—*Central Park.* Board of Water Supply B. M. 13 B. This bench mark is the top of a small rounded summit within a hollow square cut in the natural rock in the west side of Central Park. The bench mark is about on line with the north line of West Ninety-third Street and about 8 feet east of the first asphalt walk east of Central Park West. The letter B is cut in the rock on the west side of the bench.

Elevation, 114.145 feet.

B. M. 880.—*Central Park.* Board of Water Supply B. M. 13 A. A level square cut in the natural rock in the west side of Central Park. The bench mark is on a sloping ledge about on line with the south line of West Ninety-third Street and about 180 feet east of the park wall on Central Park West. The letter A is cut in the rock on the west side of the bench.

Elevation, 111.172 feet.

B. M. 881.—*West One Hundred and First Street.* A square (□) cut in the top of the northwest corner of the bluestone coping on the east side of the basement entrance to the apartment house at No. 2 West One Hundred and First Street. The bench mark is about 35 feet west of Central Park West.

Elevation, 86.955 feet.

B. M. 882.—*West One Hundred and First Street.* The top of a ¾-inch iron bolt in the top of the granite basin head on the center line of West One Hundred and First Street and on the east side of Central Park West. The bolt is 7½ inches from the west side of the basin head.

Elevation, 85.861 feet.

B. M. 883.—*Fifth Avenue.* A ⅞-inch copper bolt in the top of the north end of the lowest granite step at the main entrance to the private pavilion of Mount Sinai Hospital, on the east side of Fifth Avenue, between East One Hundredth Street and East One Hundred and First Street. The bench mark is 7 inches above the area. (Note 5; p. 89.)

Elevation, 56.915 feet.

B. M. 884.—*Central Park.* The top of a ¾-inch iron rod inside of a wrought-iron pipe casing in the east side of Central Park. The bench mark is 4 feet west of the park wall on Fifth Avenue and about 80 feet south of the center line of East One Hundred and Second Street. (Note 21, p. 91.)

Elevation, 49.612 feet.

B. M. 885.—*West One Hundred and Sixth Street.* A square (□) cut in the top of the west end of the brownstone coping at the end of the iron railing, on the south side of the apartment house at No. 1 West One Hundred and Sixth Street. The bench mark is on the east side of the brownstone balustrade on the east side of the house entrance and about 38 feet west of Central Park West.

Elevation, 80.577 feet.

B. M. 886.—*Seventh Avenue.* The bottom of a square cut in the top of the southwest corner of the foundation stone of the stone post at the west end of the Central Park wall and on line with the east side of Seventh Avenue at West One Hundred and Tenth Street. The bench mark is 11 inches above the ground. (Note 10, p. 90.)

Elevation, 35.503 feet.

B. M. 887.—*Seventh Avenue.* The top of a ⅝-inch square iron bolt in the top of the granite basin head on the northwest corner of Seventh Avenue and West One Hundred and Tenth Street. The bolt is 7 inches from the east edge of the basin head.

Elevation, 35.257 feet.

B. M. 888.—*West One Hundred and Tenth Street.* A square (□) cut in the center of the sidewalk on the north side of West One Hundred and Tenth Street, about 200 feet west of Fifth Avenue.

Elevation, 20.106 feet.

B. M. 889.—*Fifth Avenue.* A square (□) cut in the top of the northwesterly end of the long granite gate-sill between two bluestone posts, at the most easterly of the two entrances to Central Park at Fifth Avenue and West One Hundred and Tenth Street. The bench mark is about 40 feet west of the west line of Fifth Avenue.

Elevation, 18.831 feet.

B. M. 890.—*Park Avenue.* The top of a ⅝-inch iron bolt set horizontally into the brownstone on the east side of the New York Central and Hudson River Railroad at Park Avenue and East One Hundred and Tenth Street. The bolt is about on line with the north curb and 1.8 feet above the street. (Note 11, p. 90.)

Elevation, 14.991 feet.

B. M. 891.—*West One Hundred and Fifteenth Street.* The top of a galvanized iron bolt in the granite base course on the east side of the east entrance to Wadleigh High School, on the south side of West One Hundred and Fifteenth Street between Seventh Avenue and Eighth Avenue. The bolt is about 138 feet west of Seventh Avenue, or about 38 feet from the east side of the building, and 5 inches above the area. (Note 13, p. 90.)

Elevation, 34.723 feet.

B. M. 892.—*West One Hundred and Fifteenth Street.* A cross (+) cut in the granite base course on the east side of the east entrance to Wadleigh High School, on the south side of West One Hundred and Fifteenth Street between Seventh Avenue and Eighth Avenue. The cross is about 138 feet west of Seventh Avenue. (Note 15, p. 91.)

Elevation, 35.002 feet.

B. M. 893.—*West One Hundred and Fifteenth Street.* The bottom of a square cut in the top of the southwest corner of the granite water-table in front of the West One Hundred and Fifteenth Street branch of the New York Public Library. The bench mark is about 150 feet west of Seventh Avenue, is 1.1 feet east of the west side of the building and 1 foot from the face. (Note 10, p. 90.)

Elevation, 36.138 feet.

B. M. 894.—*Fifth Avenue.* The top of a 1-inch iron bolt in the top of the granite basin head on the northwest corner of Fifth Avenue and West One Hundred and Sixteenth Street. The bolt is about 2.5 feet east of the west line of Fifth Avenue and 4 inches from the south edge of the basin head.

Elevation, 21.342 feet.

B. M. 895.—*Mount Morris Park.* A ¾-inch copper bolt in the outcropping natural rock in the south side of Mount Morris Park. The bolt is in the center of a hollow square cut in the highest part of the rock, is about 70 feet west of Fifth Avenue and about 10 feet north of West One Hundred and Twentieth Street. (Note 5, p. 89.)

Elevation, 28.878 feet.

B. M. 896.—*Mount Morris Park.* The top of a ¾-inch iron rod inside of a wrought-iron pipe casing, in the south side of Mount Morris Park. The bench mark is set in the rock ledge, is about on the center line of Fifth Avenue and about 27 feet north of the north line of West One Hundred and Twentieth Street. Established by the Board of Water Supply.

Elevation, 29.922 feet.

B. M. 897.—*Madison Avenue.* A square (⊔) cut in the top of the northwest corner of the bluestone door-sill at the Madison Avenue entrance to the Pilgrim Church, on the northeast corner of Madison Avenue and East One Hundred and Twenty-first Street. The bench mark is about 10 feet north of the corner.

Elevation, 32.137 feet.

B. M. 898.—*Hancock Place.* A ¾-inch copper bolt in the top of the cone-shaped granite wagon guard on the east side of the wagon entrance to the house of Hook and Ladder Co. No. 40, on the south side of Hancock Place about 150 feet west of Hancock Square. The bolt is 1.6 feet above the sidewalk. (Note 5, p. 89.)

Elevation, 31.869 feet.

B. M. 899.—*Manhattan Street.* The bottom of a square cut in the top of the granite pedestal supporting the steel column and also the south end of the westerly steel arch of the Interborough Rapid Transit subway, at Manhattan Street. The bench mark is on the west side of the Manhattan Street subway station at Broadway, about 35 feet south of Manhattan Street. It is 4 feet above the roadway and 6 inches from the northeast corner of the stone. (Note 4, p. 89.)

Elevation, 37.988 feet.

B. M. 900.—*Manhattan Street.* A square (⊔) cut in the top of the southwest corner of the steel base of column on the east side of the Riverside Drive Viaduct on Twelfth Avenue, opposite the corner formed by the intersection of the northeasterly line of Manhattan Street and the southerly line of West One Hundred and Thirtieth Street. The column containing the bench mark is 3 feet east of the curb and about on the center line of the northeasterly sidewalk of Manhattan Street. An arrowhead cut in the westerly side of the granite base block points to the bench, which is 1.4 feet above the sidewalk.

Elevation, 13.889 feet.

B. M. 901.—*Manhattan Street.* A square (⊔) cut in the top of the southwest corner of the water-table of the two-story brick hotel on the northwest corner of Manhattan Street, West One Hundred and Thirtieth Street, and Twelfth Avenue. The bench mark is about 100 feet west of Twelfth Avenue and 3.17 feet above the sidewalk.

Elevation, 13.084 feet.

B. M. 902.—*St. Nicholas Avenue.* A horizontal line in the end of a ¾-inch copper bolt set horizontally into the granite foundation on the east side of Public School No. 157, on the west side of St. Nicholas Avenue between West One Hundred and

Twenty-sixth Street and West One Hundred and Twenty-seventh Street. The bolt is about 10 feet north of the main entrance or about 70 feet south of West One Hundred and Twenty-seventh Street. (Note 6, p. 90.)

Elevation, 36.332 feet.

B. M. 903.—*St. Nicholas Avenue.* A square (□) cut in the top of the northeast corner of the second bluestone step at the main entrance to Public School No. 157, on the west side of St. Nicholas Avenue between West One Hundred and Twenty-sixth Street and West One Hundred and Twenty-seventh Street.

Elevation, 35.616 feet.

B. M. 904.—*Madison Avenue.* A square (⊔) cut in the top of the southwest corner of the brownstone base of the brownstone balustrade on the south side of the house entrance to the five-story apartment house at No. 2033 Madison Avenue. The bench mark is about 55 feet south of East One Hundred and Twenty-ninth Street and 6 inches above the sidewalk.

Elevation, 23.176 feet.

B. M. 905.—*East One Hundred and Thirtieth Street.* A square (□) cut in the top of the south edge of the granite block at the west end of the west approach to the Third Avenue bridge over the Harlem River. The bench mark is on the south side of the approach, is about 2 feet west of the east line of Lexington Avenue and 1.7 feet above the street.*

Elevation, 14.919 feet.

B. M. 906.—*East One Hundred and Thirtieth Street.* A ⅞-inch copper bolt in the top of the easterly corner of the granite base of a steel column on the northerly side of the easterly approach to the Third Avenue bridge over the Harlem River at East One Hundred and Thirtieth Street. The bolt is at the base of the first column east of the bridge, is about 28 feet east of the east side of the bridge, and about 21 feet south of the Harlem River bulkhead. (Note 5, p. 89.)

Elevation, 10.669 feet.

B. M. 907.—*East One Hundred and Thirtieth Street.* A square (□) cut in the top of the granite base of the most easterly column on the south side of the east approach to the Third Avenue bridge over the Harlem River at East One Hundred and Thirtieth Street. The bench mark is about 55 feet east of the east line of Third Avenue and 4 inches above the sidewalk.

Elevation, 10.175 feet.

B. M. 908.—*East One Hundred and Thirtieth Street.* A square (□) cut in the top of the west side of the semicircular granite base of watering trough on the south side of the approach to the Third Avenue bridge over the Harlem River. The bench mark is about on the north line of East One Hundred and Thirtieth Street and about 5 feet west of the center line of Third Avenue.

Elevation, 10.193 feet.

B. M. 909.—*East One Hundred and Thirtieth Street.* A cross (+) cut in the dressed edge of the granite masonry at the east end of the south abutment of the Third Avenue bridge over the Harlem River at East One Hundred and Thirtieth Street. The cross is on the north face of the stone, is about 10 feet east of the east side of the bridge and about 35 feet south of the Harlem River bulkhead. It is 5 feet above the ground and the letter B is cut in the stone under the cross. (Note 15, p. 91.)

Elevation, 14.998 feet.

* Reported as having been disturbed by subway construction during 1914.

B. M. 910.—*St. Nicholas Avenue.* A square (⊡) cut in the top of the northwest corner of the brownstone coping under the iron fence on the southeast corner of St. Nicholas Avenue and West One Hundred and Thirty-fifth Street. The bench mark is 1.8 feet above the sidewalk.

Elevation, 40.060 feet.

B. M. 911.—*St. Nicholas Park.* The bottom of a 3-inch square cut in the top of the natural rock in St. Nicholas Park. The bench mark is about 100 feet west of St. Nicholas Avenue and about 10 feet north of the south line of West One Hundred and Fortieth Street. It is 4.5 feet above the ground and the letter B is cut in the vertical face of the rock on the east side of the bench.

Elevation, 55.018 feet.

B. M. 912.—*St. Nicholas Avenue.* A square (⊡) cut in the top of the southwest corner of the lowest granite step at the southeasterly entrance to the St. Nicholas Avenue Presbyterian Church, on the northwest corner of St. Nicholas Avenue and West One Hundred and Forty-first Street. The bench mark is flush with the sidewalk and about 3 feet south of the north line of West One Hundred and Forty-first Street.

Elevation, 52.057 feet.

B. M. 913.—*St. Nicholas Avenue.* The top of a ⅝-inch square iron bolt in the top of the bluestone basin head on the northeast corner of St. Nicholas Avenue and West One Hundred and Forty-fifth Street. The bolt is 9 inches from the west side of the basin head.

Elevation, 76.330 feet.

B. M. 914.—*St. Nicholas Avenue.* A cross (+) cut in the top of the northeast corner of the granite plinth of the column on the north side of the main entrance to the six-story apartment house at No. 707 St. Nicholas Avenue. The cross is about 70 feet north of West One Hundred and Forty-fifth Street. (Note 17, p. 91.)

Elevation, 80.032 feet.

B. M. 915.—*St. Nicholas Avenue.* A square (⊡) cut in the top of the southeast corner of the brownstone coping under the iron railing on the north side of the cellar entrance to the four-story brick apartment house at No. 783 St. Nicholas Avenue. The bench mark is about 25 feet north of West One Hundred and Forty-ninth Street.

Elevation, 100.372 feet.

B. M. 916.—*St. Nicholas Avenue.* The top of a ⅝-inch square iron bolt in the top of the bluestone basin head on the northwest corner of St. Nicholas Avenue and West One Hundred and Fiftieth Street. The bolt is 8 inches from the south side of the basin head.

Elevation, 105.333 feet.

B. M. 917.—*St. Nicholas Avenue.* The top of a ⅝-inch iron bolt set horizontally into the west side of the water-table of Croton Aqueduct ventilator, on the east side of St. Nicholas Avenue just north of the north line of West One Hundred and Fifty-fourth Street. The bolt is 2.5 feet above the sidewalk. (Note 11, p. 90.)

Elevation, 135.811 feet.

B. M. 918.—*West One Hundred and Fifty-fifth Street.* The top of a ⅝-inch iron bolt set horizontally into the granite base of the large ornamental stone post at the corner of Trinity Church Cemetery, on the southwest corner of West One Hundred and Fifty-fifth Street and Amsterdam Avenue. The bolt is 5½ inches above the sidewalk. (Note 11, p. 90.)

Elevation, 148.729 feet.

B. M. 919.—*St. Nicholas Avenue.* A horizontal line on the face of a brass tablet bolt 2 inches in diameter, set horizontally into the bluestone water-table on the east side of Public School No. 46, on the southwest corner of St. Nicholas Avenue and West One Hundred and Fifty-sixth Street. The bolt is 11 inches from the northeast corner of the building and 1.25 feet above the area. (Note 1, p. 89.)

Elevation, 147.949 feet.

B. M. 920.—*St. Nicholas Avenue.* A square (□) cut in the top of the northwest corner of the lowest granite step at the main entrance to the Washington Heights Free Library, at No. 922 St. Nicholas Avenue, opposite West One Hundred and Fifty-sixth Street. The bench mark is 4 inches above the sidewalk.

Elevation, 146.591 feet.

B. M. 921.—*St. Nicholas Avenue.* A ½-inch copper bolt in the marble city monument in the center of the sidewalk on the northwest corner of St. Nicholas Avenue and West One Hundred and Sixty-second Street. The top of the monument is flush with the sidewalk.

Elevation, 165.494 feet.

B. M. 922.—*West One Hundred and Sixty-ninth Street.* A square (□) cut in the top of the northeast corner of the bluestone door-sill at the boys' entrance to Public School No. 169, on the southwest corner of West One Hundred and Sixty-ninth Street and Audubon Avenue. The bench mark is on the north side of the building and about 27 feet from the east side.

Elevation, 176.288 feet.

B. M. 923.—*West One Hundred and Sixty-ninth Street.* A cross (+) cut in the granite masonry at the northeast corner of Public School No. 169, on the southwest corner of West One Hundred and Sixty-ninth Street and Audubon Avenue. The cross is on the north side of the building, is 6 inches from the corner and 1.17 feet above the sidewalk. (Note 15, p. 91.)

Elevation, 176.553 feet.

B. M. 924.—*West One Hundred and Seventieth Street.* The bottom of a square cut in the northwest corner of the granite door-sill at the east entrance to Engine House No. 67, on the south side of West One Hundred and Seventieth Street about 110 feet east of Audubon Avenue. The bench mark is on the north edge of the sloping sill and 6.8 feet from the east side of the building. (Note 10, p. 90.)

Elevation, 178.216 feet.

B. M. 925.—*West One Hundred and Seventy-fourth Street.* A square (□) cut in the top of the bluestone coping at the corner of the iron fence on the northeast corner of West One Hundred and Seventy-fourth Street and Audubon Avenue. It is 4 inches above the sidewalk.

Elevation, 185.916 feet.

B. M. 926.—*Highbridge Water Tower.* The top of a ⅝-inch iron bolt set horizontally into the granite water-table on the northwesterly side of Highbridge Water Tower, about 540 feet east of Amsterdam Avenue and about 75 feet south of West One Hundred and Seventy-fourth Street. The bolt is 5 inches above the ground. (Note 11, p. 90.)

Elevation, 202.146 feet.

B. M. 927.—*Highbridge Water Tower.* A square (⊓) cut in the top of the north edge of the granite door-sill at the entrance to Highbridge Water Tower. The bench mark is on the north side of the tower and 4 inches from the east end of the step.

Elevation, 203.246 feet.

B. M. 928.—*Highbridge Water Tower.* The top of a 2-inch hemispherical brass bolt set vertically in the stone foundation of Highbridge Water Tower. The bolt is on the northwesterly side of the tower just west of the entrance.

Elevation, 201.791 feet.

B. M. 929.—*West One Hundred and Eighty-first Street.* A square (ㅁ) cut in the top of the northwesterly corner of the bluestone basin head on the northeast corner of West One Hundred and Eighty-first Street and Amsterdam Avenue.

Elevation, 167.025 feet.

B. M. 930.—*West One Hundred and Eighty-first Street.* A ½-inch copper bolt in the marble city monument on the northeast corner of West One Hundred and Eighty-first Street and Audubon Avenue. The top of the monument is 4 inches below the sidewalk.

Elevation, 169.531 feet.

B. M. 931.—*Wadsworth Avenue.* A horizontal line on the face of a brass tablet bolt 2 inches in diameter, set horizontally into the center of the south face of the buttress on the west side of Public School No. 132, on the east side of Wadsworth Avenue between West One Hundred and Eighty-second Street and West One Hundred and Eighty-third Street. The bench mark is in the limestone, is north of the main entrance and about 40 feet south of the north side of the building. It is 4 feet above the ground. (Note 1, p. 89.)

Elevation, 188.334 feet.

B. M. 932.—*Wadsworth Avenue.* A cross (+) cut in the south face of the lowest granite block in the buttress on the west side of Public School No. 132, on the east side of Wadsworth Avenue, between West One Hundred and Eighty-second Street and West One Hundred and Eighty-third Street. The bench mark is north of the main entrance, is about 40 feet south of the north side of the building and 12 inches above the ground. (Note 15, p. 91.)

Elevation, 185.322 feet.

B. M. 933.—*Wadsworth Avenue.* A ¾-inch copper bolt in the northeast corner of the lowest bluestone step at the main entrance to Public School No. 132, on the east side of Wadsworth Avenue between West One Hundred and Eighty-second Street and West One Hundred and Eighty-third Street. The bolt is 6 inches from the north end of the step. (Note 5, p. 89.)

Elevation, 184.771 feet.

B. M. 934.—*Broadway.* A ⅝-inch copper bolt in the top of the south end of the bluestone door-sill at the house entrance to the six-story yellow brick apartment house at Nos. 4441-43 Broadway. The bench mark is on the west side of Broadway, about 56 feet north of Fairview Avenue. (Note 8, p. 90.)

Elevation, 74.842 feet.

B. M. 935.—*Broadway.* A square (ㅁ) cut in the top of the southeast corner of the bluestone door-sill at the house entrance to the six-story apartment house at Nos. 4441-43 Broadway.

Elevation, 74.828 feet.

B. M. 936.—*Broadway.* A square (ㅁ) cut in the top of the northeast corner of the bluestone door-sill at the house entrance to the three-story brick and frame hotel at No. 4601 Broadway. The bench mark is on the west side of Broadway about 400 feet north of Nagle Avenue.

Elevation, 41.668 feet.

B. M. 937.—*Dyckman Street.* A ½-inch copper bolt in the top of a marble city monument on the northeast corner of Dyckman Street and Broadway. The top of the monument is flush with the ground.

Elevation, 24.198 feet.

B. M. 938.—*Dyckman Street.* A square (□) cut in the top of the bluestone coping of the brick wall at the property line between Nos. 207 and 209 Dyckman Street. The bench mark is about 55 feet east of Broadway and on top of the wall separating the areas of both apartment houses (six-story yellow brick).

Elevation, 22.564 feet.

B. M. 939.—*Dyckman Street.* A square (□) cut in the top of the natural rock on the east side of the New York Central and Hudson River Railroad, and about 340 feet north of Dyckman Street. The bench mark is at the north end of the railroad cut and 2.6 feet above the ground.

Elevation, 11.289 feet.

B. M. 940.—*West Two Hundred and Seventh Street.* A cross (+) cut in the top of the west nozzle of the fire hydrant on the northwest corner of West Two Hundred and Seventh Street (Emerson Street) and Broadway.

Elevation, 40.971 feet.

B. M. 941.—*West Two Hundred and Seventh Street.* The bottom of a square cut in the top of the northeast corner of the granite base block at the west end of the north side of the west approach to University Bridge over the Harlem River, at West Two Hundred and Seventh Street. The bench mark is just east of the east line of Ninth Avenue and 1.6 feet above the street. (Note 4, p. 89.)

Elevation, 14.347 feet.

B. M. 942.—*West Two Hundred and Seventh Street.* The bottom of a square cut in the top of the northwest corner of the granite base block at the north end of the west abutment of University Bridge over the Harlem River, at West Two Hundred and Seventh Street. The bench mark is at the base of the abutment and 6 inches above the street. (Note 4, p. 89.)

Elevation, 9.293 feet.

B. M. 943.—*West Two Hundred and Seventh Street.* A ⅞-inch copper bolt in the top of the southwest corner of the granite water-table on top of the north end of the west abutment of University Bridge over the Harlem River, at West Two Hundred and Seventh Street. The bolt is about 400 feet east of the east line of Ninth Avenue, is on the north side of the north foot path and 6 inches from the south side of the water-table. (Note 5, p. 89.)

Elevation, 29.965 feet.

B. M. 944.—*Broadway.* A quadrant (⊾) cut in the top of the northeast corner of the bluestone basin head on the westerly side of Broadway and on line with the south curb line of West Two Hundred and Fifteenth Street.

Elevation, 32.638 feet.

B. M. 945.—*Amsterdam Avenue.* A square (⊔) cut in the top of the southwest corner of the lowest granite step at the office entrance to the old car barn of the Third Avenue Railroad Company (two-story red brick) on the east side of Tenth Avenue and about 100 feet south of West Two Hundred and Eighteenth Street. The bench mark is about 1.3 feet above the sidewalk.

Elevation, 30.945 feet.

B. M. 946.—*West Two Hundred and Sixteenth Street.* The center of the horizontal bar of a **T** cut in the granite water-table at the southwest corner of the power house of the Metropolitan Street Railway Company, on the northeast corner of West Two Hundred and Sixteenth Street and Ninth Avenue. The bench mark is on the south side of the building, is 6 feet above the sidewalk and 1 foot from the corner.

Elevation, 19.217 feet.

B. M. 947.—*West Two Hundred and Sixteenth Street.* The top of a ½-inch iron bolt set horizontally into the granite masonry at the southwest corner of the power house of the Metropolitan Street Railway Company, on the northeast corner of West Two Hundred and Sixteenth Street and Ninth Avenue. The bolt is on the west side of the building, is 3.5 feet from the south side and 9 inches above the sidewalk.

Elevation, 13.961 feet.

B. M. 948.—*U. S. Ship Canal Bridge.* A small rounded summit within a circle, cut in the top of the granite coping on the west side of the south approach to the U. S. Ship Canal Bridge on Broadway. The bench mark is 1.9 feet from the south end of the coping, is under the iron railing and about 55 feet north of the center line of Ninth Avenue.

Elevation, 33.903 feet.

B. M. 949.—*U. S. Ship Canal Bridge.* The bottom of a square cut in the top of the south end of the curved granite coping on the west side of the north approach to the U. S. Ship Canal Bridge on Broadway. The bench mark is above the west end of the bridge seat of the north abutment and about 1.3 feet below the west foot path. The letter B is cut in the coping on the north side of the point, which is about 50 feet south of the center line of West Two Hundred and Twenty-fifth Street. (Note 4, p. 89.)

Elevation, 32.604 feet.

B. M. 950.—*U. S. Ship Canal Bridge.* A ½-inch copper bolt leaded vertically in the top of the sea wall on the south side of the U. S. Ship Canal at Broadway. The bolt is set within a 3-inch square cut in the stone and the letters U. S. are cut on the south side. The bolt is 1 foot west of the west side of the U. S. Ship Canal Bridge and 5 inches south of the face of the sea wall. The top of the bolt is stamped with the figure 13.

Elevation, 7.203 feet.

B. M. 951.—*U. S. Ship Canal Bridge.* A cross (+) cut in the west side of the granite water-table at the west end of the north abutment of the U. S. Ship Canal Bridge on Broadway. The cross is about 12 feet below the roadway of the bridge and 3.3 feet from the south face of the abutment. (Note 15, p. 91.)

Elevation, 20.322 feet.

B. M. 952.—*Broadway.* A square (⊡) cut in the top of the southwest corner of the granite copingstone on the top of the west end of the south abutment of the highway bridge on Broadway, over Spuyten Duyvil Creek. The bench mark is about 78 feet south of West Two Hundred and Thirtieth Street.

Elevation, 19.582 feet.

# STANDARD ELEVATIONS OF BENCH MARKS

IN THE

# BOROUGH OF THE BRONX

AND IN

# WESTCHESTER COUNTY, N. Y.

| No. of Bench Mark. | Locality. | Elevation Above Mean Sea Level. | |
|---|---|---|---|
| | | Meters. | Feet. |
| 953 | Third Avenue Bridge | 2.7373 | 8.981 |
| 954 | " " " | 3.7424 | 12.278 |
| 955 | " " " | 3.2297 | 10.596 |
| 956 | " " " | 2.6195 | 8.594 |
| 957 | " " " | 2.8836 | 9.461 |
| 958 | Third Avenue | 3.5923 | 11.786 |
| 959 | East One Hundred and Thirty-fifth Street | 3.3696 | 11.055 |
| 960 | East One Hundred and Thirty-eighth Street | 3.5093 | 11.513 |
| 961 | Mott Avenue | 4.8138 | 15.793 |
| 962 | " " | 15.6427 | 51.321 |
| 963 | East One Hundred and Forty-fourth Street | 12.2674 | 40.247 |
| 964 | East One Hundred and Forty-sixth Street | 12.6321 | 41.444 |
| 965 | Mott Avenue | 15.2310 | 49.970 |
| 966 | Brown Place | 11.8927 | 38.953 |
| 967 | Southern Boulevard | 13.4943 | 44.273 |
| 968 | " " | 13.4675 | 44.185 |
| 969 | " " | 12.5919 | 41.312 |
| 970 | " " | 8.6807 | 28.480 |
| 971 | Whitlock Avenue | 8.6612 | 28.416 |
| 972 | Union Avenue | 10.7361 | 35.223 |
| 973 | Southern Boulevard | 9.4826 | 31.111 |
| 974 | Morris Avenue | 6.5143 | 21.372 |
| 975 | Courtlandt Avenue | 14.5673 | 47.793 |
| 976 | " " | 14.1503 | 46.425 |
| 977 | East One Hundred and Forty-ninth Street | 13.5292 | 44.387 |
| 978 | Brook Avenue | 5.2126 | 17.102 |
| 979 | Westchester Avenue | 7.1790 | 23.553 |
| 980 | " " | 17.5182 | 57.470 |
| 981 | " " | 20.2421 | 66.411 |
| 982 | Prospect Avenue | 12.6021 | 41.345 |
| 983 | " " | 12.2434 | 40.168 |
| 984 | Longwood Avenue | 14.3193 | 46.979 |
| 985 | " " | 13.7103 | 44.981 |
| 986 | " " | 15.0194 | 49.276 |
| 987 | Southern Boulevard | 8.9249 | 29.281 |
| 988 | " " | 7.3511 | 24.118 |
| 989 | Lafayette Avenue | 16.7592 | 54.984 |
| 990 | " " | 17.5251 | 57.497 |
| 991 | East One Hundred and Sixty-third Street | 14.9013 | 48.889 |
| 992 | Westchester Avenue | 10.0051 | 32.825 |
| 993 | " " | 18.8326 | 61.787 |
| 994 | " " | 18.2780 | 59.967 |
| 995 | Grand Boulevard and Concourse | 23.3592 | 76.638 |
| 996 | " " " " | 23.7300 | 77.854 |
| 997 | Sheridan Avenue | 20.6050 | 67.602 |
| 998 | Grand Boulevard and Concourse | 34.3330 | 112.641 |

| No. of Bench Mark. | Locality. | Elevation Above Mean Sea Level. | |
|---|---|---|---|
| | | Meters. | Feet. |
| 999 | Throgs Neck | 2.5184 | 8.262 |
| 1000 | " " | 5.3437 | 17.532 |
| 1001 | " " | 5.3155 | 17.439 |
| 1002 | " " | 6.3344 | 20.782 |
| 1003 | " " | 2.8165 | 9.240 |
| 1004 | " " | 4.1513 | 13.620 |
| 1005 | Fort Schuyler Road | 11.7279 | 38.477 |
| 1006 | " " " | 12.1733 | 39.939 |
| 1007 | " " " | 12.0762 | 39.620 |
| 1008 | " " " | 12.9109 | 42.359 |
| 1009 | " " " | 14.5574 | 47.760 |
| 1010 | " " " | 11.2597 | 36.941 |
| 1011 | " " " | 11.3942 | 37.383 |
| 1012 | " " " | 17.5497 | 57.578 |
| 1013 | Westchester Avenue | 10.8745 | 35.677 |
| 1014 | " " | 13.0102 | 42.684 |
| 1015 | " " | 8.9147 | 29.248 |
| 1016 | " " | 8.1787 | 26.833 |
| 1017 | " " | 2.6440 | 8.675 |
| 1018 | " " | 2.6457 | 8.680 |
| 1019 | " " | 8.1804 | 26.839 |
| 1020 | " " | 8.1831 | 26.847 |
| 1021 | " " | 8.1847 | 26.853 |
| 1022 | " " | 4.4378 | 14.560 |
| 1023 | " " | 9.8151 | 32.202 |
| 1024 | " " | 8.4555 | 27.741 |
| 1025 | Beach Avenue | 9.7390 | 31.952 |
| 1026 | " " | 9.3319 | 30.616 |
| 1027 | Westchester Avenue | 8.0454 | 26.396 |
| 1028 | Hugh J. Grant Circle | 6.4477 | 21.153 |
| 1029 | " " " | 6.4612 | 21.198 |
| 1030 | " " " | 6.4709 | 21.230 |
| 1031 | New York Catholic Protectory | 16.5375 | 54.257 |
| 1032 | " " " " | 17.5017 | 57.420 |
| 1033 | Westchester Avenue | 11.5071 | 37.753 |
| 1034 | " " | 10.9717 | 35.996 |
| 1035 | Unionport Road | 9.5479 | 31.325 |
| 1036 | Castle Hill Avenue | 12.7433 | 41.809 |
| 1037 | " " " | 15.1653 | 49.755 |
| 1038 | " " " | 16.0261 | 52.579 |
| 1039 | East One Hundred and Seventy-seventh Street | 6.2258 | 20.426 |
| 1040 | " " " " " " " | 2.7006 | 8.860 |
| 1041 | " " " " " " " | 2.6989 | 8.855 |
| 1042 | " " " " " " " | 4.4687 | 14.661 |
| 1043 | " " " " " " " | 5.5113 | 18.082 |
| 1044 | Eastern Boulevard | 4.9604 | 16.274 |
| 1045 | " " | 6.5121 | 21.365 |
| 1046 | " " | 7.6823 | 25.204 |
| 1047 | " " | 11.5746 | 37.974 |
| 1048 | " " | 15.6300 | 51.279 |
| 1049 | " " | 15.0496 | 49.375 |
| 1050 | Washington Bridge | 43.1697 | 141.633 |
| 1051 | Grand Boulevard and Concourse | 28.8907 | 94.786 |
| 1052 | " " " " | 30.5268 | 100.153 |
| 1053 | " " " " | 31.2854 | 102.642 |
| 1054 | " " " " | 31.6518 | 103.844 |
| 1055 | Tremont Avenue | 31.0607 | 101.905 |
| 1056 | " " | 28.3977 | 93.168 |
| 1057 | Anthony Avenue | 31.8634 | 104.539 |

| No. of Bench Mark. | Locality. | Elevation Above Mean Sea Level. | |
|---|---|---|---|
| | | Meters. | Feet. |
| 1058 | Grand Boulevard and Concourse | 29.9341 | 98.209 |
| 1059 | Burnside Avenue | 18.2110 | 59.747 |
| 1060 | " " | 30.6642 | 100.604 |
| 1061 | University Heights | 51.9132 | 170.319 |
| 1062 | " " | 53.4293 | 175.293 |
| 1063 | Jerome Avenue | 31.9910 | 104.957 |
| 1064 | " " | 36.1125 | 118.479 |
| 1065 | " " | 35.4443 | 116.287 |
| 1066 | University Bridge | 10.2613 | 33.666 |
| 1067 | Fordham Road | 31.9671 | 104.879 |
| 1068 | " " | 37.5233 | 123.108 |
| 1069 | " " | 37.8370 | 124.137 |
| 1070 | Fort Schuyler Road | 16.7128 | 54.832 |
| 1071 | " " " | 14.5278 | 47.663 |
| 1072 | St. Raymonds Cemetery | 16.8216 | 55.189 |
| 1073 | " " | 15.7009 | 51.512 |
| 1074 | " " | 15.8787 | 52.095 |
| 1075 | Eastern Boulevard | 10.9200 | 35.827 |
| 1076 | " " | 12.1007 | 39.700 |
| 1077 | " " | 11.1206 | 36.485 |
| 1078 | " " | 5.2270 | 17.149 |
| 1079 | St. James Park | 35.7460 | 117.277 |
| 1080 | Poe Park | 42.7407 | 140.225 |
| 1081 | Fordham Road | 38.0538 | 124.848 |
| 1082 | " " | 19.9534 | 65.464 |
| 1083 | " " | 18.1200 | 59.449 |
| 1084 | Bathgate Avenue | 25.2838 | 82.952 |
| 1085 | Crotona Avenue | 28.9705 | 95.047 |
| 1086 | " " | 29.5025 | 96,793 |
| 1087 | Bronx Park | 15.8550 | 52.018 |
| 1088 | Bronx and Pelham Parkway | 23.7998 | 78.083 |
| 1089 | " " " " | 30.3006 | 99.411 |
| 1090 | " " " " | 14.4589 | 47.437 |
| 1091 | " " " " | 13.5258 | 44.376 |
| 1092 | " " " " | 11.9917 | 39.343 |
| 1093 | " " " " | 5.1780 | 16.988 |
| 1094 | Grand Boulevard and Concourse | 43.7790 | 143.632 |
| 1095 | " " " " | 41.9474 | 137.622 |
| 1096 | " " " " | 45.3066 | 148.643 |
| 1097 | " " " " | 46.5113 | 152.596 |
| 1098 | Briggs Avenue | 41.7398 | 136.941 |
| 1099 | " " | 41.6761 | 136.732 |
| 1100 | Grand Boulevard and Concourse | 45.2162 | 148.347 |
| 1101 | " " " " | 43.7396 | 143.502 |
| 1102 | " " " " | 45.3044 | 148.636 |
| 1103 | Jerome Avenue | 43.5164 | 142.770 |
| 1104 | " " | 43.5268 | 142.804 |
| 1105 | " " | 43.5196 | 142.781 |
| 1106 | " " | 46.2914 | 151.874 |
| 1107 | Broadway | 5.9645 | 19.569 |
| 1108 | " | 5.3673 | 17.609 |
| 1109 | West Two Hundred and Thirty-eighth Street | 8.8776 | 29.126 |
| 1110 | " " " " " " " | 7.5986 | 24.930 |
| 1111 | Riverdale Avenue | 54.4051 | 178.494 |
| 1112 | " " | 53.9185 | 176.898 |
| 1113 | Van Cortlandt Park | 10.8730 | 35.672 |
| 1114 | Pelham Bridge | 9.2619 | 30.387 |
| 1115 | " " | 8.2914 | 27.203 |
| 1116 | " " | 8.0924 | 26.550 |

| No. of Bench Mark. | Locality. | Elevation Above Mean Sea Level. | |
|---|---|---|---|
| | | Meters. | Feet. |
| 1117 | Pelham Bridge Road | 6.6850 | 21.932 |
| 1118 | Prospect Avenue | 4.0712 | 13.357 |
| 1119 | " " | 4.2912 | 14.079 |
| 1120 | Broadway | 6.9984 | 22.961 |
| 1121 | " | 13.0137 | 42.696 |
| 1122 | " | 14.3260 | 47.001 |
| 1123 | Woodlawn Cemetery | 51.4794 | 168.895 |
| 1124 | Jerome Avenue | 52.4968 | 172.233 |
| 1125 | " " | 53.2970 | 174.859 |
| 1126 | Van Cortlandt Park | 58.9452 | 193.389 |
| 1127 | " " " | 32.4507 | 106.465 |
| 1128 | " " " | 16.4569 | 53.992 |
| 1129 | " " " | 14.3598 | 47.112 |
| 1130 | " " " | 12.1067 | 39.720 |
| 1131 | " " " | 16.4588 | 53.999 |
| 1132 | " " " | 12.9397 | 42.453 |
| 1133 | Broadway | 13.5950 | 44.603 |
| 1134 | Jerome Avenue | 64.2318 | 210.734 |
| 1135 | " " | 64.1235 | 210.379 |
| 1136 | Pelham Bay Park | 17.8224 | 58.472 |
| 1137 | Eastchester | 6.4855 | 21.278 |
| 1138 | " | 5.2216 | 17.131 |
| 1139 | " | 4.5931 | 15.069 |
| 1140 | Webster Avenue | 37.2201 | 122.113 |
| 1141 | " " | 37.4561 | 122.887 |
| 1142 | Carpenter Avenue | 42.5961 | 139.751 |
| 1143 | " " | 43.3751 | 142.307 |
| 1144 | " " | 36.8278 | 120.826 |
| 1145 | White Plains Road | 48.1820 | 158.077 |
| 1146 | " " " | 46.9260 | 153.956 |
| 1147 | East Two Hundred and Forty-first Street | 28.7208 | 94.228 |
| 1148 | Town of Pelham, Westchester County | 33.0118 | 108.306 |
| 1149 | " " " " " | 34.6893 | 113.810 |
| 1150 | " " " " " | 30.3056 | 99.428 |
| 1151 | City of Mount Vernon, Westchester County | 3.2112 | 10.535 |
| 1152 | " " " " " " | 28.2040 | 92.533 |
| 1153 | " " " " " " | 28.5503 | 93.669 |
| 1154 | " " " " " " | 30.0699 | 98.654 |
| 1155 | " " " " " " | 49.0596 | 160.956 |
| 1156 | " " " " " " | 46.1257 | 151.331 |
| 1157 | " " " " " " | 45.7953 | 150.247 |
| 1158 | City of Yonkers, Westchester County | 46.4727 | 152.469 |
| 1159 | " " " " " | 59.8914 | 196.494 |
| 1160 | " " " " " | 65.2826 | 214.181 |
| 1161 | " " " " " | 63.7602 | 209.187 |
| 1162 | " " " " " | 10.5054 | 34.467 |
| 1163 | " " " " " | 10.6417 | 34.914 |
| 1164 | " " " " " | 20.1310 | 66.047 |
| 1165 | " " " " " | 37.3632 | 122.582 |
| 1166 | " " " " " | 31.7295 | 104.099 |
| 1167 | " " " " " | 25.9939 | 85.282 |
| 1168 | " " " " " | 27.9407 | 91.669 |
| 1169 | Dunwoodie, Westchester County | 43.9739 | 144.271 |
| 1170 | Bryn Mawr Park, Westchester County | 70.4888 | 231.262 |
| 1171 | " " " " " | 75.0135 | 246.107 |
| 1172 | Nepperhan, Westchester County | 50.9508 | 167.161 |
| 1173 | " " " | 51.1760 | 167.900 |
| 1174 | " " " | 45.8123 | 150.302 |
| 1175 | " " " | 38.7200 | 127.034 |

| No. of Bench Mark. | LOCALITY. | Elevation Above Mean Sea Level. | |
|---|---|---|---|
| | | Meters. | Feet. |
| 1176 | Nepera Park, Westchester County.................. | 35.3563 | 115.998 |
| 1177 | " " " " .................. | 33.5991 | 110.233 |
| 1178 | Mount Hope, Westchester County.................. | 35.5837 | 116.744 |
| 1179 | " " " " .................. | 38.7153 | 127.018 |
| 1180 | Hastings-on Hudson, Westchester County........... | 59.5769 | 195.462 |
| 1181 | " " " " " ........... | 60.6359 | 198.936 |
| 1182 | " " " " " ........... | 59.2409 | 194.360 |
| 1183 | Dobbs Ferry, Westchester County.................. | 36.5223 | 119.824 |
| 1184 | " " " " .................. | 4.4799 | 14.698 |
| 1185 | " " " " .................. | 2.8781 | 9.443 |
| 1186 | " " " " .................. | 2.7598 | 9.054 |

## DESCRIPTIONS OF BENCH MARKS

IN THE

## BOROUGH OF THE BRONX

AND IN

## WESTCHESTER COUNTY, N. Y.

B. M. 953.—*Third Avenue Bridge.* A ⅞-inch copper bolt leaded vertically in a semicircular granite block on the northeasterly side of the northeasterly abutment of the Third Avenue Bridge over the Harlem River. The bolt is 3.3 feet westerly from the easterly side of the abutment, is on line with the easterly curb of Third Avenue, and about 130 feet south of the south line of East One Hundred and Thirty-third Street. An arrow cut in the abutment points down to the bench mark. (Note 5, p. 89.)

Elevation, 8.981 feet.

B. M. 954.—*Third Avenue Bridge.* A cross (+) cut in the dressed edge on the northeasterly side of the northeasterly abutment of the Third Avenue Bridge over the Harlem River. The cross is on the corner of the abutment, is on the westerly side of the easterly stairway and 3.6 feet above the sidewalk. The letter B is cut in the abutment below the bench mark. (Note 15, p. 91.)

Elevation, 12.278 feet.

B. M. 955.—*Third Avenue Bridge.* A square (⌂) cut in the top of the granite coping below the balustrade on the easterly side of the easterly stairway on The Bronx side of the Third Avenue Bridge over the Harlem River. The bench mark is about 135 feet south of the south line of East One Hundred and Thirty-third Street, is 3½ inches from the northeasterly end of the coping and 1.5 feet above the ground. An arrowhead on the easterly side of the coping points to the bench mark.

Elevation, 10.596 feet.

B. M. 956.—*Third Avenue Bridge.* A square (□) cut in the top of the northwesterly end of the granite base block at the northerly end of the arch on the westerly side of the Third Avenue Bridge over the Harlem River. The bench mark is about 50 feet north of the bulkhead on the north side of the Harlem River and the figure 11.437 is cut in the base block on the easterly side of the point.

Elevation, 8.594 feet.

B. M. 957.—*Third Avenue Bridge.* A square (⊓) lettered B. W. S., N. Y., cut in the top of the northeasterly corner of the granite base block of granite pier on the easterly side of the northerly approach to the Third Avenue Bridge over the Harlem River. The bench mark is about 6 feet north of the north side of East One Hundred and Thirty-third Street and 8 inches above the street.

Elevation, 9.461 feet.

B. M. 958.—*Third Avenue.* The top of a ¾-inch iron bolt leaded horizontally into the granite water-table on the easterly side of the five-story red brick building of the J. L. Mott Iron Works, on the westerly side of Third Avenue about 50 feet south of the prolongation of the center line of East One Hundred and Thirty-third Street. The bolt is 1.2 feet south of the north side of the building and 2.6 feet above the sidewalk. Established by the Department of Bridges in 1894.

Elevation, 11.786 feet.

B. M. 959.—*East One Hundred and Thirty-fifth Street.* A ½-inch copper bolt in the top of the marble city monument on the northeast corner of East One Hundred and Thirty-fifth Street and Exterior Street. The top of the monument is 2 inches above the ground.

Elevation, 11.055 feet.

B. M. 960.—*East One Hundred and Thirty-eighth Street.* A square (□) cut near the southeasterly corner of large platform stone at the east turnstile exit from the Mott Haven station of the New York and Harlem Railroad. The bench mark is at the foot of the southeasterly iron post supporting the shed, is on the northerly side of East One Hundred and Thirty-eighth Street and about 30 feet west of the west line of Park Avenue.

Elevation, 11.513 feet.

B. M. 961.—*Mott Avenue.* The top of a ¾-inch square iron bolt near the southwest corner of the large bluestone slab at the west entrance to the Mott Haven station (two-story red brick) of the New York and Harlem Railroad on Mott Avenue. The bolt is 4.7 feet west of the building, is 4 inches from the south edge of the slab and about 59 feet north of the extreme south end of the red brick arch facing East One Hundred and Thirty-eighth Street.

Elevation, 15.793 feet.

B. M. 962.—*Mott Avenue.* A horizontal line on the face of a brass tablet bolt 2 inches in diameter, set horizontally into the east face of the limestone pilaster on the south side of the main or east entrance to Public School No. 31, on the west side of Mott Avenue between East One Hundred and Forty-fourth Street and East One Hundred and Forty-sixth Street. (Note 1, p. 89.)

Elevation, 51.321 feet.

B. M. 963.—*East One Hundred and Forty-fourth Street.* A square (⊓) cut in the top of the second bluestone step on the east side of the stoop on the East One Hundred and Forty-fourth Street side of Public School No. 31, on the west side of Mott Avenue. The bench mark is on the east edge of the step, is 7 inches from the north end and about 54 feet west of the west line of Mott Avenue.

Elevation, 40.247 feet.

B. M. 964.—*East One Hundred and Forty-sixth Street.* A square (⊓) cut in the top of the lowest bluestone step on the east side of the stoop on the East One Hundred and Forty-sixth Street side of Public School No. 31, on the west side of Mott Avenue. The bench mark is on the east edge of the step, is 9 inches from the south end and about 64 feet west of the west line of Mott Avenue.

Elevation, 41.444 feet.

B. M. 965.—*Mott Avenue.* A square (⊔) cut in the top of the 21-inch square granite lamp base on the north side of the Mott Avenue entrance to the Interborough Rapid Transit subway at East One Hundred and Forty-ninth Street. The entrance to the subway is on the southwest corner and the bench mark is 3 feet above the sidewalk.

Elevation, 49.970 feet.

B. M. 966.—*Brown Place.* A ½-inch copper bolt in the top of the granite water-table at the southeast corner of Public School No. 43, on the northwest corner of Brown Place and East One Hundred and Thirty-fifth Street. The bolt is 3.33 feet above the area. (Note 8, p. 90.)

Elevation, 38.953 feet.

B. M. 967.—*Southern Boulevard.* A ¾-inch copper bolt in the top of the north-west corner of the bluestone door-sill at the corner or office entrance to Winter's piano factory (five-story red brick) on the northeast corner of Southern Boulevard and East One Hundred and Thirty-seventh Street. The bolt is 3¼ inches south of the base of the bluestone column on the north side of the entrance. (Note 5, p. 89.)

Elevation, 44.273 feet.

B. M. 968.—*Southern Boulevard.* A square (□) cut in the top of the bluestone coping on top of the brick balustrade on the north side of the corner or office entrance to Winter's piano factory, on the northeast corner of Southern Boulevard and East One Hundred and Thirty-seventh Street. The bench mark is 10½ inches from the east end of the coping and 2.3 feet above the ground.

Elevation, 44.185 feet.

B. M. 969.—*Southern Boulevard.* The top of a railroad spike driven in the root on the east side of an elm tree on the southeasterly side of Southern Boulevard, about 100 feet south of East One Hundred and Thirty-eighth Street.

Elevation, 41.312 feet.

B. M. 970.—*Southern Boulevard.* Public Service Commission B. M. 1,389. A square (ㅁ) cut in the sidewalk at the southerly corner of the base of southerly iron fence post on the westerly side of Southern Boulevard, about 35 feet north of East One Hundred and Forty-second Street. The bench mark is on the westerly side of the highway bridge over the Port Morris branch of the New York and Harlem Railroad.

Elevation, 28.480 feet.

B. M. 971.—*Whitlock Avenue.* A ¾-inch copper bolt set vertically in the top of the westerly end of the northerly abutment of the highway bridge on Whitlock Avenue, over the Port Morris branch of the New York and Harlem Railroad. The bolt is in the northwesterly corner of the concrete abutment, is about 110 feet east of Southern Boulevard and about 6 feet north of the south curb line of East One Hundred and Forty-second Street. It is 6 inches west of the westerly girder of the bridge. (Note 5, p. 89.)

Elevation, 28.416 feet.

B. M. 972.—*Union Avenue.* A square (ㅁ) cut in the top of the southeast corner of the lowest bluestone step at the main entrance to Public School No. 25, on the northeast corner of Union Avenue and East One Hundred and Forty-ninth Street. The bench mark is on the east side of the building and about 85 feet north of the corner.

Elevation, 35.223 feet.

B. M. 973.—*Southern Boulevard.* A square (ㅁ) cut in the top of the southeast corner of the brownstone coping under the iron railing, in front of the six-story yellow brick apartment house at No. 555 Southern Boulevard. The bench mark is on the northwesterly side of Southern Boulevard, is on the property line between No. 551 and No. 555 and about 125 feet northeasterly from East One Hundred and Forty-ninth Street.

Elevation, 31.111 feet.

B. M. 974.—*Morris Avenue.* A square (⊔) cut in the top of the northwest corner of the granite door-sill at the house entrance to the four-story yellow brick tenement at No. 560 Morris Avenue. The bench mark is about 130 feet north of East One Hundred and Forty-ninth Street.

Elevation, 21.372 feet.

B. M. 975.—*Courtlandt Avenue.* A square (⊓) cut in the top of the granite door-sill at the store entrance to the four-story red brick tenement at No. 538 Courtlandt Avenue. The bench mark is about 25 feet south of East One Hundred and Forty-ninth Street and 9 inches from the south end of the door-sill.

Elevation, 47.793 feet.

B. M. 976.—*Courtlandt Avenue.* The top of the southeast corner of the large polished granite block at the southeast corner of the Central Union Gas Company's building (five-story limestone), on the northwest corner of Courtlandt Avenue and East One Hundred and Forty-eighth Street. The bench mark is 5.87 feet above the sidewalk on the second course of polished granite, just below the curved limestone water-table. (Note 26, p. 91.)

Elevation, 46.425 feet.

B. M. 977.—*East One Hundred and Forty-ninth Street.* A ¾-inch copper bolt in the top of the northeast corner of wide granite door-sill at the store entrance to the three-story red-brick building at No. 362 East One Hundred and Forty-ninth Street. The bolt is about 145 feet east of Courtlandt Avenue, is 4.33 feet from the east side of the building and 4 inches from the east end of the door-sill. (Note 5, p. 89.)

Elevation, 44.387 feet.

B. M. 978.—*Brook Avenue.* A square (⊓) cut in the top of the west edge of the granite door-sill at the entrance to the north store at No. 562 Brook Avenue. The bench mark is about 90 feet north of East One Hundred and Fiftieth Street and near the northwest corner of the door-sill.

Elevation, 17.102 feet.

B. M. 979.—*Westchester Avenue.* A ¾-inch copper bolt in the top of the northwest corner of the granite base of pilaster on the east side of the wide entrance to the six-story brick warehouse of The Bronx Refrigerating Company, at Nos. 520-536 Westchester Avenue, between Brook Avenue and German Place. The bolt is about 220 feet east of Brook Avenue and 3 inches from the north edge of the base block. (Note 5, p. 89.)

Elevation, 23.553 feet.

B. M. 980.—*Westchester Avenue.* The bottom of a square cut in the top of the east end of the brownstone door-sill at the north entrance to store on the southwest corner of Westchester Avenue and Union Avenue. The bench mark is 2.5 feet west of the corner of the building (four-story yellow brick apartment) and six inches above the sidewalk. (Note 4, p. 89.)

Elevation, 57.470 feet.

B. M. 981.—*Westchester Avenue.* A ½-inch copper bolt in the marble city monument on the easterly corner of Westchester Avenue and Longwood Avenue. The top of the monument is flush with the sidewalk.

Elevation, 66.411 feet.

B. M. 982.—*Prospect Avenue.* A ¾-inch copper bolt in the top of the granite fender post on the south side of the wagon entrance to Engine House No. 73, on the northwest corner of Prospect Avenue and East One Hundred and Fifty-second Street.

The bolt is on the east side of the building, is 7.5 feet north of the corner and 1.2 feet above the sidewalk. (Note 5, p. 89.)

Elevation, 41.345 feet.

B. M. 983.—*Prospect Avenue.* A square (⊔) cut in the top of the northeast corner of the lowest brownstone step of stoop at the entrance to the two-story and basement red brick dwelling at No. 651 Prospect Avenue, about 29 feet south of East One Hundred and Fifty-second Street.

Elevation, 40.168 feet.

B. M. 984.—*Longwood Avenue.* A horizontal line on the face of a brass tablet bolt 2 inches in diameter, set horizontally into the center of the westerly face of limestone buttress on the front of Public School No. 39, on the northerly side of Longwood Avenue between Kelly Street and Beck Street. The bench mark is about 65 feet westerly from Beck Street and 2.5 feet above the ground. (Note 1, p. 89.)

Elevation, 46.979 feet.

B. M. 985.—*Longwood Avenue.* The center of the horizontal bar of a T cut in the bluestone base course of buttress on the front of Public School No. 39, on the northerly side of Longwood Avenue between Kelly Street and Beck Street. The bench mark is in the center of the westerly face of the buttress and 6 inches above the ground. It is about 65 feet westerly from Beck Street.

Elevation, 44.981 feet.

B. M. 986.—*Longwood Avenue.* A ¾-inch copper bolt in the top of the seventh bluestone step at the main entrance to Public School No. 39, on the north side of Longwood Avenue between Kelly Street and Beck Street. The step containing the bolt is the wide landing step. (Note 5, p. 89.)

Elevation, 49.276 feet.

B. M. 987.—*Southern Boulevard.* A ½-inch copper bolt in the top of the marble city monument on the northeasterly corner of Southern Boulevard and Tiffany Street. The top of the monument is ½ inch below the sidewalk.

Elevation, 29.281 feet.

B. M. 988.—*Southern Boulevard.* A square (⊔) cut in the top of the northeasterly corner of the bluestone coping at the corner of the iron picket fence, in front of the five-story yellow brick apartment house at No. 801 Southern Boulevard. The bench mark is about 100 feet north of Longwood Avenue.

Elevation, 24.118 feet.

B. M. 989.—*Lafayette Avenue.* A ½-inch copper bolt in the top of the marble city monument on the northeasterly corner of Lafayette Avenue and Tiffany Street. The top of the monument is flush with the sidewalk.

Elevation, 54.984 feet.

B. M. 990.—*Lafayette Avenue.* A ¾-inch copper bolt in the top of the westerly end of the bluestone door-sill at the entrance to the American Bank Note Company's building (five-story red brick), on the northeasterly corner of Lafayette Avenue and Tiffany Street. The bolt is on the Lafayette Avenue side and 5.5 feet from the Tiffany Street side. (Note 8, p. 90.)

Elevation, 57.497 feet.

B. M. 991.—*East One Hundred and Sixty-third Street.* A ½-inch copper bolt in the top of the marble city monument on the southwesterly corner of East One Hundred and Sixty-third Street (Dongan Street) and Southern Boulevard. The monument is on the south side of East One Hundred and Sixty-third Street, opposite the point of curve immediately west of Southern Boulevard.

Elevation, 48.889 feet.

B. M. 992.—*Westchester Avenue.* A square (ㅁ) cut in the top of the southeast corner of the iron bedplate under the iron column supporting the southeast corner of the three-story brick tenement on the northwest corner of Westchester Avenue and Intervale Avenue. The bench mark is at the corner entrance.

Elevation, 32.825 feet.

B. M. 993.—*Westchester Avenue.* A ¾-inch copper bolt in the top of the northeasterly corner of the large granite block on the north side of the house entrance to the Chester Hall Apartments (six-story stone and brick), on the southeast corner of Westchester Avenue and Southern Boulevard. The bolt is about 25 feet south of Westchester Avenue and 1.9 feet above the sidewalk. (Note 5, p. 89.)

Elevation, 61.787 feet.

B. M. 994.—*Westchester Avenue.* A ½-inch copper bolt in the top of the marble city monument on the southeast corner of Westchester Avenue and Southern Boulevard. The top of the monument is ½ inch below the sidewalk.

Elevation, 59.967 feet.

B. M. 995.—*Grand Boulevard and Concourse.* A ¾-inch copper bolt in the top of the northeasterly corner of the lowest granite step at the base of the bronze monument of Louis J. Heintz, on the center line of the Grand Boulevard and Concourse, about 50 feet northeasterly from the northerly line of East One Hundred and Sixty-second Street. (Note 5, p. 89.)

Elevation, 76.638 feet.

B. M. 996.—*Grand Boulevard and Concourse.* The top of a ¾-inch iron rod inside of a wrought-iron pipe casing set in the circular plot containing the white marble monument erected to the memory of Heinrich Heine, at the entrance to the Grand Boulevard and Concourse. The bench mark is about 80 feet northeasterly from the northerly line of East One Hundred and Sixty-first Street and 3.3 feet inside of the iron railing on the south side of the monument. The top of the casing is flush with the ground. (Note 21, p. 91.)

Elevation, 77.854 feet.

B. M. 997.—*Sheridan Avenue.* The center of the horizontal bar of a **T** cut in the south face of the limestone water-table of the four-story and basement brick apartment house on the northwest corner of Sheridan Avenue and East One Hundred and Sixty-third Street. The bench mark is 5.9 feet above the sidewalk and 3¼ inches from the southeast corner of the water-table.

Elevation, 67.602 feet.

B. M. 998.—*Grand Boulevard and Concourse.* The bottom of a 2-inch square cut in the top of the northeast corner of the 30" x 30½" granite copingstone on top of the east end of the south abutment of the highway bridge on the Grand Boulevard and Concourse at East One Hundred and Sixty-seventh Street. The stone containing the bench mark is also the base of the granite post at the southeast corner of the bridge. (Note 10, p. 90.)

Elevation, 112.641 feet.

B. M. 999.—*Throgs Neck.* A ⅞-inch copper bolt in the top of the southeast corner of a large copingstone on top of the sea wall on the southerly side of the military reservation at Fort Schuyler. The bolt is 3 feet south of the southeast corner of the two-story brick house occupied by Mr. Mayhew, post engineer, and about 50 feet west of the east dock. The bolt is 5 inches from the face of the sea wall and in the center of a 4-inch square cavity cut in the copingstone, which is also a part of the foundation of the post engineer's house. (Note 5, p. 89.)

Elevation, 8.262 feet.

B. M. 1000.—*Throgs Neck.* The bottom of a square cut in the projecting granite foundation stone in front of the easterly salient of the easterly bastion of old Fort Schuyler. The bench mark is 9 inches east of the east corner of the bastion and immediately west of the south end of "Battery Bell." The number of the bench mark, 1000, is cut in the top of the granite sill of the small window 2.8 feet above the point. (Note 4, p. 89.)

Elevation, 17.532 feet.

B. M. 1001.—*Throgs Neck.* A square (□) cut in the projecting foundation stone in front of the southeasterly salient of the most southeasterly bastion of old Fort Schuyler. The bench mark is near the most easterly part of Throgs Neck and about 275 feet northeast of the east dock near the post engineer's house. The number of the bench mark, 1001, is cut in the top of the granite sill of the small window 2.75 feet above the point, which is 8 inches from the corner of the bastion. This point is shown on the old plans of the fort with an elevation of 21.5 feet. Mr. Mayhew, post engineer, states that he does not know the origin of this figure, but that it is very old. The bench mark appears to have been established shortly after the fort was built, and referred to local mean low water.

Elevation, 17.439 feet.

B. M. 1002.—*Throgs Neck.* A cross (+) cut in the dressed edge of the masonry at the southeast corner of the one-story stone building below the Throgs Neck Light House, in the military reservation at Fort Schuyler. The bench mark is about 42 feet west of the west end of old Fort Schuyler, is about 130 feet north of the sea wall and 5 feet above the ground. An arrowhead cut in the stone points up to the bench mark, which is 1 inch from the corner.

Elevation, 20.782 feet.

B. M. 1003.—*Throgs Neck.* A cross (+) cut in the top of iron lug on the northeasterly side of the rim of Croton water manhole on the north side of Fort Schuyler Road, at the beginning of the causeway leading to the military reservation at Fort Schuyler. The bench mark is just south of the corner of a stone fence and 1.3 feet above the ground.

Elevation, 9.240 feet.

B. M. 1004.—*Throgs Neck.* A square (□) cut in the top of the northeasterly edge of the 10-inch square granite monument at the base of the stone fence on the southwesterly side of Fort Schuyler Road, alongside of the one-story brick dairy on the Havemeyer estate. The bench mark is about 1 mile west of Fort Schuyler, is about 45 feet south of a private road to the estate and nearly opposite a fire hydrant on the easterly side of Fort Schuyler Road. The top of the monument is 2 feet above the ground and an arrowhead on the easterly side of the monument points to the bench mark.

Elevation, 13.620 feet.

B. M. 1005.—*Fort Schuyler Road.* A square (⊓) cut in the top of the west edge of the 12-inch square granite monument in the corner of the stone fence on the southeast corner of Fort Schuyler Road and a private road leading southerly to the Havemeyer estate. The top of the monument is 1.2 feet above the ground and contains the letter X. It is about 1,000 feet west of Pennyfield Road.

Elevation, 38.477 feet.

B. M. 1006.—*Fort Schuyler Road.* The center of a square (□) cut in the top of the center of the large granite gate post on the east side of the entrance to the Havemeyer estate, on the south side of Fort Schuyler Road about 1.3 miles southeasterly from Eastern Boulevard and about 600 feet south of Ocean Avenue (Pennyfield Avenue). The bench mark is on the most easterly of four granite gate posts and 7 feet above the ground.

Elevation, 39.939 feet.

B. M. 1007.—*Fort Schuyler Road.* A square (⊓) cut in the top of the east edge of the large cubical granite block on the south side of the south gate to the Morris estate, on the west side of Fort Schuyler Road. The bench mark is about 9 feet north of the south line of Ocean Avenue (Pennyfield Avenue), is 2.2 feet above the ground and 4 inches from the north edge of the stone. An arrowhead cut in the east face of the stone points to the bench mark.

Elevation, 39.620 feet.

B. M. 1008.—*Fort Schuyler Road.* The bottom of a square cut in the top of the north end of the projecting stone at the bottom of the stone fence on the easterly side of Fort Schuyler Road, in front of the Jackson estate. The bench mark is about 800 feet north of Ocean Avenue (Pennyfield Avenue), is about 20 feet north of the lodge gate and 1 foot above the ground. The letter B is cut in the stone, at the south side of the point. (Note 20, p. 91.)

Elevation, 42.359 feet.

B. M. 1009.—*Fort Schuyler Road.* The bottom of a square cut in the top of the south corner of the bluestone door-sill at the entrance to the four-story red brick water tower on the Jackson estate. The bench mark is about 800 feet north of Ocean Avenue (Pennyfield Avenue) and about 270 feet east of Fort Schuyler Road. (Note 4, p. 89.)

Elevation, 47.760 feet.

B. M. 1010.—*Fort Schuyler Road.* A square (⊓) cut in the top of the northeast corner of the stone wall on the south side of the gate at the entrance to Colford's pasture for horses, on the west side of Fort Schuyler Road about 150 feet south of Burdett Avenue. The bench mark is 3.3 feet above the ground.

Elevation, 36.941 feet.

B. M. 1011.—*Fort Schuyler Road.* A square (⊓) cut in the top of the west edge of the bluestone sill of the south gate at the entrance to the Montgomery estate, on the east side of Fort Schuyler Road about 50 feet south of the south side of Roosevelt Avenue. The bench mark is 4 inches from the south end of the sill.

Elevation, 37.383 feet.

B. M. 1012.—*Fort Schuyler Road.* A ½-inch copper bolt in the top of the marble city monument marking the U. S. C. & G. S. triangulation station Crosby. This point marks the eastern end of the Unionport Base Line and is located in Hoffman's Picnic Grounds, formerly the Kelley estate. The monument is set in concrete 3′ x 3′ x 4′, is about 225 feet west of Fort Schuyler Road and about 830 feet south of Eastern Boulevard. The top is flush with the ground.

Elevation, 57.578 feet.

B. M. 1013.—*Westchester Avenue.* A 1¼-inch iron bolt set vertically in the center of the 2.5-foot square top of the south wing wall of the east abutment of the highway bridge on Westchester Avenue, over the New York, New Haven and Hartford Railroad. The bolt is 4.7 feet south of the south side of the bridge and level with the sidewalk.

Elevation, 35.677 feet.

B. M. 1014.—*Westchester Avenue.* The center of the horizontal bar of a **T** cut in the brick on the northeasterly side of the three-story terra cotta brick hotel on the northwesterly corner of Westchester Avenue and Freeman Street. The bench mark is on the Freeman Street side of the hotel, is on the nineteenth course of brick above the sidewalk and 14 feet from the corner. The brick containing the bench mark is in the first course above the second course of limestone, and 5.5 feet above the sidewalk.

Elevation, 42.684 feet.

B. M. 1015.—*Westchester Avenue.* A square (□) cut in the top of the extreme northwest corner of the white copingstone under the iron railing on the south side of Westchester Avenue, about 33 feet west of the southwest corner of the highway bridge over the Bronx River. Established by the Public Service Commission.

Elevation, 29.248 feet.

B. M. 1016.—*Westchester Avenue.* A ¾-inch copper bolt in the top of the white stone coping on the top of the concrete abutment at the east end of the highway bridge on Westchester Avenue, over the Bronx River. The bolt is 5 inches south of the south side of the bridge and 5 inches east of the west side of the copingstone. (Note 5, p. 89.)

Elevation, 26.833 feet.

B. M. 1017.—*Westchester Avenue.* A square (□) lettered U. S., cut in the top of the southwest corner of the white stone coping of the sea wall on the east side of the Bronx River at Westchester Avenue. The bench mark is 6 inches northeast of the curved corner of the copingstone.

Elevation, 8.675 feet.

B. M. 1018.—*Westchester Avenue.* A ¾-inch copper bolt in the top of the large copingstone on the southwest corner of the sea wall on the east side of the Bronx River at Westchester Avenue. The bolt is 1.33 feet from the curved corner of the copingstone. (Note 5, p. 98.)

Elevation, 8.680 feet.

B. M. 1019.—*Westchester Avenue.* A 1-inch square (□) cut in the top of the white stone coping on the top of the concrete abutment at the east end of the highway bridge on Westchester Avenue, over the Bronx River. The bench mark is 10 inches south of the south side of the bridge and 2 inches from the west edge of the copingstone.

Elevation, 26.839 feet.

B. M. 1020.—*Westchester Avenue.* A square (□) cut in the top of the east edge of the white stone coping on the top of the concrete abutment at the east end of the highway bridge on Westchester Avenue over the Bronx River. The bench mark is 7½ inches north of the south side of the bridge.

Elevation, 26.847 feet.

B. M. 1021.—*Westchester Avenue.* The surface between two parallel lines (| |) cut in the top of the white stone coping on the top of the concrete abutment at the east end of the highway bridge on Westchester Avenue over the Bronx River. The bench mark is 9 feet south of the south side of the bridge and 5 inches from the east edge of the copingstone.

Elevation, 26.853 feet.

B. M. 1022.—*Westchester Avenue.* The top of a railroad spike driven horizontally into the east side of a large sycamore tree about 75 feet south of Westchester Avenue and about 50 feet west of Old Clason Point Road. The spike is 8 inches above the ground.

Elevation, 14.560 feet.

B. M. 1023.—*Westchester Avenue.* The bottom of a square cut in the top of the rough stone water-table at the southeast corner of Public School No. 15, on the north side of Westchester Avenue at St. Lawrence Avenue (East One Hundred and Seventy-second Street). The bench mark is 4.7 feet above the area. (Note 10, p. 90.)

Elevation, 32.202 feet.

B. M. 1024.—*Westchester Avenue.* A cross (+) cut in the top of the west end of the bluestone sill of east basement window on the south side of Public School No. 15, on the north side of Westchester Avenue at St. Lawrence Avenue (East One Hundred and Seventy-second Street). The cross is on the south wing of the school, at the small window with iron bars.

Elevation, 27.741 feet.

B. M. 1025.—*Beach Avenue.* A square (□) cut in the top of the northwest corner of the door-sill at the west store entrance to the three-story brick apartment house on the southeast corner of Beach Avenue and Westchester Avenue. The bench mark is on the west side of the building, about 25 feet south of Westchester Avenue.

Elevation, 31.952 feet.

B. M. 1026.—*Beach Avenue.* A ½-inch copper bolt in the top of the marble city monument on the southeast corner of Beach Avenue and Westchester Avenue. The top of the monument is flush with the sidewalk.

Elevation, 30.616 feet.

B. M. 1027.—*Westchester Avenue.* The top of a railroad spike driven horizontally into the northerly side of an elm tree on the southerly side of Westchester Avenue about 75 feet westerly from the center line of White Plains Road. The spike is 6 inches above the ground.

Elevation, 26.396 feet.

B. M. 1028.—*Hugh J. Grant Circle.* A square (□) cut in the top of the southeast corner of the granite basin head on the east side of Hugh J. Grant Circle and on the north side of Tremont Avenue.

Elevation, 21.153 feet.

B. M. 1029.—*Hugh J. Grant Circle.* A square (□) cut in the top of the west end of the straight curb adjoining the basin head on the east side of Hugh J. Grant Circle and on the south side of Westchester Avenue.

Elevation, 21.198 feet.

B. M. 1030.—*Hugh J. Grant Circle.* The surface between two parallel lines (| |) cut in the top of the granite basin head on the east side of Hugh J. Grant Circle and on the south side of Westchester Avenue. The bench mark is near the east end of the basin head.

Elevation, 21.230 feet.

B. M. 1031.—*New York Catholic Protectory.* A square (⊓) cut in the top of the center of the west edge of the bluestone sill of blind window on the west side of the one-story red brick boiler house in the grounds of the New York Catholic Protectory. The bench mark is about 425 feet south of Unionport Road, is 4 feet south of the north side of the building and 2 feet above the ground.

Elevation, 54.257 feet.

B. M. 1032.—*New York Catholic Protectory.* A square (⊔) cut in the top of the corner of the limestone water-table at the northwest corner of the two-story and basement red brick cooking school and laundry building, in the grounds of the New York Catholic Protectory. The bench mark is about 350 feet south of Unionport Road, is at the top of the steps and 3.5 feet above the ground.

Elevation, 57.420 feet.

B. M. 1033.—*Westchester Avenue.* A square (⊔) cut in the top of the southeast corner of the marble city monument on the northeast corner of Westchester Avenue and Unionport Road. The top of the monument is flush with the ground.

Elevation, 37.753 feet.

B. M. 1034.—*Westchester Avenue.* A small cross (+) cut in the top of the southwest corner of the bluestone sill of cellar window on the west side of the two-story and attic frame house on the northeast corner of Westchester Avenue and Purdy Street. The cross is about 18 feet north of the north line of Westchester Avenue and 1 foot above the ground. This point is also a bench mark of the Public Service Commission. The Bronx Highway Bureau, and the Department of Docks and Ferries.

Elevation, 35.996 feet.

B. M. 1035.—*Unionport Road.* A ½-inch copper bolt in the top of the marble city monument on the southeast corner of Unionport Road and Starling Avenue. The top of the monument is ½ inch above the ground.

Elevation, 31.325 feet.

B. M. 1036.—*Castle Hill Avenue.* A ½-inch copper bolt in the top of the granite wagon guard on the south side of the wagon entrance to Engine House No. 64, on the east side of Castle Hill Avenue between Ellis Avenue and Gleason Avenue. The bolt is about 65 feet south of Ellis Avenue and 2.33 feet above the sidewalk. (Note 8, p. 90.)

Elevation, 41.809 feet.

B. M. 1037.—*Castle Hill Avenue.* A square (⊓) cut in the top of the north edge of the bluestone sill of basement window on the north side of Public School No. 36, on the east side of Castle Hill Avenue between Watson Avenue and Blackrock Avenue. The bench mark is near the center of the sill and 12.2 feet east of the west side of the building.

Elevation, 49.755 feet.

B. M. 1038.—*Castle Hill Avenue.* A horizontal line on the face of a brass tablet bolt 2 inches in diameter, set horizontally into the masonry on the west side of Public School No. 36, on the east side of Castle Hill Avenue between Watson Avenue and

Blackrock Avenue. The bolt is about 50 feet north of the south side of the building and 5 feet above the sidewalk. It is 1 foot north of the north side of the third basement window from Blackrock Avenue. (Note 1, p. 89.)

Elevation, 52.579 feet.

B. M. 1039.—*East One Hundred and Seventy-seventh Street.* A ¾-inch copper bolt in the top of the bluestone coping at the corner entrance to Kaiser's Hotel (two-story terra cotta brick), on the southeast corner of East One Hundred and Seventy-seventh Street and Zerega Avenue. The entrance to the building is at the southwest corner, there being a triangular space or vestibule before the door. The bolt is in the northwest corner of this space, 4.5 feet from the corner of the building. It is 10 inches above the sidewalk. (Note 5, p. 89.)

Elevation, 20.426 feet.

B. M. 1040.—*East One Hundred and Seventy-seventh Street.* A square (ㄣ) cut in the top of the northeast corner of the granite copingstone on top of the north end of the bridge seat of west abutment of the old highway bridge on East One Hundred and Seventy-seventh Street, over Westchester Creek. The bench mark is 3.7 feet below the roadway of the bridge. Established by the United States Engineers.

Elevation, 8.860 feet.

B. M. 1041.—*East One Hundred and Seventy-seventh Street.* A quadrant (ㄣ) cut in the top of the southeast corner of the granite copingstone on the top of the south end of the bridge seat of the west abutment of the old highway bridge on East One Hundred and Seventy-seventh Street, over Westchester Creek. The bench mark is 3.7 feet below the roadway of the bridge. Established by The Bronx Highway Bureau.

Elevation, 8.855 feet.

B. M. 1042.—*East One Hundred and Seventy-seventh Street.* A square (□) cut in the top of the bluestone sill of the west door on the north side of the two-story red brick pottery building on the south side of East One Hundred and Seventy-seventh Street, about 400 feet east of Westchester Creek. The bench mark is about 25 feet from the west end of the building and 6 inches from the east end of the door-sill.

Elevation, 14.661 feet.

B. M. 1043.—*East One Hundred and Seventy-seventh Street.* A square (□) cut near the west end of a large partly buried boulder just west of the east corner of East One Hundred and Seventy-seventh Street and Eastern Boulevard. The bench mark is west of the gutter and about 90 feet north of the center line of East One Hundred and Seventy-seventh Street. The letter B is cut in the top of the boulder just east of the bench mark.

Elevation, 18.082 feet.

B. M. 1044.—*Eastern Boulevard.* A square (ㄣ) cut in the top of the northeast corner of the marble city monument on the northwest corner of Eastern Boulevard and Swinton Street. The top of the monument is flush with the ground.

Elevation, 16.274 feet.

B. M. 1045.—*Eastern Boulevard.* A square (ㄣ) cut in the top of the southwest corner of the marble city monument on the northwest corner of Eastern Boulevard and Quincy Street. The top of the monument is 3 inches above the ground.

Elevation, 21.365 feet.

B. M. 1046.—*Eastern Boulevard.* The center of the horizontal bar of a **T** cut in the side of a large foundation stone on the west side of the two-story frame dwelling on the north side of Eastern Boulevard about 100 feet east of Swinton Street. The bench mark is 2.75 feet north of the front of the house and 3.6 feet above the ground.

Elevation, 25.204 feet.

B. M. 1047.—*Eastern Boulevard.* A square (⊔) cut in the top of the northwest corner of the stone wall on the easterly side of the gate at the west entrance to Hoffman's Picnic Grounds (formerly the Kelley estate). The bench mark is on the southeasterly side of Eastern Boulevard, is about 240 feet westerly from Fort Schuyler Road and 2.3 feet above the ground. Established by The Bronx Highway Bureau.

Elevation, 37.974 feet.

B. M. 1048.—*Eastern Boulevard.* A square (⊓) cut in the top of the stone wall on the southwesterly corner of Eastern Boulevard and Throgs Neck Road. The bench mark is on the Eastern Boulevard side of the corner and 3 feet above the ground. Established by The Bronx Highway Bureau.

Elevation, 51.279 feet.

B. M. 1049.—*Eastern Boulevard.* A ½-inch copper bolt in the top of the marble city monument on the southeasterly corner of Eastern Boulevard and Fort Schuyler Road. The top of the monument is flush with the ground.

Elevation, 49.375 feet.

B. M. 1050.—*Washington Bridge.* A square (⊔) cut in the top of the extreme northeast corner of the large granite copingstone at the southeast corner of the east approach to Washington Bridge. The copingstone containing the bench mark is also the base of the massive granite block 4 feet high at the east end of the south railing, and is on the west side of University Avenue (Aqueduct Avenue).

Elevation, 141.633 feet.

B. M. 1051.—*Grand Boulevard and Concourse.* A ½-inch copper bolt in the top of the marble city monument on the northeast corner of the Grand Boulevard and Concourse and East One Hundred and Seventy-third Street. The top of the monument is flush with the sidewalk.

Elevation, 94.786 feet.

B. M. 1052.—*Grand Boulevard and Concourse.* A ¾-inch copper bolt in the extreme northeasterly corner of the second course of granite in the granite post at the extreme southeasterly corner of the viaduct on the Grand Boulevard and Concourse over Morris Avenue. The bench mark is on the easterly side of the Concourse, is about 125 feet northeasterly from the northerly line of East One Hundred and Seventy-fourth Street and 2.3 feet above the ground. (Note 5, p. 89.)

Elevation, 100.153 feet.

B. M. 1053.—*Grand Boulevard and Concourse.* A square (⊔) cut in the top of the extreme northwesterly corner of the second course of granite in the large granite pier on the southeasterly side of the viaduct on the Grand Boulevard and Concourse over Morris Avenue. The bench mark is about on line with the west line of Morris Avenue and 2.2 feet above the sidewalk.

Elevation, 102.642 feet.

B. M. 1054.—*Grand Boulevard and Concourse.* A square (⊔) cut in the top of the extreme northwesterly corner of the second course of granite in the granite post at the extreme northwesterly corner of the viaduct on the Grand Boulevard and Concourse over Morris Avenue. The bench mark is on the westerly side of the Concourse at its intersection with the west line of Morris Avenue and just west of East One Hundred and Seventy-fifth Street.

Elevation, 103.844 feet.

B. M. 1055.—*Tremont Avenue.* The bottom of a 2-inch square cut in the top of the northeast corner of the 30-inch square granite copingstone on top of the east end of the south abutment of highway bridge on Tremont Avenue at Mount Hope Avenue. The stone containing the bench mark is also the base of the granite post at the southeast corner of the bridge. (Note 10, p. 90.)

Elevation, 101.905 feet.

B. M. 1056.—*Tremont Avenue.* A square (□) cut in the top of the north end of the lowest bluestone step above the bluestone platform on the east side of the stoop at the boys' entrance to Public School No. 28, on the southwest corner of Tremont Avenue and Anthony Avenue. The bench mark is on the north side of the building, is about 21 feet west of Anthony Avenue and 10 inches above the sidewalk.

Elevation, 93.168 feet.

B. M. 1057.—*Anthony Avenue.* A horizontal line on the face of a brass tablet bolt 2 inches in diameter set horizontally into the limestone water-table on the north side of the buttress near the southeast corner of Public School No. 28, on the west side of Anthony Avenue between Mount Hope Place and Tremont Avenue. The bolt is on the east side of the building and 1.5 feet from the east face of the buttress. It is about 35 feet north of Mount Hope Place and 3.5 feet above the sidewalk. (Note 1, p. 89.)

Elevation, 104.539 feet.

B. M. 1058.—*Grand Boulevard and Concourse.* The bottom of a 2-inch square cut in the top of the southwest corner of the 30-inch square granite copingstone on top of the west end of the north abutment of highway bridge on the Grand Boulevard and Concourse at Burnside Avenue. The stone containing the bench mark is also the base of the granite post at the northwest corner of the bridge. (Note 10, p. 90.)

Elevation, 98.209 feet.

B. M. 1059.—*Burnside Avenue.* A square (ㅂ) cut in the top of the northwest corner of the bluestone coping on the west side of the four-story brick apartment house on the northeast corner of Burnside Avenue and Jerome Avenue. The bench mark is on line with the north side of the building.

Elevation, 59.747 feet.

B. M. 1060.—*Burnside Avenue.* A small rounded summit within a semicircle cut in the top of the granite foundation stone under the east retaining wall of the Croton Aqueduct. The bench mark is level with the ground and 1.2 feet north of the north side of the east wing wall on the north side of Burnside Avenue. The letters B. M. are cut in the face of the retaining wall, above the point. Established by The Bronx Highway Bureau.

Elevation, 100.604 feet.

B. M. 1061.—*University Heights.* The center of the horizontal bar of a T cut in the granite foundation at the northwest corner of Gould Hall, in the grounds of the New York University, near the southwest corner of University Avenue and Hall of Fame Terrace. The bench mark is on the east side of the athletic field, is on the north side of the buttress at the corner of the building, is 8 inches from the extreme west face and 2.6 feet above the ground.

Elevation, 170.319 feet.

B. M. 1062.—*University Heights.* The center of the top of a brass tablet bolt 2 inches in diameter, set in the top of the northeasterly corner of the plinth of the most northerly limestone column on the easterly side of the "Library of New York University." This building is in the grounds of the New York University, on the westerly side of the campus. The bolt is 4.6 feet above the ground and 7 inches from the extreme northeasterly corner of the plinth. (Note 2, p. 89.)

Elevation, 175.293 feet.

B. M. 1063.—*Jerome Avenue.* A square (⊔) cut in the top of the granite wagon guard on the north side of the north wagon entrance to Engine House No. 75 (two-story brick), on the west side of Jerome Avenue about 90 feet north of West One Hundred and Eighty-third Street.

Elevation, 104.957 feet.

B. M. 1064.—*Jerome Avenue.* A square (⊓) cut in the top of the west edge of the seventh bluestone step at the playground entrance to Public School No. 33, on the east side of Jerome Avenue between East One Hundred and Eighty-fourth Street and Fordham Road. The bench mark is about 65 feet north of the south side of the building, is about 4 feet above the sidewalk and 3½ inches from the south end of the step.

Elevation, 118.479 feet.

B. M. 1065.—*Jerome Avenue.* A horizontal line in the end of a ¾-inch copper bolt set horizontally into the granite foundation at the northwest corner of Public School No. 33, on the east side of Jerome Avenue between East One Hundred and Eighty-fourth Street and Fordham Road. The bolt is on the Jerome Avenue side of the building, is about 275 feet south of Fordham Road and 6 inches above the sidewalk. It is 6 inches from the north side of the building. (Note 6, p. 90.)

Elevation, 116.287 feet.

B. M. 1066.—*University Bridge.* A square (□) cut in the top of the southeast corner of the granite coping under the east end of the iron railing on the north side of the north sidewalk of University Bridge, over the Harlem River at West One Hundred and Eighty-fourth Street. The bench mark is about 175 feet west of the west line of Harlem River Terrace and 5½ inches above the sidewalk.

Elevation, 33.666 feet.

B. M. 1067.—*Fordham Road.* A square (⊓) cut in the top of the southwest corner of the granite door-sill at the entrance to the one-story stone gate house on the north side of Fordham Road at Loring Place. Established by The Bronx Highway Bureau.

Elevation, 104.879 feet.

B. M. 1068.—*Fordham Road.* A square (⊓) cut in the top of the northeast corner of the brownstone door-sill at the house entrance to the four-story brick apartment house on the southeast corner of Fordham Road and Jerome Avenue.

Elevation, 123.108 feet.

B. M. 1069.—*Fordham Road.* A ½-inch copper bolt in the top of the southeast corner of the bluestone basin head on the north side of Fordham Road and on the west side of the center driveway of the Grand Boulevard and Concourse. The bolt is about 3 feet north of the north curb line of Fordham Road, is 5 inches from the east edge of the basin head and 4 inches from the south edge. (Note 8, p. 90.)

Elevation, 124.137 feet.

B. M. 1070.—*Fort Schuyler Road.* A square (☐) cut in the top of the southwest corner of the bluestone door-sill at the northwesterly entrance to the two-story red brick building on the northeasterly corner of Fort Schuyler Road and Meyers Street (Bohn Green). The bench mark is on the westerly side of the building and about 17 feet northwesterly from Meyers Street. It is 5 inches from the southerly end of the door-sill and 1.2 feet above the ground.

Elevation, 54.832 feet.

B. M. 1071.—*Fort Schuyler Road.* A square (⊔) cut in the top of the southeast corner of the granite base block of the large iron gate post on the east side of the coach entrance to St. Raymonds Cemetery, on the south side of Fort Schuyler Road opposite Mayflower Avenue. Established by The Bronx Highway Bureau.

Elevation, 47.663 feet.

B. M. 1072.—*St. Raymonds Cemetery.* The bottom of a square cut in the top of the northwest corner of the bluestone coping on the granite wall on the east side of the most northerly entrance to St. Raymonds Cemetery, on the southerly side of Fort Schuyler Road about 30 feet east of Whittemore Avenue. The bench mark is about 28 feet east of the east side of Pilgrim Avenue, is on the west end of the curved wall and 3.1 feet east of the large granite gate post on the east side of the coach entrance. It is 3.4 feet above the ground. Established by the Department of Docks and Ferries. (Note 10, p. 90.)

Elevation, 55.189 feet.

B. M. 1073.—*St. Raymonds Cemetery.* A ½-inch copper bolt in the top of the northeast corner of the granite door-sill at the entrance to the mausoleum of the Ismay family, in the most northerly part of St. Raymond's Cemetery opposite Pilgrim Avenue. The bolt is about 90 feet south of the intersection of the south side of Fort Schuyler Road and the southeasterly side of Whittemore Avenue, is about 60 feet southeasterly from Whittemore Avenue and about 45 feet west of the center line of the cemetery road running southerly from the entrance. It is 1.29 feet from the extreme east end of the door-sill. (Note 8, p. 90.)

Elevation, 51.512 feet.

B. M. 1074.—*St. Raymonds Cemetery.* This point is the top of the northeast corner of the 15½-inch square granite plinth of the most easterly of four granite columns in front of the mausoleum of James T. Barry, in the most northerly part of St. Raymonds Cemetery opposite Pilgrim Avenue. The bench mark is about 70 feet south of the intersection of the south side of Fort Schuyler Road and the southeasterly side of Whittemore Avenue, is about 35 feet southeasterly from Whittemore Avenue and about 60 feet west of the center line of the cemetery road running southerly from the entrance. The bench mark is 1 foot above the ground and is marked by a small hole.

Elevation, 52.095 feet.

B. M. 1075.—*Eastern Boulevard.* The bottom of a square cut in the top of the northeast corner of the water-table on the northeast corner of Public School No. 14 (one-story stone), on the west side of Eastern Boulevard about 350 feet north of the center line of Fairmount Avenue. The bench mark is 5 feet above the area and an arrowhead cut in the west side of the water-table points to the spot. (Note 10, p. 90.)

Elevation, 35.827 feet.

B. M. 1076.—*Eastern Boulevard.* A horizontal line in the end of a ¾-inch copper bolt set horizontally into a large foundation stone on the north side of the two-story and attic stone and frame house on the southwesterly corner of Eastern Boulevard and Morris Park Avenue (Willow Lane). The bolt is in the northeasterly corner of the foundation of the bay window, is 8 feet west of the front of the house and 1.2 feet above the ground. (Note 6, p. 90.)

Elevation, 39.700 feet.

B. M. 1077.—*Eastern Boulevard.* A ½-inch copper bolt in the top of a marble city monument on the southwesterly corner of Eastern Boulevard and Morris Park Avenue (Willow Lane). The monument is about 75 feet south of the south side of Arnow Avenue, and the top is flush with the ground.

Elevation, 36.485 feet.

B. M. 1078.—*Eastern Boulevard.* A 2½-inch cross (X) cut in the top of the northeasterly lug on the iron rim of water supply manhole, on the northwest corner of Eastern Boulevard and Fordham Road. The bench mark is in the center of the north sidewalk and about 80 feet west of the center line of Eastern Boulevard.

Elevation, 17.149 feet.

B. M. 1079.—*St. James Park.* The top of a ¾-inch iron rod inside of a wrought-iron pipe casing in the westerly side of St. James Park. The bench mark is 20 feet east of the iron fence on the east side of Jerome Avenue and about 55 feet south of the center line of West One Hundred and Ninety-second Street. The top of the casing is 3 inches above the ground. (Note 21, p. 91.)

Elevation, 117.277 feet.

B. M. 1080.—*Poe Park.* The center of the top of a granite monument 10″ x 10″ x 4′ set in concrete 3′ x 3′ x 4′ in the west side of Poe Park. The bench mark is about on line with the center line of East One Hundred and Ninety-third Street and 6.8 feet east of the iron fence on the east side of the Grand Boulevard and Concourse. The top of the monument is ½ inch above the ground.

Elevation, 140.225 feet.

B. M. 1081.—*Fordham Road.* A ½-inch copper bolt in the top of the marble city monument on the northeast corner of Fordham Road and the Grand Boulevard and Concourse. The top of the monument is flush with the sidewalk.*

Elevation, 124.848 feet.

B. M. 1082.—*Fordham Road.* A ½-inch copper bolt in the top of the marble city monument on the southeast corner of Fordham Road and Webster Avenue. The top of the monument is flush with the sidewalk.

Elevation, 65.464 feet.

B. M. 1083.—*Fordham Road.* A ¾-inch copper bolt in the top of the west end of the sill of the most southerly window on the south side of the one-story and attic stone house in the southwest corner of the grounds of Fordham University. The bolt is on the north side of Fordham Road, is about 21 feet east of the east line of Third Avenue and 6 inches from the west end of the window-sill. (Note 5, p. 89.)

Elevation, 59.449 feet.

B. M. 1084.—*Bathgate Avenue.* A square (⊓) cut in the top of the north end of the lowest brownstone step of stoop at the main entrance to the three-story and basement red brick tenement at No. 2530 Bathgate Avenue. The bench mark is on the east side of Bathgate Avenue and about 100 feet north of Fordham Road.

Elevation, 82.952 feet.

* This monument may have been disturbed during 1914 on account of the widening of Fordham Road.

B. M. 1085.—*Crotona Avenue.* A square (⊔) cut in the top of the extreme southwest corner of the bluestone coping about 14 feet south of the south side of the main entrance to Fordham Hospital, on the west side of Crotona Avenue, at its intersection with the westerly side of Southern Boulevard. The bench mark is about 450 feet north of Fordham Road and at the west end of the low iron fence where it abuts against the limestone front of the building.

Elevation, 95.047 feet.

B. M. 1086.—*Crotona Avenue.* A horizontal line on the face of a brass tablet bolt 2 inches in diameter set horizontally into the second course of limestone on the east side of Fordham Hospital, on the west side of Crotona Avenue at its intersection with the westerly side of Southern Boulevard. The bolt is 9½ inches south of the south side of the main entrance and 1.75 feet above the vault light platform. (Note 1, p. 89.)

Elevation, 96.793 feet.

B. M. 1087.—*Bronx Park.* A square (⊔) cut in the top of the northwest corner of the granite base block in the center of the driveway at the north entrance to the Zoological Gardens in Bronx Park. This entrance is on the south side of Fordham Road, is about 150 feet west of the west side of the bridge over the Bronx River and is known as the "Concourse Entrance." The bench mark is 3 feet north of the iron gate and 10 inches above the roadway.

Elevation, 52.018 feet.

B. M. 1088.—*Bronx and Pelham Parkway.* This is the center of the top of the capstone on the 24-inch square stone post on the north side of the north bicycle path on the Bronx and Pelham Parkway, about 185 feet west of the west side of White Plains Avenue. The post containing the bench mark is one of twelve stone posts crossing the parkway in one line, and is the fifth from the north end. The bench mark is about 19 feet north of the north side of the center driveway of the parkway and about 8 feet west of the west curb of Old Bear Swamp Road. A 3½-inch square (□) was cut in the top of the capstone surrounding the bench mark, which is 4.5 feet above the ground.

Elevation, 78.083 feet.

B. M. 1089.—*Bronx and Pelham Parkway.* A ¾-inch copper bolt in the center of the top of the east end of the bluestone basin head on the north side of the Bronx and Pelham Parkway, about 435 feet west of the center line of Williamsbridge Road. The bolt is 4 inches from the east end of the basin head. (Note 5, p. 89.)

Elevation, 99.411 feet.

B. M. 1090.—*Bronx and Pelham Parkway.* A square (⊔) cut in the top of the northwest corner of the granite coping on the north parapet wall of culvert crossing the Bronx and Pelham Parkway, about 1,000 feet west of the center line of Eastchester Road. The bench mark is 2.5 feet above the sidewalk. Established by The Bronx Highway Bureau.

Elevation, 47.437 feet.

B. M. 1091.—*Bronx and Pelham Parkway.* The bottom of a square cut in the top of the projecting granite base course on the south side of the north parapet wall of culvert crossing the Bronx and Pelham Parkway, about 1,000 feet west of the center line of Eastchester Road. The bench mark is 4.8 feet from the east end of the wall and 5½ inches from the south edge of the base block. An arrow points to the spot. (Note 4, p. 89.)

Elevation, 44.376 feet.

B. M. 1092.—*Bronx and Pelham Parkway.* This is the head of a rivet in the top of the northwest corner of the plate girder on the south side of the highway bridge on the Bronx and Pelham Parkway over the New York, New Haven and Hartford Railroad. The bench mark is on the south side of the parkway, is 2.3 feet above the sidewalk and 4 inches from the extreme northwest corner of the girder. The head of the rivet is filed flat.

Elevation, 39.343 feet.

B. M. 1093.—*Bronx and Pelham Parkway.* The top of a 1-inch copper bolt set in a granite monument marking the U. S. C. & G. S. triangulation station Baychester. The bolt marks the south end of the old Baychester Base Line, is about 30 feet south of the south side of the Bronx and Pelham Parkway and about 350 feet east of the southeast corner of the highway bridge over the New York, New Haven and Hartford Railroad. The top of the monument is exposed and 3 feet below the surrounding ground surface.

Elevation, 16.988 feet.

B. M. 1094.—*Grand Boulevard and Concourse.* The bottom of a 2-inch square cut in the top of the southwest corner of the 30-inch square granite copingstone on top of the west end of the north abutment of highway bridge on the Grand Boulevard and Concourse at Kingsbridge Road. The stone containing the bench mark is also the base of the granite post at the northwest corner of the bridge. (Note 10, p. 90.)

Elevation, 143.632 feet.

B. M. 1095.—*Grand Boulevard and Concourse.* A ½-inch copper bolt in the top of the marble city monument on the northwest corner of the Grand Boulevard and Concourse and East One Hundred and Ninety-eighth Street. The top of the monument is flush with the sidewalk.

Elevation, 137.622 feet.

B. M. 1096.—*Grand Boulevard and Concourse.* A ½-inch copper bolt in the top of the marble city monument on the northwest corner of the Grand Boulevard and Concourse and East One Hundred and Ninety-ninth Street. The top of the monument is flush with the sidewalk.

Elevation, 148.643 feet.

B. M. 1097.—*Grand Boulevard and Concourse.* A ½-inch copper bolt in the top of the marble city monument on the northwest corner of the Grand Boulevard and Concourse and Bedford Park Boulevard. The top of the monument is flush with the sidewalk.

Elevation, 152.596 feet.

B. M. 1098.—*Briggs Avenue.* A ¾-inch copper bolt in the top of the granite wagon guard on the north side of the north wagon entrance to Engine House No. 79 (two-story yellow brick), on the east side of Briggs Avenue about 125 feet south of Bedford Park Boulevard. (Note 5, p. 89.)

Elevation, 136.941 feet.

B. M. 1099.—*Briggs Avenue.* A cross (+) cut in the granite base course on the west side of Engine House No. 79 (two-story yellow brick), on the east side of Briggs Avenue about 125 feet south of Bedford Park Boulevard. The cross is 5 inches from the north side of the building and 1.75 feet above the sidewalk. (Note 15, p. 91.)

Elevation, 136.732 feet.

B. M. 1100.—*Grand Boulevard and Concourse.* A square (□) cut in the top of the northeast corner of a bluestone copingstone on the north side of the stone basement steps at the southeast corner of the stone church on the west side of the Grand Boulevard and Concourse opposite East Two Hundred and Second Street.

Elevation, 148.347 feet.

B. M. 1101.—*Grand Boulevard and Concourse.* The bottom of a 2-inch square cut in the top of the northeast corner of the 30-inch square granite copingstone on top of the east end of the south abutment of the highway bridge on the Grand Boulevard and Concourse at East Two Hundred and Fourth Street. The stone containing the bench mark is also the base of the granite post at the southeast corner of the bridge. (Note 10, p. 90.)

Elevation, 143.502 feet.

B. M. 1102.—*Grand Boulevard and Concourse.* A ½-inch copper bolt in the top of the marble city monument on the northwest corner of the Grand Boulevard and Concourse and East Two Hundred and Fifth Street. The top of the monument is flush with the sidewalk.

Elevation, 148.636 feet.

B. M. 1103.—*Jerome Avenue.* A ⅞-inch copper bolt in the top of the granite water-table at the extreme southeast corner of the Jerome Avenue pumping station of the Department of Water Supply, Gas & Electricity. The building is a long one-story red brick structure, on the west side of Jerome Avenue between Van Cortlandt Avenue and Mosholu Parkway. The bolt is in a hollow square cut in the southwest corner of the water-table, is 4 inches from the corner of the building and 2.6 feet above the area. (Note 5, p. 89.)

Elevation, 142.770 feet.

B. M. 1104.—*Jerome Avenue.* A square (ㅂ) cut in the top of the northeast corner of a granite block forming part of the coping of the low wall in front of the Jerome Avenue pumping station on the west side of Jerome Avenue between Van Cortlandt Avenue and Mosholu Parkway. The bench mark is 33 feet south of the south side of the building, is 7.1 feet north of the 26-inch square granite post at the south end of the wall and 2.9 feet above the ground. Established by The Bronx Highway Bureau.

Elevation, 142.804 feet.

B. M. 1105.—*Jerome Avenue.* A square (ㅂ) lettered B. W. S., N. Y., cut in the top of the extreme northeast corner of the granite water-table on the south side of the entrance to the boiler room of the Jerome Avenue pumping station, on the west side of Jerome Avenue between Van Cortlandt Avenue and Mosholu Parkway. The bench mark is about 10 feet north of the north side of the water tower and 3.5 feet above the red brick pavement.

Elevation, 142.781 feet.

B. M. 1106.—*Jerome Avenue.* A ¾-inch copper bolt in the top of the concrete coping on the east side of the reinforced concrete bridge on Jerome Avenue, over Mosholu Parkway. The bolt is 10 inches from the east face of the coping, is level with the east sidewalk and 3.6 feet from the extreme south end of the bridge. (Note 5, p. 89.)

Elevation, 151.874 feet.

B. M. 1107.—*Broadway.* A square (⊡) cut in the top of the northwest corner of the granite copingstone on top of the west end of the north abutment of the highway bridge on Broadway, over Spuyten Duyvil Creek. The bench mark is about 30 feet south of West Two Hundred and Thirtieth Street. Established by The Bronx Highway Bureau.

Elevation, 19.569 feet.

B. M. 1108.—*Broadway.* A ½-inch copper bolt in the top of the marble city monument on the southwest corner of Broadway and Kingsbridge Avenue. The top of the monument is ½ inch above the sidewalk.

Elevation, 17.609 feet.

B. M. 1109.—*West Two Hundred and Thirty-eighth Street.* The center of a circle cut in the center of the top of the semicircular granite wagon guard on the top of the west abutment of highway bridge over the New York and Putnam Railway, and on the center line of West Two Hundred and Thirty-eighth Street. The bench mark is at the west end of the center girder of the bridge. Established by The Bronx Highway Bureau.

Elevation, 29.126 feet.

B. M. 1110.—*West Two Hundred and Thirty-eighth Street.* A ⅞-inch copper bolt in the top of the south end of the bridge seat of the west abutment of highway bridge over the New York and Putnam Railway at West Two Hundred and Thirty-eighth Street. The bolt is 1 foot south of the south girder, is 1 foot west of the east face of the abutment and 4 feet below the roadway. (Note 5, p. 89.)

Elevation, 24.930 feet.

B. M. 1111.—*Riverdale Avenue.* A square (⊡) cut in the top of the north end of the granite base of watering trough on the northeast corner of Riverdale Avenue and Spuyten Duyvil Parkway. The bench mark is at the east end of the reentrant curve and 1.2 feet above the ground.

Elevation, 178.494 feet.

B. M. 1112.—*Riverdale Avenue.* A square (⊡) cut in the top of the southwest corner of the granite door-sill at the wagon entrance to Engine House No. 52 (three-story and attic frame), on the east side of Riverdale Avenue about 500 feet north of Spuyten Duyvil Parkway.

Elevation, 176.898 feet.

B. M. 1113.—*Van Cortlandt Park.* A ¾-inch copper bolt in the top of the northwest corner of the lowest granite step at the base of the bronze monument of Major General Porter, in Van Cortlandt Park. The monument is about opposite West Two Hundred and Forty-sixth Street and about 675 feet east of Broadway. The bolt is about 16 feet north of the one-story and attic stone annex of the Cortlandt House, is 1 foot from the extreme northwest corner of the step and 6 inches above the ground. (Note 5, p. 89.)

Elevation, 35.672 feet.

B. M. 1114.—*Pelham Bridge.* The bottom of a square cut in the top of the southwest corner of the square granite capstone at the extreme southwest corner of Pelham Bridge, on Pelham Bridge Road, over Eastchester Bay. The capstone containing the bench mark supports the most southerly lamp-post on the west side of the bridge and is 3.5 feet above the sidewalk. (Note 10, p. 90.)

Elevation, 30.387 feet.

B. M. 1115.—*Pelham Bridge.* A cross (+) cut in the west face of the large dressed granite block at the extreme northwest corner of Pelham Bridge, on Pelham Bridge Road, over Eastchester Bay. The block containing the bench mark supports the most northerly lamp-post on the west side of the bridge. The bench mark is about 4 inches above the water-table. (Note 15, p. 91.)

Elevation, 27.203 feet.

B. M. 1116.—*Pelham Bridge.* The top of a galvanized iron bolt in the west side of the granite water-table on the west side of Pelham Bridge, on Pelham Bridge Road, over Eastchester Bay. The bolt is at the northwest corner of the north abutment, is 6 inches from the north end of the water-table and about level with the sidewalk. (Note 13, p. 90.)

Elevation, 26.550 feet.

B. M. 1117.—*Pelham Bridge Road.* The bottom of a square cut in the top of the northeast corner of the bluestone platform step at the east entrance to the one-story and attic red brick dwelling on the west side of Pelham Bridge Road, about 450 feet south of City Island Road. The building has a two-story brick tower over the entrance, was erected in 1858 and was formerly occupied by the Park Department Engineers. The bench mark is at the base of the northeast column supporting the tower. (Note 10, p. 90.)

Elevation, 21.932 feet.

B. M. 1118.—*Prospect Avenue.* The bottom of a square cut in the top of the northwest corner of the lowest stepped stone at the west end of the west wing wall of the south abutment of the New York, New Haven and Hartford Railroad bridge over Prospect Avenue (Eastchester Road). The bench mark is 4.5 feet above the ground. (Note 10, p. 90.)

Elevation, 13.357 feet.

B. M. 1119.—*Prospect Avenue.* A cross (+) cut in the granite masonry at the west end of the north abutment of the New York, New Haven and Hartford Railroad bridge over Prospect Avenue (Eastchester Road). The cross is 4 inches from the south face of the abutment and 5 feet above the ground. (Note 15, p. 91.)

Elevation, 14.079 feet.

B. M. 1120.—*Broadway.* A square (⊡) cut in the top of the southeast corner of the iron base of the most southerly of two small elevated railroad columns supporting the station platform, on the west side of Broadway at Spuyten Duyvil Parkway. The bench mark is on line with the center line of the sidewalk on the west side of Broadway, is about 32 feet north of the south curb of Spuyten Duyvil Parkway and 7 inches above the roadway.

Elevation, 22.961 feet.

B. M. 1121.—*Broadway.* A ½-inch copper bolt in the top of the marble city monument on the northwest corner of Broadway and West Two Hundred and Fifty-third Street. The top of the monument is flush with the ground.

Elevation, 42.696 feet.

B. M. 1122.—*Broadway.* A square (□) cut in the top of the west end of the third cement step on the north side of the stoop leading to the Mosholu Inn (three-story frame), on the west side of Broadway opposite Mosholu Avenue.

Elevation, 47.001 feet.

B. M. 1123.—*Woodlawn Cemetery.* A ¾-inch copper bolt in the top of the granite coping of low wall on the north side of the southwest entrance to Woodlawn Cemetery, on the easterly side of Jerome Avenue just north of Bainbridge Avenue (Woodlawn Road). The bolt is in that portion of the wall running east and west, is 8½ inches west of the point of curve at the east end of the wall and 8½ inches from the south side. It is 3.4 feet west of the extreme east end of the iron fence and 2 feet above the ground. (Note 5, p. 89.)

Elevation, 168.895 feet.

B. M. 1124.—*Jerome Avenue.* The top of a ¾-inch iron rod inside of a wrought-iron pipe casing in the top of the rock ledge on the west side of Jerome Avenue, about 125 feet north of the north side of the southwest entrance to Woodlawn Cemetery. The bench mark is opposite New York Telephone Company pole No. 304 and 3.5 feet above the sidewalk. Established by the Board of Water Supply.

Elevation, 172.233 feet.

B. M. 1125.—*Jerome Avenue.* The top of a railroad spike driven horizontally into the west side of a chestnut tree on the west side of Jerome Avenue, about 500 feet north of the southwest entrance to Woodlawn Cemetery. Established by The Bronx Highway Bureau.

Elevation, 174.859 feet.

B. M. 1126.—*Van Cortlandt Park.* A cross (+) cut in the top of the fixed rim of the fire hydrant on the south side of Mosholu Avenue about 35 feet west of the west curb of Jerome Avenue. The cross is on the east side of the hinge and 2.6 feet above the ground.

Elevation, 193.389 feet.

B. M. 1127.—*Van Cortlandt Park.* A square (□) cut in the top of a large boulder on the easterly side of Mosholu Avenue, about 3,000 feet northwesterly from Jerome Avenue, measured along Mosholu Avenue. The boulder containing the bench mark is about 8 feet in diameter and the top is 1.5 feet above the ground. The bench mark is about 10 feet north of lamp-post No. 27 and about 300 feet south of a culvert crossing Mosholu Avenue. The letter B is cut in the west side of the boulder near the point.

Elevation, 106.465 feet.

B. M. 1128.—*Van Cortlandt Park.* A square (□) cut in the top of a partly buried boulder 1.5 feet in diameter, on the southwesterly side of Mosholu Avenue about 500 feet east of the Putnam Division of the New York Central and Hudson River Railroad and about 30 feet southerly from lamp-post No. 40.

Elevation, 53.992 feet.

B. M. 1129.—*Van Cortlandt Park.* The bottom of a square cut in the top of the highest point of the outcropping natural rock about 10 feet long and 3 feet high, in the golf field in Van Cortlandt Park. The bench mark is about 450 feet east of the Putnam Division of the New York Central and Hudson River Railroad and about 250 feet northeast of Mosholu Avenue. (Note 4, p. 89.)

Elevation, 47.112 feet.

B. M. 1130.—*Van Cortlandt Park.* The bottom of a square cut in the top of the west end of a partly buried boulder 1.5 feet in diameter, in the gutter on the westerly side of Mosholu Avenue about 1,000 feet west of the Putnam Division of the New York Central and Hudson River Railroad and about 60 feet south of lamp-post No. 53.

Elevation, 39.720 feet.

B. M. 1131.—*Van Cortlandt Park.* On the Yonkers Rapid Transit Railroad bridge over Mosholu Avenue, at the Mosholu station in Van Cortlandt Park. The bench mark is a 3-inch square (□) cut in the top of the easterly end of the bridge seat of the southerly abutment.

Elevation, 53.999 feet.

B. M. 1132.—*Van Cortlandt Park.* A square (ㄴ) cut in the top of the southwest corner of the third stepped stone of the westerly wing wall of the northerly abutment of the Yonkers Rapid Transit Railroad bridge over Mosholu Avenue. The bench mark is 3.3 feet above the ground. Established by The Bronx Highway Bureau.

Elevation, 42.453 feet.

B. M. 1133.—*Broadway.* A square (ㄴ) cut in the top of the southwest corner of the granite basin head on the northeast corner of Broadway and Mosholu Avenue.

Elevation, 44.603 feet.

B. M. 1134.—*Jerome Avenue.* The bottom of a square cut in the top of the natural rock on the west side of Jerome Avenue, about 52 feet south of East Two Hundred and Thirty-third Street. Established by the Board of Water Supply.

Elevation, 210.734 feet.

B. M. 1135.—*Jerome Avenue.* A ½-inch copper bolt in the top of the marble city monument on the northeast corner of Jerome Avenue and East Two Hundred and Thirty-third Street. The top of the monument is flush with the ground.

Elevation, 210.379 feet.

B. M. 1136.—*Pelham Bay Park.* The bottom of a 3½-inch square cut in the southeasterly corner of the north half of "Split Rock," near the west side of Prospect Avenue (Eastchester Road or Split Rock Road), in the northwest corner of Pelham Bay Park. The bench mark is about 700 feet south of the city line, is about 30 feet west of Prospect Avenue and 3 feet above the ground. An arrow cut in the east side of the rock points up to the bench mark. (Note 20, p. 91.)

Elevation, 58.472 feet.

B. M. 1137.—*Eastchester.* The center of a brass bolt marking the U. S. C. & G. S. triangulation station Maguire. The bench mark is on the land of Paul Kohn, about 500 feet southeasterly from Boston Road and about 500 feet westerly from Eastchester Creek. The bolt is about 150 feet south of the house and flush with the ground. (Note 25, p. 91.)

Elevation, 21.278 feet.

B. M. 1138.—*Eastchester.* The bottom of a square cut in the top of the easterly corner of the coping of the parapet wall at the south end of the west abutment of the old highway bridge on Boston Road, over Eastchester Creek (Hutchinson River). Established by The Bronx Highway Bureau. (Note 10, p. 90.)

Elevation, 17.131 feet.

B. M. 1139.—*Eastchester.* A square (□) cut in the top of the southwest corner of the bluestone door-sill of wide door at the southeast corner of the two-story and attic red brick gas purification building of the Westchester Lighting Company, about 1,000 feet northwesterly from Boston Road and about 1,000 feet east of Eastchester Creek (Hutchinson River). The bench mark is about 50 feet north of storage holder No. 1 and flush with the ground.

Elevation, 15.069 feet.

B. M. 1140.—*Webster Avenue.* The bottom of a square cut in the projecting foundation stone at the northwest corner of R. H. Macy's two-story brick and frame stable on the east side of Webster Avenue about 125 feet south of McLean Avenue. The bench mark is on the west side of the building, is 7 inches from the north side of the foundation and 9 inches above the sidewalk. (Note 20, p. 91.)

Elevation, 122.113 feet.

B. M. 1141.—*Webster Avenue.* A cross (+) cut in the west side of the bluestone water-table on the west side of R. H. Macy's two-story brick and frame stable, on the east side of Webster Avenue about 125 feet south of McLean Avenue. The cross is 7 inches from the north end of the water-table and 1.5 feet above the sidewalk. (Note 15, p. 91.)

Elevation, 122.887 feet.

B. M. 1142.—*Carpenter Avenue.* The top of a galvanized iron bolt in the granite base course on the west side of Public School No. 16, on the east side of Carpenter Avenue between East Two Hundred and Thirty-ninth Street and East Two Hundred and Fortieth Street. The bolt is on the north side of the south buttress and 1 foot above the area. (Note 13, p. 90.)

Elevation, 139.751 feet.

B. M. 1143.—*Carpenter Avenue.* A horizontal line on the face of a brass tablet bolt 2 inches in diameter set horizontally into the limestone on the west side of Public School No. 16, on the east side of Carpenter Avenue between East Two Hundred and Thirty-ninth Street and East Two Hundred and Fortieth Street. The bench mark is on the north side of the south buttress, is about 32 feet north of the south end of the building, is 6 inches from the west side and 3.5 feet above the area. (Note 1, p. 89.)

Elevation, 142.307 feet.

B. M. 1144.—*Carpenter Avenue.* A ½-inch copper bolt in the top of the marble city monument on the southeast corner of Carpenter Avenue and Baychester Avenue. The monument is about 10 feet south of the south house line and the top is flush with the ground.

Elevation, 120.826 feet.

B. M. 1145.—*White Plains Road.* A square (□) cut in the top of the northwest corner of the bluestone door-sill at the house entrance to the three-story red brick tenement at No. 4640 White Plains Road, about 100 feet south of Baychester Avenue.

Elevation, 158.077 feet.

B. M. 1146.—*White Plains Road.* A ½-inch copper bolt in the top of the marble city monument on the southeast corner of White Plains Road and Baychester Avenue. The top of the monument is flush with the ground.

Elevation, 153.956 feet.

B. M. 1147.—*East Two Hundred and Forty-first Street.* The bottom of a square cut in the south edge of the granite bridge seat of the north abutment of the New York, New Haven and Hartford Railroad bridge over East Two Hundred and Forty-first Street (Baychester Avenue). The bench mark is about 200 feet west of the west side of Bronx Boulevard and an arrow cut in the course below the bridge seat points to the spot.* (Note 10, p. 90.)

Elevation, 94.228 feet.

*Reported as being partially covered by a new course of concrete.

B. M. 1148.—*Town of Pelham, Westchester County.* The center of the brass bolt marking the U. S. C. & G. S. triangulation station Prospect. The bench mark is in a vacant field, is about 37 feet east of the center line of Peace Street and about 205 feet south of the center line of Washington Avenue. The bolt is flush with the ground. (Note 25, p. 91.)

Elevation, 108.306 feet.

B. M. 1149.—*Town of Pelham, Westchester County.* A square (□) cut in the top of the southeast corner of the bluestone sill in front of the most southerly window on the north side of the east entrance to the wide two-story and attic stone and frame house on the westerly side of Peace Street (No. 957) between Jackson Avenue and Prospect Avenue. The bench mark is 2.25 feet above the floor of the porch.

Elevation, 113.810 feet.

B. M. 1150.—*Town of Pelham, Westchester County.* A square (ㅂ) cut in the top of the northeasterly corner of the 7-inch square granite monument on the northeasterly corner of Boston Road and Wolfs Lane. The top of the monument is 8 inches above the ground.

Elevation, 99.428 feet.

B. M. 1151.—*City of Mount Vernon, Westchester County.* The bottom of a square cut in the top of the southwest corner of the bluestone coping at the west end of the south abutment of the concrete arch highway bridge on East Sixth Street, over the Hutchinson River. The copingstone containing the bench mark supports the iron post at the west end of the iron railing. The bench mark is flush with the sidewalk. (Note 10, p. 90.)

Elevation, 10.535 feet.

B. M. 1152.—*City of Mount Vernon, Westchester County.* A square (ㅂ) cut in the top of the southeast corner of the concrete door-sill at the entrance to the store on the west side of the passenger entrance to the New York, Westchester and Boston Railway station (one-story concrete), on the north side of East Sixth Street between Franklin Avenue and South Fulton Avenue. The bench mark is about 27 feet east of Franklin Avenue, is 3.5 feet west of the west side of the passenger entrance and 2 inches above the sidewalk.

Elevation, 92.533 feet.

B. M. 1153.—*City of Mount Vernon, Westchester County.* A ¾-inch copper bolt in the top of the south end of the west abutment of the highway bridge on East Sixth Street over the New York, Westchester and Boston Railway. The bolt is 6 inches south of the south side of the bridge. (Note 5, p. 89.)

Elevation, 93.669 feet.

B. M. 1154.—*City of Mount Vernon, Westchester County.* A cross (+) cut in the top of the 4-inch iron ball on top of the iron post at the west end of the iron fence on the south side of the highway bridge on East Sixth Street, over the New York, Westchester and Boston Railway. The top of the ball is 4 feet above the sidewalk.

Elevation, 98.654 feet.

B. M. 1155.—*City of Mount Vernon, Westchester County.* A cross (+) cut in the dressed edge of the granite cornerstone inscribed 1897, at the southwest corner of the old Mount Vernon High School, on the east side of South Fourth Avenue, about 115 feet north of East Fourth Street. The cross is on the south side of the building, is 1¼ inches from the corner and 4 feet above the ground. (Note 15, p. 91.)

Elevation, 160.956 feet.

B. M. 1156.—*City of Mount Vernon, Westchester County.* A ¾-inch copper bolt set horizontally into the bluestone sill of the first basement window south of the east or main entrance to Public School No. 3, on the west side of South Tenth Avenue between West First Street and West Second Street. The bolt is about 16 feet from the south side of the building, is 4 inches from the south end of the sill and 9 inches above the ground. (Note 6, p. 90.)

Elevation, 151.331 feet.

B. M. 1157.—*City of Mount Vernon, Westchester County.* A square (□) cut in the top of the northeast corner of the large granite copingstone at the southwest corner of the highway bridge on South Tenth Avenue, over the New York, New Haven and Hartford Railroad. The bench mark is at the west end of the south abutment and level with the pavement of West First Street.

Elevation, 150.247 feet.

B. M. 1158.—*City of Yonkers, Westchester County.* A cross (+) cut in the bluestone water-table on the northeasterly corner of the three-story brick apartment house at No. 819 McLean Avenue, at the intersection of McLean Avenue and East Two Hundred and Forty-second Street. The cross is on the north side of the building, is 3 inches from the east side and 8 inches above the sidewalk. (Note 15, p. 91.)

Elevation, 152.469 feet.

B. M. 1159.—*City of Yonkers, Westchester County.* A square (□) cut in the top of a partly buried boulder 2.5 feet in diameter, on the southerly side of McLean Avenue about 1,700 feet east of Jerome Avenue. The bench mark is opposite the easterly end of McLean's Pond, is about 20 feet east of lamp-post No. 2801 and 7 feet south of the south rail of the trolley line.

Elevation, 196.494 feet.

B. M. 1160.—*City of Yonkers, Westchester County.* Aqueduct Commission B. M. 47. The top of a ¾-inch iron bolt in the top of the natural rock on the east side of Jerome Avenue, about 225 feet south of McLean Avenue. The bolt is about 10 feet east of the east rail of the trolley line and about 3.5 feet above the ground.

Elevation, 214.181 feet.

B. M. 1161.—*City of Yonkers, Westchester County.* A square (□) cut in the top of the southeast corner of the bluestone door-sill on the east side of the one-story stone head house of the Croton Aqueduct near the northwest corner of McLean Avenue and Jerome Avenue. The bench mark is about 60 feet north of McLean Avenue and about 115 feet west of Jerome Avenue.

Elevation, 209.187 feet.

B. M. 1162.—*City of Yonkers, Westchester County.* The center of the horizontal bar of a **T** cut in the north face of the east pier of highway bridge on McLean Avenue, over the Putnam Division of the New York Central and Hudson River Railroad. The bench mark is on the west side of the railroad, is 2 feet from the west side of the pier and 3 feet above the ground.

Elevation, 34.467 feet.

B. M. 1163.—*City of Yonkers, Westchester County.* The bottom of a square cut in the east face of the east pier of highway bridge on McLean Avenue, over the Putnam Division of the New York Central and Hudson River Railroad. The bench mark is on the west side of the railroad, is 7 feet north of the south end of the pier and 3 feet above the ground. An arrowhead cut in the third course of masonry points down to the bench mark. (Note 21, p. 91.)

Elevation, 34.914 feet.

B. M. 1164.—*City of Yonkers, Westchester County.* The bottom of a square cut in the top of the northwest corner of the projecting granite copingstone on top of the north side of the east pier of highway bridge on McLean Avenue, over the Putnam Division of the New York Central and Hudson River Railroad. The copingstone projects 11 inches north of the north side of the bridge and is flush with the roadway. It is about 350 feet west of Tibbets road. (Note 10, p. 90.)

Elevation, 66.047 feet.

B. M. 1165.—*City of Yonkers, Westchester County.* A square (ㅂ) cut in the top of the northwest corner of the capstone of the large stone post on the southeast corner of Caryl Avenue and Broadway. The bench mark is 4.3 feet above the sidewalk.

Elevation, 122.582 feet.

B. M. 1166.—*City of Yonkers, Westchester County.* A ¾-inch copper bolt in the top of the northeast corner of the large granite block on the north side of the house entrance to the five-story brick apartment house at No. 506 South Broadway, opposite Valentine Lane. The bolt is 3 inches from the north side of the block, is 3½ inches west of the base of the limestone pilaster and 1.8 feet above the sidewalk. (Note 5, p. 89.)

Elevation, 104.099 feet.

B. M. 1167.—*City of Yonkers, Westchester County.* The bottom of a square cut in the top of the southeast corner of the lowest step of the east wing wall of the north abutment of the Yonkers Rapid Transit Railroad at Lawrence Street (Lowerre station). The bench mark is 2 feet above the sidewalk and an arrowhead cut in the stone points to the spot.

Elevation, 85.282 feet.

B. M. 1168.—*City of Yonkers, Westchester County.* The center of the horizontal bar of a **T** cut in the stone foundation at the southwesterly corner of Public School No. 13 (four-story and attic stone and brick) on the easterly side of McLean Avenue opposite Carrol Street. The bench mark is on the McLean Avenue side of the building, is 10 inches from the southerly side and 5 feet above the area.

Elevation, 91.669 feet.

B. M. 1169.—*Dunwoodie, Westchester County.* The bottom of a square cut in the top of the northwest corner of the granite door-sill on the north side of the Dunwoodie blow-off chamber, on the line of the Croton Aqueduct. The bench mark is about 800 feet north of Yonkers Avenue and about 300 feet east of the Putnam Division of the New York Central and Hudson River Railroad. (Note 4, p. 89.)

Elevation, 144.271 feet.

B. M. 1170.—*Bryn Mawr Park, Westchester County.* The center of the horizontal bar of a **T** cut in the southeasterly side of the stone railroad station on the northwesterly corner of Palmer Avenue and the Putnam Division of the New York Central and Hudson River Railroad. The bench mark is in the vertical face of a large stone block in the southeast corner of the building, is 1.5 feet east of the south window and 3 feet above the cement platform.

Elevation, 231.262 feet.

B. M. 1171.—*Bryn Mawr Park, Westchester County.* The top of a 1-inch iron bolt in the top of the natural rock on the south side of Palmer Avenue, about 700 feet east of the Putnam Division of the New York Central and Hudson River Railroad.

The bolt is on line with the prolongation of the center line of Maple Avenue and 5 feet south of the stone fence on the south side of Palmer Avenue. Established by the Board of Water Supply.

Elevation, 246.107 feet.

B. M. 1172.—*Nepperhan, Westchester County.* The bottom of a square cut in the top of the southeast corner of the granite water-table of gate house on the line of the Croton Aqueduct. The bench mark is about 100 feet north of Tuckahoe Road and about 50 feet east of the Putnam Division of the New York Central and Hudson River Railroad.*

Elevation, 167.161 feet.

B. M. 1173.—*Nepperhan, Westchester County.* The center of the horizontal bar of a T cut in the granite water-table of gate house on the Croton Aqueduct line. The bench mark is about 100 feet north of Tuckahoe Road and about 50 feet east of the Putnam Division of the New York Central and Hudson River Railroad. It is on the north side of the building, is 6 inches from the northeast corner and 6 inches above the ground.

Elevation, 167.900 feet.

B. M. 1174.—*Nepperhan, Westchester County.* This point is on the gray marble cornerstone in the southwesterly corner of the small stone and frame church on the northwest corner of Tuckahoe Road and Buckingham Road, about 700 feet west of the Putnam Division of the New York Central and Hudson River Railroad. The bench mark is the center of the horizontal bar of the final T of the word Methodist, in the inscription "MEMORIAL METHODIST EPISCOPAL CHURCH," which appears on the cornerstone.

Elevation, 150.302 feet.

B. M. 1175.—*Nepperhan, Westchester County.* A square (□) cut in the top of the northwesterly abutment of the railroad bridge over Sawmill River Road, about 2,600 feet north of Tuckahoe Road. The bench mark is 4.3 feet west of the west rail of the Putnam Division of the New York Central and Hudson River Railroad and 4.3 feet from the southwesterly end of the abutment.

Elevation, 127.034 feet.

B. M. 1176.—*Nepera Park, Westchester County.* The bottom of a square cut in the top of the northeast corner of the granite copingstone on top of the north end of the west abutment of the highway bridge on Hearst Street, over the Sawmill River. The bench mark is about 300 feet west of the Putnam Division of the New York Central and Hudson River Railroad and about 500 feet south of the Nepera Park railroad station. It is 4 inches east of the iron post at the corner of the iron railing and 4 inches above the sidewalk. (Note 4, p. 89.)

Elevation, 115.998 feet.

B. M. 1177.—*Nepera Park, Westchester County.* The center of the horizontal bar of a T cut in the northeasterly face of the west abutment of the highway bridge on Hearst Street, over the Sawmill River. The bench mark is about 300 feet west of the Putnam Division of the New York Central and Hudson River Railroad and about 500 feet south of the Nepera Park railroad station. It is about 5.8 feet below the coping and 3 feet northwest of the northwest corner of the bridge.

Elevation, 110.233 feet.

*Reported as destroyed in 1914.

B. M. 1178.—*Mount Hope, Westchester County.* A square (□) cut in the top of the east end of the north abutment of railroad bridge on the Putnam Division of the New York Central and Hudson River Railroad over the Sawmill River. The bench mark is about 600 feet south of Farragut Road.*

Elevation, 116.744 feet.

B. M. 1179.—*Mount Hope, Westchester County.* The bottom of a square cut in the top of the outcropping natural rock in the lawn on the west side of Ravensdale Avenue, about 200 feet west of the west side of the Putnam Division of the New York Central and Hudson River Railroad. The bench mark is about 25 feet north of the north side of Ravensdale Avenue and about 125 feet west of the west side of Stanley Avenue. An arrow cut in the south side of the rock points to the bench mark. (Note 4, p. 89.)

Elevation, 127.018 feet.

B. M. 1180.—*Hastings-on-Hudson, Westchester County.* The center of the top of the 6½" x 6½" granite property monument on the east side of Farragut Road about 60 feet south of the south side of the Union Free School of the Village of Hastings. The monument marks the south line of the school property and the top is 5 inches above the ground.

Elevation, 195.462 feet.

B. M. 1181.—*Hastings-on-Hudson, Westchester County.* This point is on the cornerstone of the Union Free School of the Village of Hastings, on the east side of Farragut Road, opposite Olinda Park. The building is a two-story and attic red brick structure and the cornerstone is in the northwest corner. The cornerstone contains the inscription, "UNION FREE SCHOOL, DISTRICT NO. 4, TOWN OF GREENBURGH," and the bench mark is the center of the horizontal bar of the final letter H in Greenburgh. It is 5.5 feet above the ground and 3.17 feet from the north side of the building.

Elevation, 198.936 feet.

B. M. 1182.—*Hastings-on-Hudson, Westchester County.* A square (ㅌ) cut in the top of the bluestone water-table at the southwest corner of the Union Free School of the Village of Hastings, on the east side of Farragut Road opposite Olinda Park. The bench mark is 5 inches above the ground.

Elevation, 194.360 feet.

B. M. 1183.—*Dobbs Ferry, Westchester County.* A square (ㅌ) cut in the top of the southwest corner of the monolithic granite base of the Washington-Rochambeau monument, on the east side of Broadway about 325 feet north of the center line of Colonial Avenue. The bench mark is just east of the Mackenzie Boarding School and 1.5 feet above the ground.

Elevation, 119.824 feet.

B. M. 1184.—*Dobbs Ferry, Westchester County.* Coast and Geodetic Survey B. M. U. A punch hole in the center of the end of a ½-inch copper bolt lettered U. S. B. M., leaded horizontally into the west side of the stone railroad station. The bolt is in the fourth stone above the door-sill, is 10 inches south of the door to the baggage room, is 4.4 feet above the cement platform and 4.3 feet north of the southwest corner of the building. Established in 1902.

Elevation, 14.698 feet.

* In 1913 it was reported that this bridge was to be re-constructed.

B. M. 1185.—*Dobbs Ferry, Westchester County.* Coast and Geodetic Survey B. M. V. The intersection of a cross (+) cut in the end of a ½-inch copper bolt leaded horizontally into the brick on the south side of the two-story red brick building occupied by Besson & Company's (formerly Taylor's) lumber office, about 400 feet south of the railroad station and about 100 feet east of the Hudson River. The bolt is 10 inches above the ground and 3.7 feet west of the southeast corner of the building. Established in August, 1887.

Elevation, 9.443 feet.

B. M. 1186.—*Dobbs Ferry, Westchester County.* Coast and Geodetic Survey B. M. W. On the west side of the most southerly building (two-story brick engine room and ice house) of the Manila Anchor Brewing Company's plant, on the east side of the Hudson Division of the New York Central and Hudson River Railroad, about 1,800 feet north of the station. The bench mark is the bottom of a square lettered U. S. B. M., cut in the top of the projecting foundation stone 16.3 feet from the north end of the building and 8 inches above the ground. Established in 1902.

Elevation, 9.054 feet.